AF572407

Residential Construction Academy

Electrical Principles

Residential Construction Academy

Electrical Principles

Stephen L. Herman

THOMSON
DELMAR LEARNING

Australia Canada Mexico Singapore Spain United Kingdom United States

Residential Construction Academy: Electrical Principles
Stephen L. Herman

Executive Director:
Alar Elken

Acquisitions Editor:
Christopher Will and Mark Huth

Executive Editor:
Sandy Clark

Development Editor:
Monica Ohlinger

Editorial Assistant:
Jennifer Luck

Executive Marketing Manager:
Maura Theriault

Marketing Coordinator:
Brian McGrath

Executive Production Manager:
Mary Ellen Black

Production Manager:
Andrew Crouth

Senior Project Editor:
Christopher Chien

Senior Art/Design Coordinator:
Mary Beth Vought

Full Production Services:
Carlisle Publishers Services

Printed in the United States of America
4 5 XXX 06 05

For more information contact:
Delmar Learning
5 Maxwell Dr.
Clifton Park, NY 12065

Or find us on the World Wide Web at
http://www.delmarlearning.com

Library of Congress Cataloging-in-Publication Data

Residential Construction Academy: Electrical Principles
Stephen L. Herman
Library of Congress Cataloging-in-Publication Data
Herman, Stephen L.
Residential construction academy : electrical principles / Stephen Herman.
p. cm.
Includes bibliographical references and index.
ISBN 1-4018-1294-5
1. Electric engineering—Safety measures. 2. Dwellings—Electric equipment. I. Title.
TK152.H44 2003
621.319—dc21
2002031320

NOTICE TO THE READER

Table of Contents

SECTION 2 Alternating Current 145

Preface

Home Builders Institute Residential Construction Academy: Electrical Principles

About the Residential Construction Academy

One of the most pressing problems confronting the building industry today is the shortage of skilled labor. It is estimated that the construction industry must recruit 200,000 to 250,000 new craft workers each year to meet future needs. This shortage is expected to continue well into the next decade because of projected job growth and a decline in the number of available workers. At the same time, a lack of training opportunities for available labor is causing increasing concern throughout the country. It has resulted in a shortage of 65,000 to 80,000 skilled workers per year. The crisis is affecting all construction trades and is threatening the ability of builders to produce quality homes.

These are the reasons for the creation of the innovative *Residential Construction Academy Series.* This series is the perfect way to introduce people of all ages to the building trades while guiding them in the development of essential workplace skills, including carpentry, electrical, HVAC, plumbing, and facilities maintenance. The products and services offered through the *Residential Construction Academy* are the result of cooperative planning and rigorous joint efforts by industry and education. The program was originally conceived by the National Association of Home Builders (NAHB)—the premier association of more than 200,000 member groups in the residential construction industry—and its workforce development arm, the Home Builders Institute (HBI).

For the first time, construction professionals and educators created national standards for the construction trades. In the summer of 2001, the NAHB, through HBI, began the process of developing residential craft standards in five trades. They were carpentry, electrical wiring, HVAC, plumbing, and facilities maintenance. Groups of electrical employers from across the country met with an independent research and measurement organization to begin developing new craft training standards. The guidelines from the National Skills Standards Board were followed in developing the new standards. In addition, the process met or exceeded the American Psychological Association standards for occupational credentialing.

Then, through a partnership between HBI and Delmar Learning, textbooks, videos, and instructor's curriculum and teaching tools were created to effectively teach these standards. A foundational tenet of this series is that students *learn by doing.* Integrated into this colorful, highly illustrated text are *examples* and *practice problems* that apply the theoretical concepts of electricity to the practical applications that an electrician will encounter. There are also *lab exercises,* found in the instructor's package, that are designed to help students apply information through hands-on, active application. A constant focus of the *Residential Construction Academy* is teaching the skills needed to be successful in the construction industry and applying them to real-world applications.

Perhaps most exciting to learners and industry is the creation of a national registry of students who have successfully completed courses in the *Residential Construction Academy Series.* This registry or transcript service provides easy access to verification of skills and competencies achieved. The registry links construction industry employers and qualified potential employees in an online database facilitating student job search and the employment of skilled workers.

About This Book

Electrical Principles is intended to present the concepts of electricity to learners interested in the field of residential wiring. It assumes that the learner has no knowledge of electricity and presents the subject in a very practical, straightforward manner. The book begins with electrical safety and progresses through basic electrical theory; electrical quantities and Ohm's Law; series, parallel, and combination circuits; measuring instruments; wire tables; magnetic induction; alternating current; transformers; and motors. All mathematical calculations are explained in step-by-step detail. This book is an excellent choice for anyone entering the electrical field.

Electrical Principles is a blend of practical and theoretical. The text not only explains the different concepts relating to electrical theory, it also provides practical examples of how to do many of the common tasks the electrician must perform. It was the author's intention to make this book one that an electrician would want to keep in his toolbox for a reference. Although the text covers many areas of both direct and alternating current theory, the math level has been kept to basic algebra and basic trigonometry.

The subject matter for *Electrical Principles* has been divided into separate stand-alone chapters. Although the chapters have been divided in a logical progression, the stand-alone concept permits the information to be presented in almost any sequence the instructor desires. In addition, the student or electrician does not have to search the entire text to find information concerning a particular subject; it is located within a certain chapter.

Features

This innovative series was designed with input from educators and industry and informed by the curriculum and training objectives established by the Standards Committee. The following features aid learning:

Learning Features such as the **introduction, objectives,** and **glossary** set the stage for the coming body of knowledge and help the learner identify key concepts and information. These learning features serve as a road map for continuing through the chapter. The learner also may use them as a reference later.

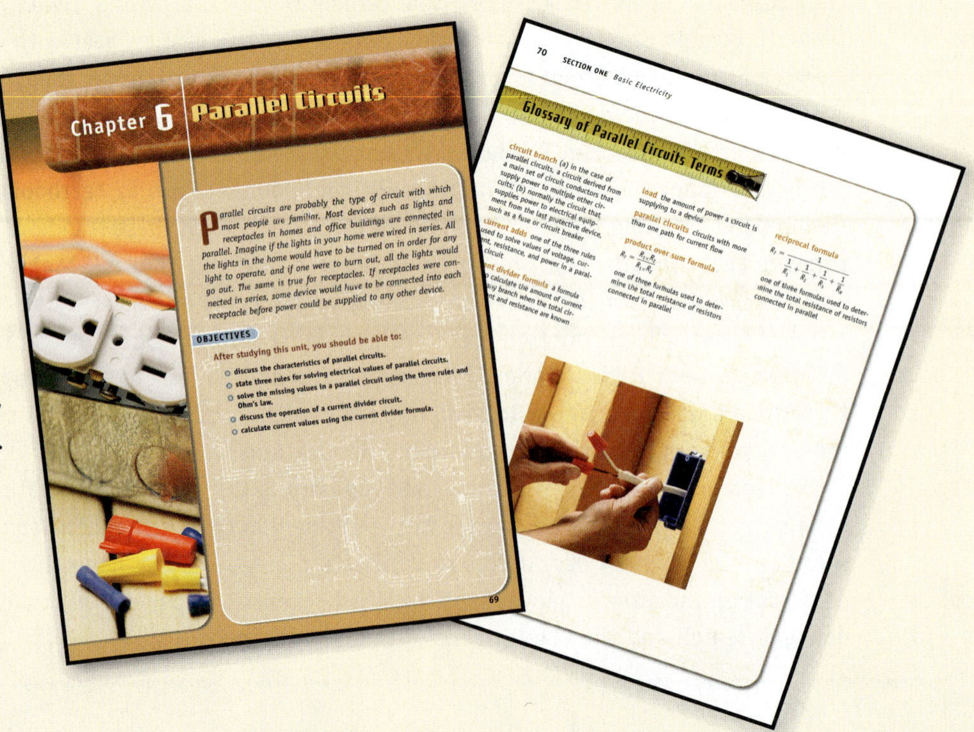

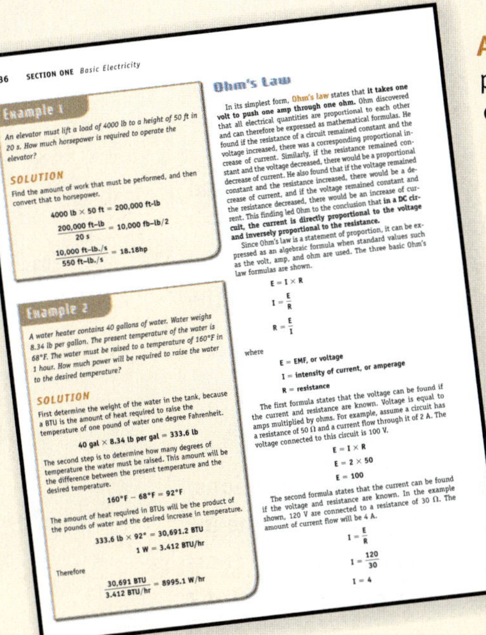

36 SECTION ONE *Basic Electricity*

Example 1

An elevator must lift a load of 4000 lb to a height of 50 ft in 20 s. How much horsepower is required to operate the elevator?

SOLUTION

Find the amount of work that must be performed, and then convert that to horsepower.

$$4000 \text{ lb} \times 50 \text{ ft} = 200{,}000 \text{ ft-lb}$$

$$\frac{200{,}000 \text{ ft-lb}}{20 \text{ s}} = 10{,}000 \text{ fb-lb/2}$$

$$\frac{10{,}000 \text{ ft-lb./s}}{550 \text{ ft-lb./s}} = 18.18\text{hp}$$

Example 2

A water heater contains 40 gallons of water. Water weighs 8.34 lb per gallon. The present temperature of the water is 68°F. The water must be raised to a temperature of 160°F in 1 hour. How much power will be required to raise the water to the desired temperature?

SOLUTION

First determine the weight of the water in the tank, because a BTU is the amount of heat required to raise the temperature of one pound of water one degree Fahrenheit.

$$40 \text{ gal} \times 8.34 \text{ lb per gal} = 333.6 \text{ lb}$$

The second step is to determine how many degrees of temperature the water must be raised. This amount will be the difference between the present temperature and the desired temperature.

$$160°\text{F} - 68°\text{F} = 92°\text{F}$$

The amount of heat required in BTUs will be the product of the pounds of water and the desired increase in temperature.

$$333.6 \text{ lb} \times 92° = 30{,}691.2 \text{ BTU}$$

$$1 \text{ W} = 3.412 \text{ BTU/hr}$$

Therefore

$$\frac{30{,}691 \text{ BTU}}{3.412 \text{ BTU/hr}} = 8995.1 \text{ W/hr}$$

Ohm's Law

In its simplest form, Ohm's law states that **it takes one volt to push one amp through one ohm.** Ohm discovered that all electrical quantities are proportional to each other and can therefore be expressed as mathematical formulas. He found if the resistance of a circuit remained constant and the voltage increased, there was a corresponding proportional increase of current. Similarly, if the resistance remained constant and the voltage decreased, there would be a proportional decrease of current. He also found that if the voltage remained constant and the resistance increased, there would be a decrease of current, and if the voltage remained constant and the resistance decreased, there would be an increase of current. This finding led Ohm to the conclusion that **in a DC circuit, the current is directly proportional to the voltage and inversely proportional to the resistance.**

Since Ohm's law is a statement of proportion, it can be expressed as an algebraic formula when standard values such as the volt, amp, and ohm are used. The three basic Ohm's law formulas are shown.

$$E = I \times R$$

$$I = \frac{E}{R}$$

$$R = \frac{E}{I}$$

where

E = EMF, or voltage

I = intensity of current, or amperage

R = resistance

The first formula states that the voltage can be found if the current and resistance are known. Voltage is equal to amps multiplied by ohms. For example, assume a circuit has a resistance of 50 Ω and a current flow through it of 2 A. The voltage connected to this circuit is 100 V.

$$E = I \times R$$

$$E = 2 \times 50$$

$$E = 100$$

The second formula states that the current can be found if the voltage and resistance are known. In the example shown, 120 V are connected to a resistance of 30 Ω. The amount of current flow will be 4 A.

$$I = \frac{E}{R}$$

$$I = \frac{120}{30}$$

$$I = 4$$

Application of Theory is a core concept of this text. This volume explains concepts relating to electrical theory, but also provides practical examples of how to do many of the common tasks the electrician must perform. Information is heavily illustrated to show concepts and hardware encountered by the learner. An *example* section integrated throughout each chapter takes the information presented and immediately applies it so theory is linked to the practical. The overall effect is a clear link between theory and practice, making learning easier and more meaningful.

Safety is featured at the beginning of the text to instill safety as an "attitude" among learners. Learners will come to appreciate that safety is a blend of ability, skill, and knowledge that should be continuously applied to all they do in the construction industry.

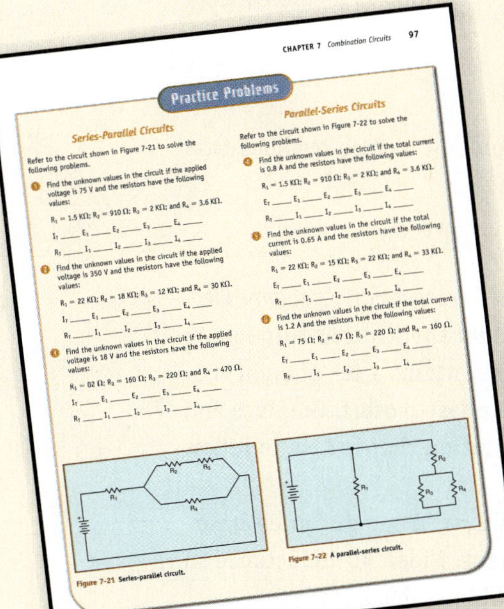

CHAPTER 7 *Combination Circuits* 97

Practice Problems

Series-Parallel Circuits

Refer to the circuit shown in Figure 7-21 to solve the following problems.

1. Find the unknown values in the circuit if the applied voltage is 75 V and the resistors have the following values:

   R_1 = 1.5 KΩ; R_2 = 910 Ω; R_3 = 2 KΩ; and R_4 = 3.6 KΩ.

   I_T ___ E_1 ___ E_2 ___ E_3 ___ E_4 ___

   R_T ___ I_1 ___ I_2 ___ I_3 ___ I_4 ___

2. Find the unknown values in the circuit if the applied voltage is 350 V and the resistors have the following values:

   R_1 = 22 KΩ; R_2 = 18 KΩ; R_3 = 12 KΩ; and R_4 = 30 KΩ.

   I_T ___ E_1 ___ E_2 ___ E_3 ___ E_4 ___

   R_T ___ I_1 ___ I_2 ___ I_3 ___ I_4 ___

3. Find the unknown values in the circuit if the applied voltage is 18 V and the resistors have the following values:

   R_1 = 02 Ω; R_2 = 160 Ω; R_3 = 220 Ω; and R_4 = 470 Ω.

   I_T ___ E_1 ___ E_2 ___ E_3 ___ E_4 ___

   R_T ___ I_1 ___ I_2 ___ I_3 ___ I_4 ___

Parallel-Series Circuits

Refer to the circuit shown in Figure 7-22 to solve the following problems.

4. Find the unknown values in the circuit if the total current is 0.8 A and the resistors have the following values:

   R_1 = 1.5 KΩ; R_2 = 910 Ω; R_3 = 2 KΩ; and R_4 = 3.6 KΩ.

   E_T ___ E_1 ___ E_2 ___ E_3 ___ E_4 ___

   R_T ___ I_1 ___ I_2 ___ I_3 ___ I_4 ___

5. Find the unknown values in the circuit if the total current is 0.65 A and the resistors have the following values:

   R_1 = 22 KΩ; R_2 = 15 KΩ; R_3 = 22 KΩ; and R_4 = 33 KΩ.

   E_T ___ E_1 ___ E_2 ___ E_3 ___ E_4 ___

   R_T ___ I_1 ___ I_2 ___ I_3 ___ I_4 ___

6. Find the unknown values in the circuit if the total current is 1.2 A and the resistors have the following values:

   R_1 = 75 Ω; R_2 = 47 Ω; R_3 = 220 Ω; and R_4 = 160 Ω.

   E_T ___ E_1 ___ E_2 ___ E_3 ___ E_4 ___

   R_T ___ I_1 ___ I_2 ___ I_3 ___ I_4 ___

Figure 7-21 Series-parallel circuit.

Figure 7-22 A parallel-series circuit.

Practice Problems provide practice with applied math concepts. Placing mathematics in context and applying it helps the student better grasp the material and makes academics meaningful. For example, while students are learning about series-parallel circuits, they will be asked to calculate unknown values when given voltage and resistor values. Students are thus given an opportunity to practice challenging math concepts in an applied and meaningful context.

Review Questions complete each chapter. These are designed to reinforce the information learned in the chapter and to give learners the opportunity to think about what they have learned and what they have accomplished.

Turnkey Curriculum and Teaching Material Package

We understand that a text is only one part of a complete, turnkey educational system. We also understand that instructors want to spend their time on teaching, not preparing to teach. The *Residential Construction Academy Series* is committed to providing thorough curriculum and prepatory materials to aid instructors and alleviate some of their heavy preparation commitments. An integrated teaching solution is ensured with the text, Instructor's e.resource™, print instructor's resource guide, and student videos.

e.resource™

Delmar Learning's e.resource™ is a complete guide to classroom management. The CD-ROM contains lecture outlines; notes to instructors providing teaching hints, cautions, and answers to review questions; and other aids for the instructor using this series. Designed as a complete and integrated package, the instructor is also provided with suggestions for when and how to use the accompanying **PowerPoint, computerized test bank, lab exercises,** and **video** package components. There is also a print **instructor's resource guide** available.

Lab Exercises

The author has compiled a number of lab exercises designed to address hands-on application of electrical theory concepts presented in the text. The labs can be used as is or modified by instructors in their Microsoft Word™ format. These hands-on activities reflect the fact that active learning is a core concept of the *Residential Construction Academy Series.*

PowerPoint

The author has created a series of PowerPoint presentations providing thorough lecture outlines that can be used to teach the course.

Videos

The *Electrical Principles Video Series* is an integrated part of the *Residential Construction Academy Electricity* package. This video series introduces the viewer to electrical theory and its applications through an overview of important concepts, equations, and practical examples. The series contains a set of eight, 20-minute videos covering direct and alternating current, and introducing single phase motors. Graphics and animations illustrate important concepts, creating a visually appealing presentation that is easy to follow and understand.

The complete set includes the following: Video #1 Basic Electricity & Ohms' Law. Video #2 Series & Parallel Circuits. Video #3 Combination Circuits. Video #4 Basic Alternating Current. Video #5 Inductance in AC Circuits. Video #6 Capacitors. Video #7 Capacitors in AC Circuits. Video #8 Single-Phase Motors.

CD Courseware

This package also includes computer-based training that uses video, animation, and testing to introduce, teach, or remediate the concepts covered in the videos. Students will be pretested on the material and then, if needed, provided with suitable remediation to ensure understanding of the concepts. Final posttesting will be administered to ensure that the student has gained mastery of all material.

Online Companion

The Online Companion is an excellent supplement for students. It features many useful resources to support the *Electrical Principles* book, videos, and CDs. Linked from the Student Materials section of www.residentialacademy.com, the Online Companion includes chapter quizzes, an online glossary, product updates, related links, and more. Visit; http://www.delmarlearning.com/companions/index.asp?isbn=1401812945.

About the Author

Stephen L. Herman has been both a teacher of industrial electricity and an industrial electrician for many years. His formal training was obtained at Catawba Valley Technical College in Hickory, North Carolina. Mr. Herman has worked as a maintenance electrician for Superior Cable Corp. and as a class "A" electrician for National Liberty Pipe and Tube Co. During those years of experience, Mr. Herman learned to combine his theoretical knowledge of electricity with practical application. The books he has authored reflect his strong belief that a working electrician must have a practical knowledge of both theory and experience to be successful.

Mr. Herman was the Electrical Installation and Maintenance instructor at Randolph Technical College in Asheboro, North Carolina for nine years. After a return to industry, he became the lead instructor of the Electrical Technology Curriculum at Lee College in Baytown, Texas. He retired from Lee College after 20 years of service and, at present, resides in Pittsburg, Texas, with his wife. He continues to stay active in industry, write, and update his books.

Compliance with Apprenticeship, Training, Employer, and Labor Services (ATELS)

These materials are in full compliance with the Apprenticeship, Training, Employer, and Labor Services (ATELS) requirements for classroom training.

Acknowledgments

Electrical National Skill Standards

The NAHB and the HBI would like to thank the many individual members and companies that participated in the creation of the Electrical National Skill Standards. Special thanks are extended to the following individuals and companies:

Robert W. Baird
Independent Electrical Contractors

John Gaddis
Golconda Job Corps Center

Steve Herman
Lee College

Roy Hogue
TruRoy Electrical

Fred Humphreys
Home Builders Institute

Ray Mullin
Delmar Learning Author

Ron Rodgers
Wasdyke Associates/Employment Research

Jack Sanders
Home Builders Institute

Gordon Stewart
Joe Swartz Electric Company

Roy Stoner
Maryland State Department of Education

Clarence Tibbs
STE Electrical Systems, Inc.

Ray Wasdyke
Wasdyke Associates

In addition to the standard's committee, there were many other people who contributed their time and expertise to the project. They have spent hours attending focus groups, reviewing, and contributing to the work. Delmar Learning and the author extend our sincere gratitude to the following:

Ray Pearce
Mohave Community College

David Gehlauf
Tri-County Vocational School

Orville Lake
Augusta Technical College

Ralph B. Hathaway
Dean Institute of Technology

Section ONE

Basic Electricity

SECTION ONE

BASIC ELECTRICITY

- **Chapter 1:** *Safety*
- **Chapter 2:** *Basic Electrical Theory*
- **Chapter 3:** *Electrical Quantities and Ohm's Law*
- **Chapter 4:** *Resistors*
- **Chapter 5:** *Series Circuits*
- **Chapter 6:** *Parallel Circuits*
- **Chapter 7:** *Combination Circuits*
- **Chapter 8:** *Measuring Instruments*
- **Chapter 9:** *Using Wire Tables and Determining Conductor Sizes*
- **Chapter 10:** *Magnetic Induction*

Chapter 1 Safety

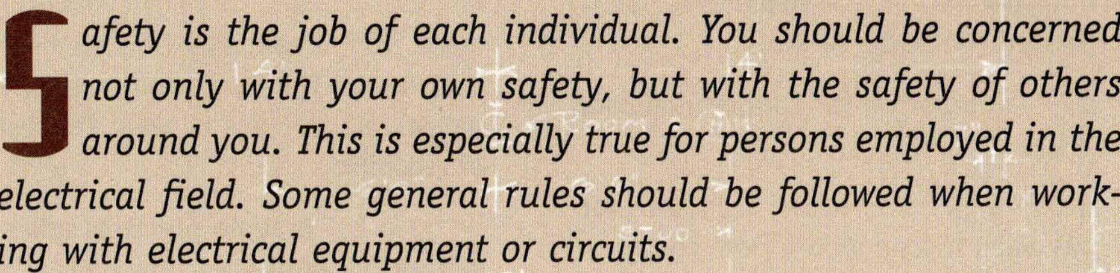

Safety is the job of each individual. You should be concerned not only with your own safety, but with the safety of others around you. This is especially true for persons employed in the electrical field. Some general rules should be followed when working with electrical equipment or circuits.

OBJECTIVES

After completing this unit, the student should be able to:

- state basic safety rules.
- describe the effects of electric current on the body.
- discuss the origin and responsibilities of OSHA.
- discuss material safety data sheets.
- discuss lock out and tag out procedures.
- discuss types of protective clothing.
- explain how to properly place a straight ladder against a structure.
- discuss different types of scaffolds.
- discuss classes of fires.
- discuss ground fault circuit interrupters.
- discuss the importance of grounding.

Glossary of Safety Terms

artificial respiration a method of providing oxygen to a person unable to breath on his or her own

confined space a space that has limited entrances or exits

CPR an acronym for cardiopulmonary resuscitation

de-energized circuit a circuit that has been disconnected from the power supply

disconnect to turn off or remove power from a circuit

egress an entrance or exit into a structure

energized circuit a circuit connected to the power supply; a voltage or potential exists between conductors

fibrillation a condition in which the heart starts to quiver and stops its pumping action

fire retardant clothing clothing made from a specially treated material that resists ignition; many industries require electricians to wear this type of clothing

horseplay playing practical jokes or exhibiting adolescent behavior in an inappropriate place

idiot proofing designing a piece of equipment so it cannot be misused by an experienced person

lock out and tag out a procedure to prevent a circuit from being energized accidentally

meter an instrument used to test for some electrical quantity such as voltage, current, or resistance

milliamperes one milliampere equals 1/1000 of an amp (0.001)

MSDS an acronym for material safety data sheets

OSHA an acronym for Occupational Safety and Health Administration

scaffold a platform for working in high places, generally designed so it can be taken apart and transported to a different job site or area

General Safety Rules

Never Work on an Energized Circuit If the Power can Be Disconnected

When possible, use the following three-step check to make certain that power is turned off:

1. Test the **meter** on a known live circuit to make sure the meter is operating.
2. Test the circuit that is to become the **de-energized circuit** with the meter.
3. Test the meter on the known live circuit again to make certain the meter is still operating.

Install a warning tag at the point of disconnection so people will not restore power to the circuit. If possible, use a lock to prevent anyone from turning the power back on.

Think

Of all the rules concerning safety, this one is probably the most important. No amount of safeguarding or **idiot proofing** a piece of equipment can protect a person as well as taking time to think before acting. Many technicians have been killed by supposedly "dead" circuits. Do not depend on circuit breakers, fuses, or someone else to open a circuit. Test it yourself before you touch it. If you are working on high voltage equipment, use insulated gloves and meter probes to measure the voltage being tested. **Think** before you touch something that could cost you your life.

Avoid Horseplay

Jokes and **horseplay** have a time and place, but not when someone is working on an electric circuit or a piece of moving machinery. Do not be the cause of someone being injured or killed and do not let someone else be the cause of your being injured or killed.

Do Not Work Alone

This is especially true when working in a hazardous location or on a live circuit. Have someone with you who can turn off the power or give **artificial respiration** and/or **CPR.** Possible effects of several electrical shocks is breathing difficulties and causing the heart to go into fibrillation.

Work with One Hand When Possible

The worst kind of electrical shock occurs when the current path is from one hand to the other, which permits the current to pass directly through the heart. A person can survive a severe shock between the hand and foot that would cause death if the current path was from one hand to the other.

Learn First Aid

Anyone working on electrical equipment, especially those working with voltages greater than 50 volts, should make an effort to learn first aid. A knowledge of first aid, especially CPR, might save your own or someone else's life.

Effects of Electric Current on the Body

Most people have heard that its not the voltage that kills, but the current. This is true, but do not be mislead into thinking voltage cannot harm you. Voltage is the force that pushes the current though the circuit. It can be compared to the pressure that pushes water through a pipe. The more pressure available, the greater the volume of water flowing through the pipe. Students often ask how much current will flow through the body at a particular voltage. There is no easy answer to this question. The amount of current that can flow at a particular voltage is determined by the resistance of the current path. Different people have different resistances. A body has less resistance on a hot day when sweating, because salt water is a very good conductor. What one eats and drinks for lunch can have an effect on the body's resistance as can the length of the current path. Is the current path between two hands or from one hand to one foot? All of these factors affect body resistance.

The chart (Fig. 1-1) illustrates the effects of different amounts of current on the body. This chart is general—some people may have less tolerance to electricity and others may have a greater tolerance.

A current of 2 to 3 **milliamperes** (0.002 to 0.003 amp) usually causes a slight tingling sensation, which increases as current increases and becomes very noticeable at about 10 (0.010 amp) milliamperes. The tingling sensation is very painful at about 20 milliamperes. Currents between 20 and 30 milliamperes cause a person to seize the line and be unable to let go of the circuit. Currents between 30 and 40 milliamperes cause muscular paralysis, and those between 40 and 60 milliamperes cause breathing difficulty. When the current increases to about 100 milliamperes, breathing is extremely difficult. Currents from 100 to 200 milliamperes generally cause death because the heart usually goes into **fibrillation,** a condition in which the heart begins to "quiver" and the pumping action stops. Currents above 200 milliamperes cause the heart to squeeze shut. When the current is removed, the heart usually returns to a normal pumping action. This is the operating principle of a defibrillator. The voltage considered to be the most dangerous to work with is 120 volts, because that generally causes a current flow of between 100 and 200 milliamperes through most people's bodies. Large amounts of current can cause severe electrical burns that are often very serious because they occur on the inside of the body. The exterior of the body may not look seriously burned, but the inside may be severely burned.

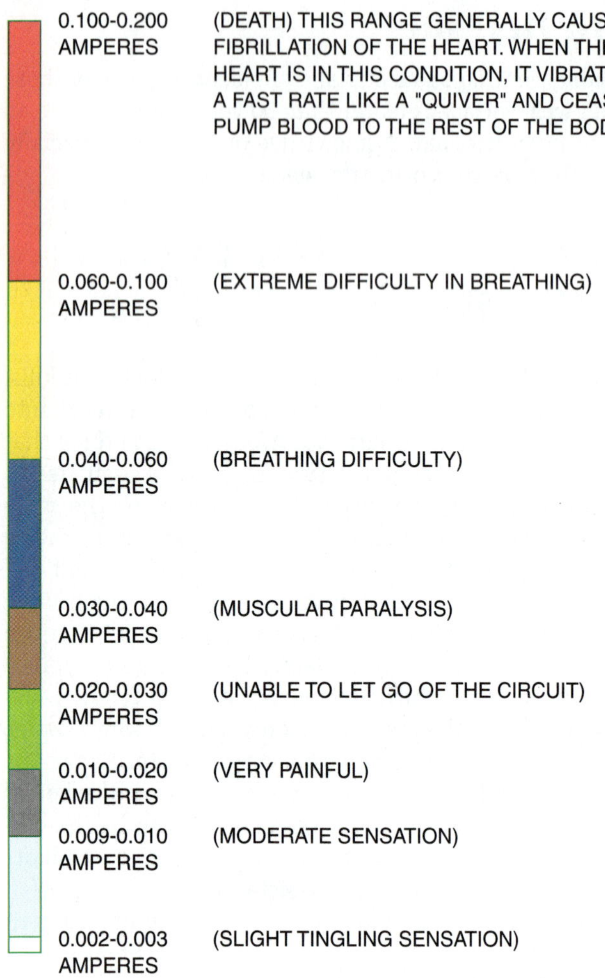

Figure 1-1 The effects of electric current on the body.

OSHA

OSHA is an acronym for Occupational Safety and Health Administration, U.S. Department of Labor. Created by congress in 1971, its mission is to ensure safe and healthful workplaces in the United States. Since its creation, workplace fatalities have been cut in half, and occupational injury and illness rates have declined by 40%. Enforcement of OHSA regulations is the responsibility of the Secretary of Labor.

OSHA standards cover many areas, such as the handling of hazardous materials, fall protection, protective clothing, and hearing and eye protection. Part 1910 subpart S deals mainly with the regulations concerning electrical safety. These regulations are available in books and can be accessed at the OSHA website on the Internet at *www.osha.org*.

Hazard Materials

It may become necessary to deal with some type of hazardous material. A hazardous material or substance is any substance which exposure to may result in adverse effects on the health or safety of employees. Hazardous materials may be chemical, biological, or nuclear. OSHA sets standards for dealing with many types of hazardous materials. The required response is determined by the type of hazard associated with the material. Hazardous materials are required to be listed as such. Much information concerning hazardous materials is generally found on Material Safety Data Sheets (MSDS). (A sample MSDS is included at the end of the chapter.) If you are working in an area that contains hazardous substances, always read any information concerning the handling of the material and any safety precautions that should be observed. Waiting until a problem exists in not the time to start looking for information on what to do. Some hazardous materials require a HAZMAT (Hazardous Materials Response Team) to handle any problems. A HAZMAT means any group of employees designated by the employer that are expected to handle and control an actual or potential leak or spill of a hazardous material. They are expected to work in close proximity to the material. A HAZMAT team is not a fire brigade and a fire brigade may not necessarily be a HAZMAT team. A HAZMAT team may be part of a fire brigade or fire department.

Employer Responsibilities

Section 5(a)1 of the occupational safety and health act basically states that each employer must furnish each of his employees a place of employment that is free of recognized hazards that are likely to cause death or serious injury. This places the responsibility for compliance on the employer. The employer must identify hazards or potential hazards within the work site and eliminate them, control them, or provide the employee with suitable protection from them. It is the employee's responsibility to follow the safety procedures setup by the employer.

To help facilitate these safety standards and procedures, OSHA requires that an employer have a competent person oversee implementation and enforcement of these standards and procedures. This person must be able to recognize unsafe or dangerous conditions and have the authority to correct or eliminate them. He also has the authority to stop work or shut down a job site until safety regulations are met.

MSDS

MSDS stands for material safety data sheets, which are provided with many products. They generally warn users of any hazards associated with the product. They outline the physical and chemical properties of the product, list precautions that should be taken when using the product, list any potential health hazards, storage consideration, flammability, reactivity, and in some instances radioactivity. They sometimes list the name, address, and telephone number of the manufacturer, MSDS date, and emergency telephone numbers, and usually provide information on first aid procedures to use if the product is swallowed or comes in contact with the skin. Material safety data sheets can be found on many home products such as cleaning products, insecticides, and flammable liquids.

Alcohol and Drugs

The use of alcohol and drugs has no place on a construction site. It is not only dangerous to users and those who work around them, it also costs industry millions of dollars a year. Alcohol and drug abusers kill thousands of people on the highways each year, and are just as dangerous on a work site as they are behind the wheel of a vehicle. Many industries have instituted testing policies to screen for alcohol and drugs. A person who tests positive generally receives a warning the first time and is fired the second time.

Trenches

It is often necessary to dig trenches to bury conduit. Under some conditions, these trenches can be deep enough to bury a person if a cave-in should occur. Safety regulations for the shoring of trenches is found in OSHA Standard 1926 Subpart P App C titled "Timber Shoring for Trenches." These procedures and regulations are federally mandated and must be followed. Some general safety rules should be followed also:

1. Don't walk close to trenches unless it is necessary. This can cause the dirt to loosen and increase the possibility of a cave-in.
2. Don't jump over trenches if it is possible to walk around them.
3. Place barricades around trenches (Fig. 1-2).
4. Use ladders to enter and exit trenches.

Figure 1-2 Place a barricade around a trench and use a ladder to enter and exit the trench.

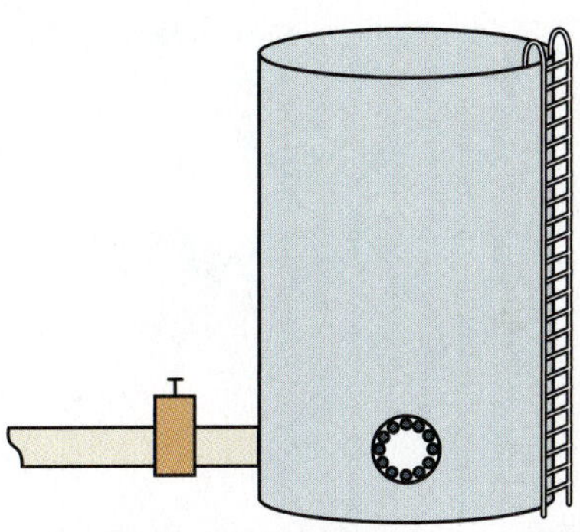

Figure 1-3 A confined space is any space having a limited means of egress.

Confined Spaces

Confined spaces have a limited means of **egress** (Fig. 1-3). They can be very hazardous workplaces, often containing atmospheres that are extremely harmful or deadly. Confined spaces are very difficult to ventilate because of their limited openings. It is often necessary for a workman to wear special clothing and use a separate air supply to work there. OSHA Section 12: "Confined Space Hazards," lists rules and regulations for working in a confined space. In addition, many industries have written procedures that must be followed when working in confined spaces. Some general rules include the following:

1. Have a person stationed outside the confined space to watch the person or persons working inside. The outside person should stay in voice or visual contact with the inside workers at all times. He should check air sample readings, and monitor oxygen and explosive gas levels.
2. The outside person should never enter the space, even in an emergency, but should contact the proper emergency personnel. If he or she should enter the space and become incapacitated, there would be no one available to call for help.
3. Use only electrical equipment and tools that are approved for the atmosphere found inside the confined area. It may be necessary to obtain a burning permit to operate tools that have open brushes and that spark when they are operated.
4. As a general rule, a person working in a confined space should wear a harness with a lanyard that extends to the outside person, so the outside person could pull him or her to safety if necessary.

Lock Out and Tag Out Procedures

Lock out and tag out procedures are generally employed to prevent someone from energizing a piece of equipment by mistake. This could apply to switches, circuit breakers, or valves. Most industries have their own internal policies and procedures. Some require that a tag similar to the one shown in Figure 1-4 be placed on the piece of equipment being serviced; some also require that the equipment be locked out with a padlock. The person performing the work places the lock on the equipment and keeps the key in his or her possession. A device that permits the use of multiple padlocks and a safety tag is shown in Figure 1-5. This is used when more than one person is working on the same piece of equipment. Violating lock and tag out procedures is considered an extremely serious offense in most industries and often results in immediate termination of employment. As a general rule, there are no first time warnings.

Figure 1-4 Safety tag used to tag-out equipment.

Figure 1-5 The equipment can be locked out by several different people.

Protective Clothing

Maintenance and construction workers alike are usually required to wear certain articles of protective clothing, dictated by the environment of the work area and the job being performed.

Head Protection

Some type of head protection is required on almost any construction site or in any industrial location. A typical electrician's hard hat, made of non-conductive plastic, is shown in Figure 1-6. It has a pair of safety goggles attached that can be used when desired or necessary.

Eye Protection

Eye protection is another piece of safety gear required on almost all construction sites and industrial locations. Eye protection can come in different forms ranging from the goggles shown in Figure 1-6 to the safety glasses with side shields shown in Figure 1-7. Common safety glasses may or may not be prescription glasses, but almost all provide side protection (Fig. 1-7). Sometimes a full face shield may be required.

Hearing Protection

Section III Chapter 5 of the OSHA Technical Manual includes requirements concerning hearing protection. The need for hearing protection is based on the ambient sound level of the construction site or the industrial location. It is not unusual for workers to be required to wear some type of hearing protection when working in certain areas, usually in the form of ear plugs or ear muffs.

Fire Retardant Clothing

Special clothing made of fire retardant material is required in some areas, generally certain industries as opposed

Figure 1-6 Typical electrician's hard hat with attached safety goggles.

Figure 1-7 Safety glasses provide side protection.

Figure 1-8 Leather gloves with rubber inserts.

to new construction sites. **Fire retardant clothes** are often required for maintenance personnel who work with high power sources such as transformer installations and motor control centers. An arc-flash in a motor control center can easily catch a person's clothes on fire. The typical motor control center can produce enough energy during an arc-flash to kill a person 30 feet away.

Gloves

Another common article of safety clothing is gloves. Electricians often wear leather gloves with rubber inserts when it is necessary to work on energized circuits (Fig. 1-8). These gloves are usually rated for a certain amount of voltage. They should be inspected for holes or tears before they are used. Kevlar gloves (Fig. 1-9) help protect against cuts when stripping cable with a sharp blade.

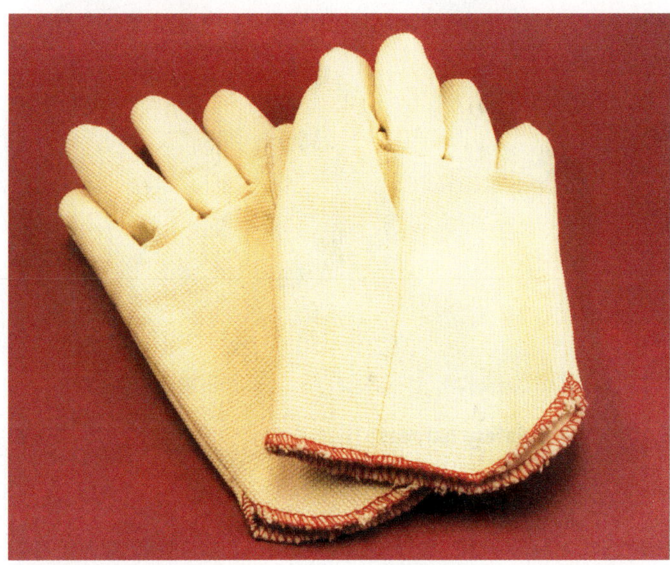

Figure 1-9 Kevlar gloves protect against cuts.

Safety Harness

Safety harnesses provide protection from falling. They buckle around the upper body with leg, shoulder, and chest straps, and the back has a heavy metal D-ring (Fig. 1-10). A section of rope approximately 6 feet in length, called a lanyard, is attached to the D-ring and secured to a stable structure above the worker. If the worker falls, the lanyard limits the distance he can drop. A safety harness should be worn

1. when working more than 6 feet above the ground or floor.
2. when working near a hole or drop-off.
3. when working on high scaffolding

A safety harness is shown in Figure 1-10.

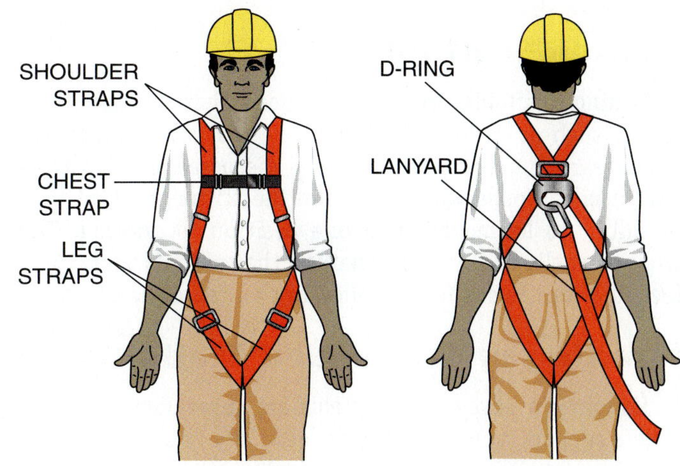

Figure 1-10 Typical safety harness.

Ladders and Scaffolds

It is often necessary to work in an elevated location. When this is the case, ladders or scaffolds are employed. **Scaffolds** generally provide the safest elevated working platforms. They are commonly assembled on the job site

Figure 1-11 Safety harness.

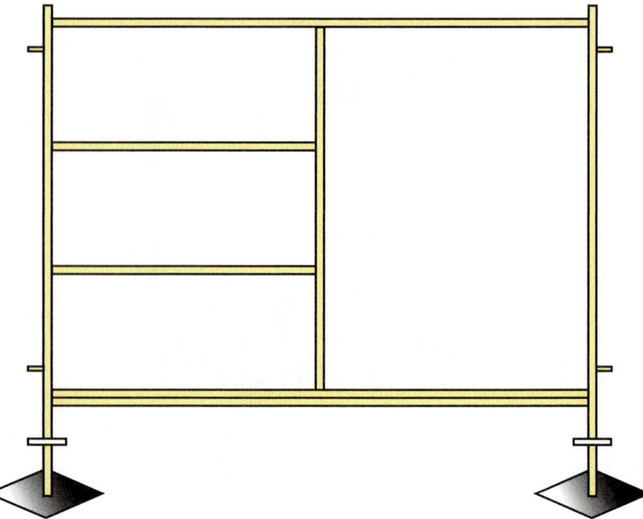

Figure 1-12 Typical section of scaffolding.

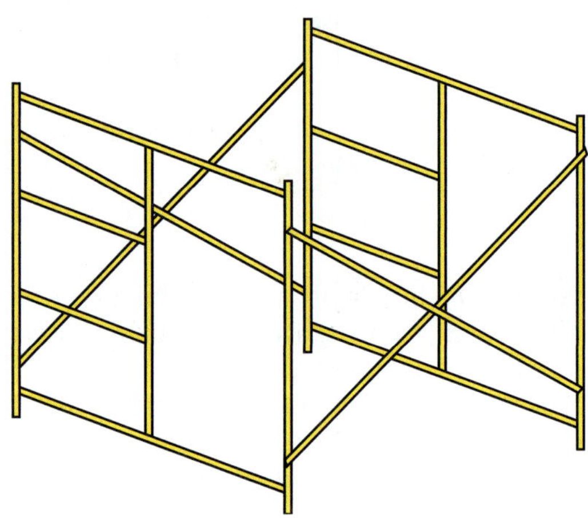

Figure 1-13 X braces connect scaffolding sections together.

from standard sections (Fig. 1-12). The bottom sections usually contain adjustable feet that can be used to level the section. Two end sections are connected by X braces that form a rigid work platform (Fig. 1-13). Sections of scaffolding are stacked on top of each other to reach the desired height.

Rolling Scaffolds

Rolling scaffolds are used in areas that contain level floors, such as inside a building. The major difference between a rolling scaffold and those discussed previously is that it is equipped with wheels on the bottom section that permit it to be moved from one position to another. The wheels usually contain a mechanism that permits them to be locked after the scaffold is rolled to the desired location.

Hanging or Suspended Scaffolds

Hanging or suspended scaffolds are suspended by cables from a support structure. They are generally used on the sides of buildings to raise and lower workers by using hand cranks or electric motors.

Ladders

Ladders can be divided into two main types, straight and step. Straight ladders are constructed by placing rungs between two parallel rails (Fig. 1-14). They generally contain safety feet on one end that helps prevent the ladder from slipping. Ladders used for electrical work are usually wood or fiberglass; aluminum ladders are avoided because they conduct electricity. Regardless of the type of ladder used, you should check its load capacity before using it. This information is found on the side of the ladder. Load capacities of 200 lbs., 250 lbs., and 300 lbs. are common. Do not use a ladder that does not have enough load capacity to support your weight plus the weight of your tools and the weight of any object you are taking up the ladder with you.

Straight ladders should be placed against the side of a building or other structure at an angle of approximately 76° (Fig. 1-15). This can be accomplished by moving the base of the ladder away from the structure a distance equal to one fourth the height of the ladder. If the ladder is 20 feet high,

Figure 1-14 **Straight ladder.**

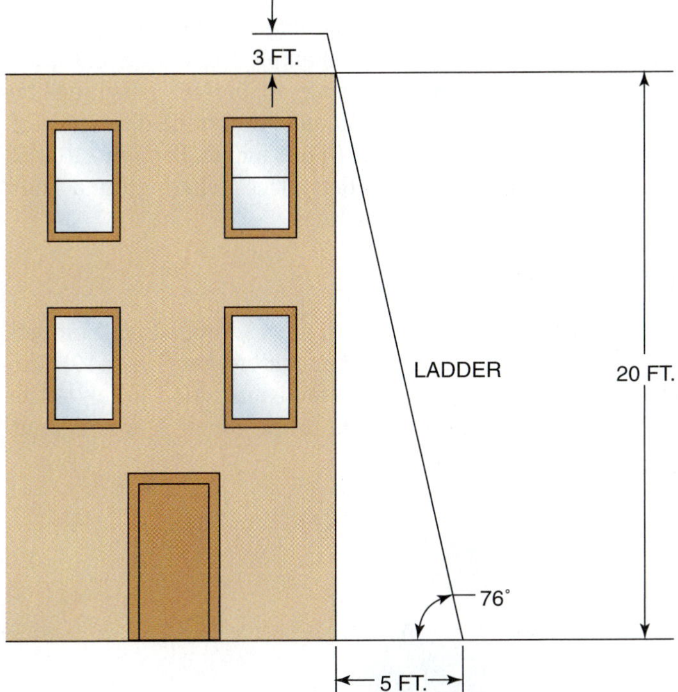

Figure 1-15 **A ladder should be placed at an angle of approximately 76°.**

it should be placed 5 feet from the base of the structure. If the ladder is to provide access to the top of the structure, it should extend 3 feet above the structure.

Step Ladders

Step ladders are self-supporting, constructed of two sections hinged at the top (Fig. 1-16). The front section has two rails and steps; the rear portion two rails and braces. Like straight ladders, step ladders are designed to withstand a certain load capacity. Always check the load capacity before using a ladder. As a general rule, ladder manufacturers recommend that the top step not be used because of the danger of becoming unbalanced and falling. Many people mistakenly think the top step is the top of the ladder, but it is actually the last step before the ladder top.

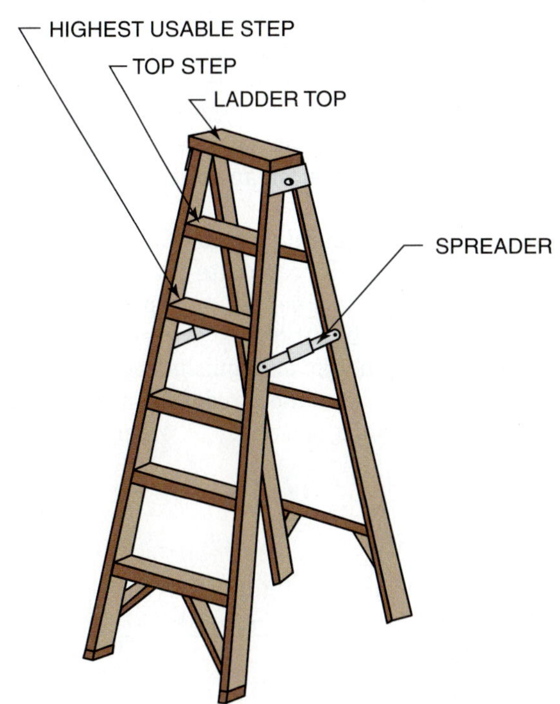

Figure 1-16 **Typical step ladder.**

Fires

For a fire to burn it must have three things: fuel, heat, and oxygen. Fuel is anything that can burn, including materials such as wood, paper, cloth, combustible dusts, and even some metals. Different materials require different amounts of heat for combustion to take place. If the temperature of any material is below its combustion temperature, it will not burn. Oxygen must be present for combustion to take place. If a fire is denied oxygen, it will extinguish.

Fires are divided into four classes: A, B, C, and D. Class A fires involve common combustible materials such as wood or paper. They are often extinguished by lowering the temperature of the fuel below the combustion temperature. Class A fire extinguishers often use water to extinguish a fire. A fire extinguisher listed as class A only should never be used on an electrical fire.

Class B fires involve fuels such as grease, combustible liquids, or gases. A class B fire extinguisher generally employs carbon dioxide (CO_2), which greatly lowers the temperature of the fuel and deprives the fire of oxygen. Carbon dioxide extinguishers are often used on electrical fires, because they do not destroy surrounding equipment by coating it with a dry powder.

Class C fires involve energized electrical equipment. A class C fire extinguisher usually uses a dry powder to smother the fire. Many fire extinguishers can be used on multiple types of fires; for example, an extinguisher labeled ABC could be used on any of the three classes of fire. The important thing to remember is never to use an extinguisher on a fire for which it is not rated. Using a class A extinguisher filled with water on an electrical fire could be fatal.

Class D fires consist of burning metal. Spraying water on some burning metals actually can cause the fire to increase. Class D extinguishers place a powder on top of the burning metal that forms a crust to cut off the oxygen supply to the metal. Some metals cannot be extinguished by placing powder on them, in which case, the powder should be used to help prevent the fire from spreading to other combustible materials.

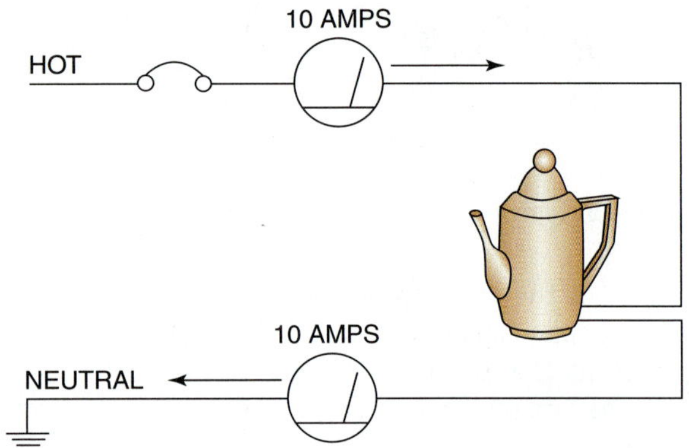

Figure 1-17 The current in both the hot and neutral conductors should be the same but flowing in opposite directions.

Ground Fault Circuit Interrupters

Ground fault circuit interrupters (GFCI) are used to prevent people from being electrocuted. They work by sensing the amount of current flow on both the ungrounded (hot) and grounded (neutral) conductors supplying power to a device. In theory, the amount of current in both conductors should be equal, but opposite in polarity (Fig. 1-17). In this example, a current of 10 amperes flows in both the hot and neutral conductors.

A ground fault occurs when a path to ground, other than the intended path, is established (Fig. 1-18). Assume that a person comes in contact with a defective electrical appliance. If the person is grounded, a current path can be established through the person's body. In the example shown in Figure 1-18, it is assumed that a current of 0.1 amps is flowing through the person. This means that the hot conductor now has a current of 10.1 amps, but the neutral conductor has a current of only 10 amps. The GFCI is designed to detect this current difference to protect personnel by opening the circuit when it detects a current difference of approximately 5 milliamperes (0.005 amps). Section 210.8 of the National Electrical Code lists places where ground fault protection is required in dwellings.

GFCI Devices

Several devices can be used to provide ground fault protection, including the ground fault circuit breaker (Fig. 1-19). The circuit breaker provides ground fault protection for an entire circuit, so any device connected to the circuit is ground fault

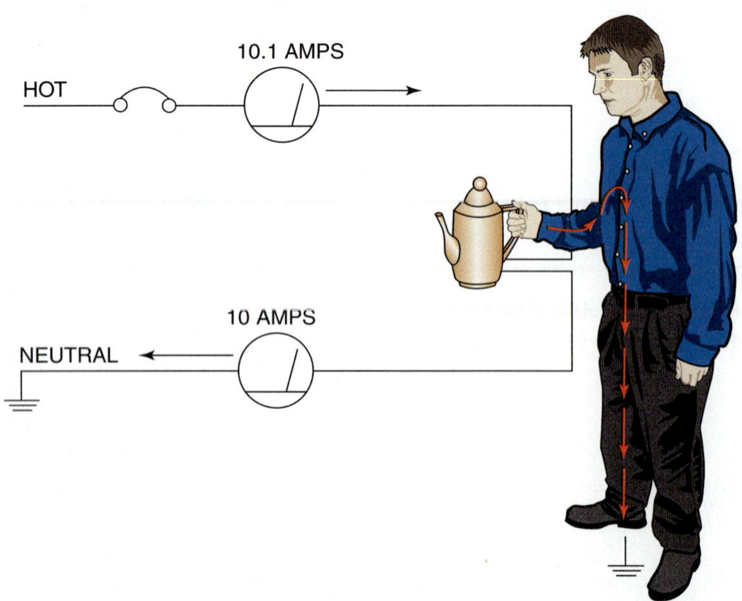

Figure 1-18 A ground fault occurs when a path to ground other than the intended path is established.

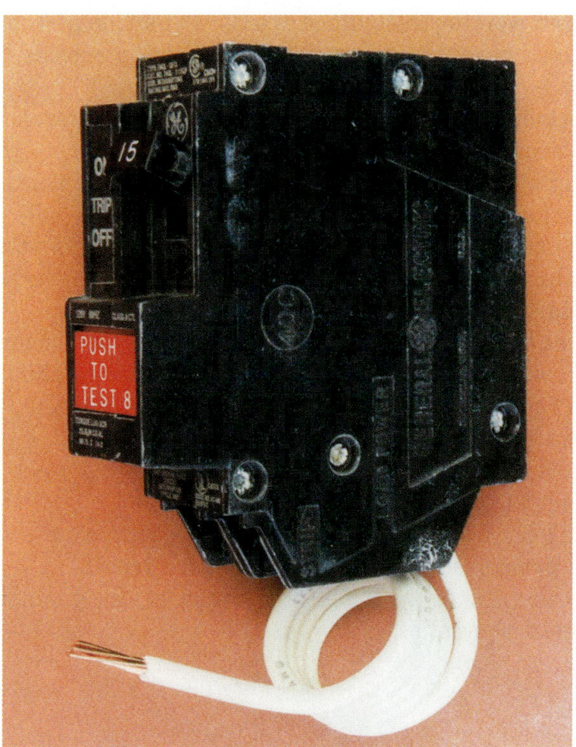

Figure 1-19 Ground fault circuit breaker.

Figure 1-20 Ground fault receptacle.

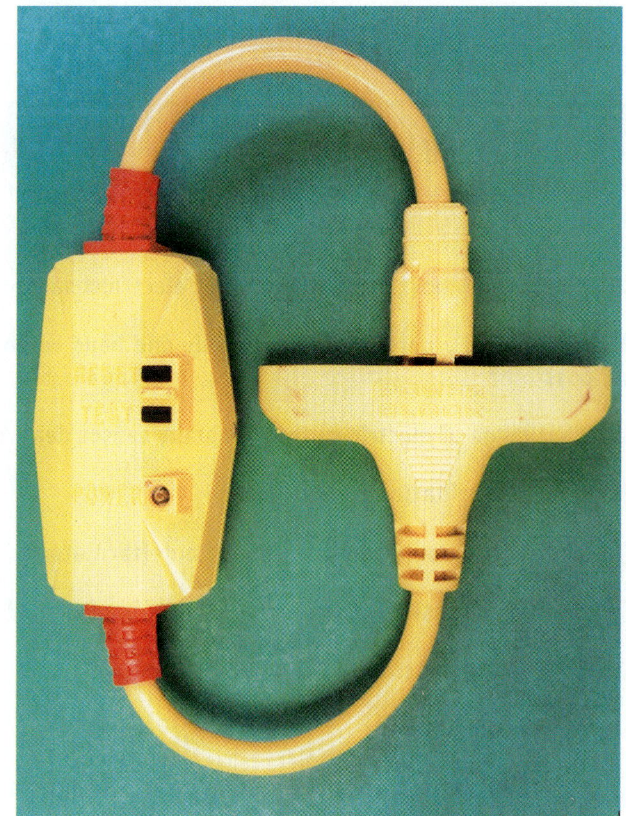
Figure 1-21 Ground fault extension.

protected. A second method of protection, ground fault receptacles (Fig. 1-20), provide protection at the point of attachment. They have some advantages over the GFCI circuit breaker. They can be connected so that they protect only the devices connected to them and do not protect any other outlets on the same circuit, or they can be connected so they provide protection to other outlets. Another advantage is that, since they are located at the point of attachment for the device, there is no stray capacitance loss between the panel box and the equipment being protected. Long wire runs often cause nuisance tripping of GFCI circuit breakers. A third ground fault protective device is the GFCI extension cord (Fig. 1-21). It can be connected into any standard electrical outlet, and any devices connected to it are then ground fault protected.

Grounding

Grounding is one the most important safety considerations in the electrical field. Grounding provides a low resistance path to ground to prevent conductive objects from existing at a high potential. Many electrical appliances are provided with a three-wire cord. The third prong is connected to the case of the appliance and forces the case to exist at ground potential. If an ungrounded conductor comes in contact with the case, the grounding conductors conduct the current directly to ground. The third prong on a plug should never be cut off or defeated. Grounding requirements are far too numerous to list in this chapter, but Section 250 of the National Electrical Code (NEC) covers the requirements for the grounding of electrical systems.

Summary

- Never work on an **energized circuit** if the power can be **disconnected.**
- The most important rule of safety is to think.
- Avoid horseplay.
- Do not work alone.
- Work with one hand when possible.
- Learn first aid and CPR.
- A current of 100 to 200 milliamperes passing through the heart generally causes death.
- The mission of OSHA is to ensure safe and healthy work-places in the United States.
- Avoid using alcohol and drugs in the work place.
- Don't walk close to trenches unless it is necessary.
- Don't jump over trenches if it is possible to walk around them.
- Place barricades around trenches.
- Use ladders to enter and exit trenches.
- When working in confined spaces, an outside person should keep in constant contact with people inside the space.
- Lock out and tag out procedures are used to prevent someone from energizing a circuit by mistake.
- Scaffolds generally provide the safest elevated working platforms.
- The bottom of a straight ladder should be placed at a distance from the wall that is equal to one fourth the height of the ladder.
- Fires can be divided into four classes: class A is common items such as wood and paper; class B is grease, liquids, and gases; class C is energized electrical equipment; and class D is metals.
- Ground fault circuit interrupters are used to protect people from electrical shock.
- GFCI protectors open the circuit when approximately 5 milliamperes of ground fault current is sensed.
- Section 250 of the NEC lists requirements for grounding electrical systems.

Review Questions

1. **What is the most important rule of electrical safety?**
2. **Why should a person work with only one hand when possible?**
3. **What range of electric current generally causes death?**
4. **What is fibrillation of the heart?**
5. **What is the operating principle of a defibrillator?**
6. **Who is responsible for enforcing OSHA regulations?**
7. **What is the mission of OSHA?**
8. **What is an MSDS?**
9. **A padlock is used to lock out a piece of equipment. Who should have the key?**
10. **A ladder is used to reach the top of a building 16 feet tall. What distance should the bottom of the ladder be placed from the side of the building?**
11. **What is a ground fault?**
12. **What is the approximate current at which a ground fault detector will open the circuit?**
13. **Name three devices used to provide ground fault protection.**
14. **What type of fire is class B?**
15. **What section of the National Electrical Code covers grounding?**

HEAVY DUTY CLEAR LO-VOC PVC CEMENT

SECTION 1	IDENTITY OF MATERIAL
Trade Name	OATEY HEAVY DUTY CLEAR LO-VOC PVC CEMENT
Product Numbers	31850, 31851, 31853, 31854
Formula	PVC Resin in Solvent Solution
Synonyms	PVC Plastic Pipe Cement
Firm Name & Mailing Address	OATEY CO., 4700 West 160th Street, P.O. Box 35906 Cleveland, Ohio 44135, U.S.A. http://www.oatey.com
Oatey Phone Number	1-216-267-7100
Emergency Phone Numbers	For Emergency First Aid call 1-303-623-5716 COLLECT. For chemical transportation emergencies ONLY, call Chemtrec at 1-800-424-9300
Prepared By	Charles N. Bush, Ph.D.

SECTION 2	HAZARDOUS INGREDIENTS		
INGREDIENTS	***%***	***CAS NUMBER***	***SEC 313***
Acetone	0–5%	67-64-1	No
Amorphous Fumed Silica (Non-Hazardous)	1–3%	112945-52-5	No
Proprietary (Non-Hazardous)	5–15%	N/A	No
PVC Resin (Non-Hazardous)	10–16%	9002-86-2	No
Cyclohexanone	5–15%	108-94-1	No
Tetrahydrofuran (See SECTION 11)	30–50%	109-99-9	No
Methyl Ethyl Ketone	20–35%	78-93-3	Yes

SECTION 3	KNOWN HAZARDS UNDER U.S. 29 CFR 1910.1200				
HAZARDS	***YES***	***NO***	***HAZARDS***	***YES***	***NO***
Combustible Liquid		x	Skin Hazard	x	
Flammable Liquid	x		Eye Hazard	x	
Pyrophoric Material		x	Toxic Agent	x	
Explosive Material		x	Highly Toxic Agent		x
Unstable Material		x	Sensitizer		x
Water Reactive Material		x	Kidney Toxin	x	
Oxidizer		x	Reproductive Toxin	x	
Organic Peroxide		x	Blood Toxin		x
Corrosive Material		x	Nervous System Toxin	x	
Compressed Gas		x	Lung Toxin	x	
Irritant	x		Liver Toxin	x	
Carcinogen NTP/IARC/OSHA (see SECTION 11)		x			

SECTION 4	EMERGENCY AND FIRST AID PROCEDURES – CALL 1-303-623-5716 COLLECT
Skin	If irritation arises, wash thoroughly with soap and water. Seek medical attention if irritation persists. Remove dried cement with Oatey Plumber's Hand Cleaner or baby oil.
Eyes	If material gets into eyes or if fumes cause irritation, immediately flush eyes with water for 15 minutes. If irritation persists, seek medical attention.
Inhalation	Move to fresh air. If breathing is difficult, give oxygen. If not breathing, give artificial respiration. Keep victim quiet and warm. Call a poison control center or physician immediately. If respiratory irritation occurs and does not go away, seek medical attention.
Ingestion	**DO NOT INDUCE VOMITING.** This product may be aspirated into the lungs and cause chemical pneumonitis, a potentially fatal condition. Drink water and call a poison control center or physician immediately. Avoid alcoholic beverages. Never give anything by mouth to an unconscious person.

(continued)

Courtesy of Oatey Co.

HEAVY DUTY CLEAR LO-VOC PVC CEMENT *(continued)*

SECTION 5	FIRE FIGHTING MEASURES
Precautions	Do not use or store near heat, sparks, or flames. Do not smoke when using. Vapors may accumulate in low places and may cause flash fires.
Special Fire Fighting Procedure	For Small Fires: Use dry chemical, C02, water or foam extinguisher. For Large Fires: Evacuate area and call Fire Department immediately.

SECTION 6	ACCIDENTAL RELEASE MEASURES
Spill or Leak Procedures	Remove all sources of ignition and ventilate area. Stop leak if it can be done without risk. Personnel cleaning up the spill should wear appropriate personal protective equipment, including respirators if vapor concentrations are high. Soak up spill with absorbent material such as sand, earth or other non-combusting material. Put absorbent material in covered, labeled metal containers. Contaminated absorbent material may pose the same hazards as the spilled product. See Section 13 for disposal information.

SECTION 7	HANDLING AND STORAGE
Precautions	**HANDLING & STORAGE:** Keep away from heat, sparks and flames; store in cool, dry place. **OTHER:** Containers, even empties will retain residue and flammable vapors.

SECTION 8	EXPOSURE CONTROLS/PERSONAL PROTECTION
Protective Equipment Types	**EYES:** Safety glasses with side shields. **RESPIRATORY:** NIOSH-approved canister respirator in absence of adequate ventilation. **GLOVES:** Rubber gloves are suitable for normal use of the product. For long exposures to pure solvents chemical resistant gloves may be required. OTHER: Eye wash and safety shower should be available.
Ventilation	**LOCAL EXHAUST:** Open doors & windows. Exhaust ventilation capable of maintaining emissions at the point of use below PEL. If used in enclosed area, use exhaust fans. Exhaust fans should be explosion-proof or set up in a way that flammable concentrations of solvent vapors are not exposed to electrical fixtures or hot surfaces.

SECTION 9	PHYSICAL AND CHEMICAL PROPERTIES			
NFPA Hazard Signal	Health 2	Stability 1	Flammability 3	Special None
HMIS Hazard Signal	Health 3	Stability 1	Flammability 4	Special None
Boiling Point	151 Degrees F / 66 C			
Melting Point	N/A			
Vapor Pressure	145 mmHg @ 20 Degrees C			
Vapor Density (Air = 1)	2.5			
Volatile Components	70–80%			
Solubility In Water	Negligible			
PH	N/A			
Specific Gravity	0.95 +/− 0.015			
Evaporation Rate	(BUAC = 1) = 5.5 − 8.0			
Appearance	Clear Liquid			
Odor	Ether-Like			
Will Dissolve In	Tetrahydrofuran			
Material Is	Liquid			

(continued)

HEAVY DUTY CLEAR LO-VOC PVC CEMENT *(continued)*

SECTION 10	STABILITY AND REACTIVITY
Flammability	LEL = 1.8% Volume, UEL = 11.8% Volume
Flashpoint And Method Used	0–5 Degrees F./PMCC
Stability	Stable. CONDITIONS TO AVOID: Heat, sparks and open flame. HAZARDOUS DECOMPOSITION PRODUCTS: Carbon monoxide/carbon dioxide/hydrogen chloride/smoke.
Hazardous Polymerization	Will Not Occur. CONDITIONS TO AVOID: None
Incompatibility/ Materials To Avoid	Acids, oxidizing materials, alkalis, chlorinated inorganics (potassium, calcium and sodium hypochlorite).

SECTION 11	TOXICOLOGICAL INFORMATION			
Entry Route	Inhale – Yes	Ingest – Yes	Skin – Yes	Eye – Yes
Target Organs	Eye, Skin, Kidney, Lung, Liver, Central Nervous System			
Inhalation	**Avoid breathing vapors.** High vapor concentrations may cause irritation of mucous membranes, nose & throat, headache, dizziness, nausea, numbness of the extremities and narcosis in high concentrations. Some solvents at concentrations much higher than the OSHA PEL's have caused CNS depression & liver damage in animals and/or retardation of fetal development in rats. Vapor concentrations above OSHA PEL's will generally not occur unless the product is being used in an enclosed area with no ventilation. Exposure to vapors could aggravate pre-existing respiratory conditions. There have been reports in the medical literature of epileptic seizures after exposure to solvents. These are rare and links to the solvents in this product are not proven. See Section 8.			
Tetrahydro-furan Warning	The National Toxicology Program has reported that exposure of mice and rats to Tetrahydrofuran (THF) vapor levels up to 1800 ppm 6 hr/day, 5 days/week for their lifetime caused an increased incidence of kidney tumors in male rats and liver tumors in female mice. The significance of these findings for human health are unclear at this time, and may be related to "species specific" effects. Elevated incidences of tumors in humans have not been reported for THF. THF is not listed as a carcinogen by NTP, IARC, or OSHA. One THF vendor has recommended a reduction in the "acceptable exposure limit" from 200 ppm to 25 ppm, 8 and 12 hour time weighted average.			
Skin	Chronic contact may lead to irritation & dermatitis. Chronic exposure to vapors of high concentration may cause dermatitis. May possibly be absorbed through the skin.			
Eye	Vapors or direct contact may cause irritation.			
Ingestion	See Section 4.			

SECTION 12	ECOLOGICAL INFORMATION
VOC Information	This product emits VOC's (volatile organic compounds) in its use. Make sure that use of this product complies with local VOC emission regulations, where they exist.
VOC Level	510 g/l per SCAQMD Test Method 316A.

SECTION 13	DISPOSAL INFORMATION
Waste Disposal	Product and absorbent material containing product is considered to be hazardous waste. Dispose of according to local, state, and federal regulations. Residual material after solvent evaporation is non-hazardous. Empty cans are also considered non-hazardous.

SECTION 14	TRANSPORTATION INFORMATION
DOT Proper Shipping Name	CONSUMER COMMODITY ORM-D; For Gallons: Adhesives, 3, UN1133, PG II
DOT Hazard Class	Class 3 Flammable Liquid
Shipping ID Number	UN 1133 (Gallons Only)
RCRA Hazardous Waste Number	U002, U057, U159, U213
EPA Hazardous Waste ID Number	D001, D035, F003, F005
EPA Hazard Waste Class	Ignitable Waste. Toxic Waste (Methyl Ethyl Ketone content)

(continued)

HEAVY DUTY CLEAR LO-VOC PVC CEMENT *(continued)*

SECTION 15		REGULATIONS		
CHEMICAL	***TLV (TWA)***	***PEL***	***STEL***	***HAZARD ACTION LEVEL***
Methyl Ethyl Ketone	200 ppm	200 ppm	300 ppm	N/A
	590 mg/cu m	590 mg/cu m	885 mg/cu m	
Tetrahydrofuran	200 ppm	200 ppm	250 ppm	N/A
	590 mg/cu m	590 mg/cu m	735 mg/cu m	
Cyclohexanone	25 ppm	50 ppm	N/A	N/A
	100 mg/cu m	200 mg/cu m		
Amorphous Fumed Silica	10 mg/cu m	20 mppcf	N/A	N/A
Acetone	500 ppm	1000 ppm	750 ppm	N/A
	1200 mg/cu m	2400 mg/cu m	1800 mg/cu m	
California Proposition 65	This product contains trace amounts of chemicals known to the State of California to cause cancer. Under normal use conditions, exposures to these chemicals at levels above the State of California "No Significant Risk Level" (NSRL) are unlikely. Oatey strongly encourages the use of proper personal protective equipment (PPE) and ventilation guidelines noted in Section 8 to minimize exposure to these chemicals.			

SECTION 16	DISCLAIMER

The information herein has been compiled from sources believed to be reliable, up-to-date, and is accurate to the best of our knowledge. However, Oatey cannot give any guarantees regarding information from other sources, and expressly does not make warranties, nor assumes any liability for its use.

Chapter 2 Basic Electrical Theory

Electricity is the driving force that provides most of the power for the industrialized world. It is used to light homes, cook meals, heat and cool buildings, drive motors, and supply the ignition for most automobiles. The technician who understands electricity can seek employment in almost any part of the world.

OBJECTIVES

After completing this unit, the student should be able to:

- **list the three major parts of an atom.**
- **state the law of charges.**
- **discuss the law of centrifugal force.**
- **discuss the differences between conductors and insulators.**

Glossary of Basic Electrical Theory Terms

alternating current current that reverses its direction of flow at periodic intervals

atom the smallest part of an element, atoms are composed of three primary sub atomic particles; the proton, the neutron, and the electron

atomic number the number of protons located in the nucleus of an atom

attraction generally the force that exists between objects that contain opposite charges

bidirectional a device that can conduct current in either direction, or current that flows in two directions

centrifugal force the force that causes a spinning object to pull away from its axis point

conductors materials that conduct electricity easily

direct current (DC) current that flows in only one direction

electron a sub atomic particle that orbits the nucleus of an atom and contains a negative charge

electron orbit the space in which an electron circles the nucleus of an atom

element a material, the smallest part of which would be an atom

insulators materials that hinder the flow of electricity

matter any material composed of atoms

molecules a combination of atoms that form the smallest part of a compound

negative a charge formed by an excess of electrons

neutron one of the principal parts of an atom. The neutron has no charge and is part of the nucleus.

nucleus the center or core of an object

positive a charge formed by a lack of electrons

proton a sub-atomic particle located in the nucleus of an atom; the proton has a positive charge

repulsion generally the force that exists between two objects that contain like electrical charges

unidirectional something that moves in only one direction

valence electrons electrons located in the outermost orbit or shell of an atom

Electrical sources are divided into two basic types, **direct current** (DC) and **alternating current** (AC). Direct current is **unidirectional,** which means that it flows in only one direction. The first part of this text will be devoted mainly to the study of direct current. Alternating current is **bidirectional,** which means that it reverses its direction of flow at regular intervals. The latter part of this text is devoted mainly to the study of alternating current.

Early History of Electricity

Although the practical use of electricity has become common only within the last hundred years, it has been known as a force for much longer. The Greeks were the first to discover electricity, about 2500 years ago. They noticed that when amber was rubbed with other materials it became charged with an unknown force that had the power to attract objects such as dried leaves, feathers, bits of cloth, or other lightweight materials. The Greeks called amber *elektron*. The word *electric* was derived from it, meaning "to be like amber" or to have the ability to attract other objects.

This mysterious force remained little more than a curious phenomenon until about 2000 years later, when other people began to conduct experiments. In the early 1600s, William Gilbert discovered that amber was not the only material that could be charged to attract other objects. He called materials that could be charged *electriks* and materials that could not be charged *nonelektriks*.

About 300 years ago, a few men began to study the behavior of various charged objects. In 1733, a Frenchman named Charles DuFay found that a piece of charged glass would repel some charged objects and attract others. These men soon learned that the force of **repulsion** was just as important as the force of **attraction.** From these experiments, two lists were developed (Fig. 2-1). It was determined that any material in list A would attract any material in list B, and that all materials in list A would repel each other and all materials in list B would repel each other (Fig. 2-2). Various names were suggested for the materials in lists A and B. Any opposite-sounding names could have been chosen, such as east and west, north and south, or male and female. Benjamin Franklin named the materials in list A **positive** and the materials in list B **negative.** These names are still used today. The first item in each list was used as a standard for determining if a charged object was positive or negative. Any object repelled by a piece of glass rubbed on silk would have a positive charge and any item repelled by a hard rubber rod rubbed on wool would have a negative charge.

LIST A	LIST B
GLASS (RUBBED ON SILK)	HARD RUBBER (RUBBED ON WOOL)
GLASS (RUBBED ON WOOL OR COTTON)	BLOCK OF SULFUR (RUBBED ON WOOL OR FUR)
MICA (RUBBED ON CLOTH)	MOST KINDS OF RUBBER (RUBBED ON CLOTH)
ASBESTOS (RUBBED ON CLOTH OR PAPER)	SEALING WAX (RUBBED ON SILK, WOOL, OR FUR)
STICK OF SEALING WAX RUBBED ON WOOL)	GLASS OR MICA (RUBBED ON DRY WOOL)
	AMBER (RUBBED ON CLOTH)

Figure 2-1 List of charged materials.

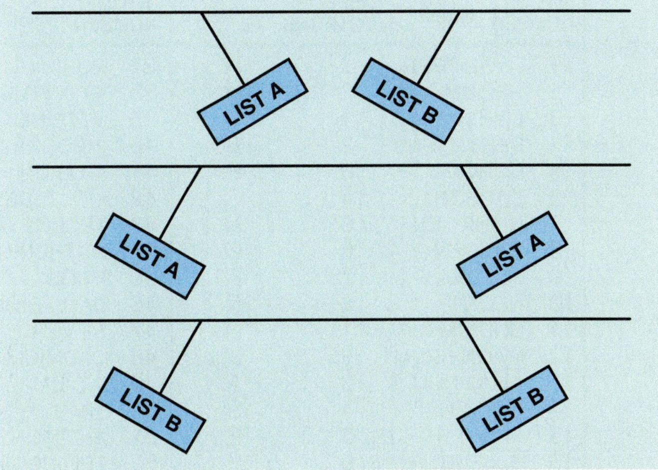

Figure 2-2 Unlike charges attract and like charges repel.

Atoms

To understand electricity, it is necessary to start with the study of atoms. The **atom** is the basic building block of the universe. All **matter** is made from a combination of atoms. Matter is any substance that has mass and occupies space. It can exist in any of three states: solid, liquid, or gas. Water, for example, can exist as a solid in the form of ice, as a liquid, or as a gas in the form of steam (Fig. 2-3). An **element** is a substance that cannot be chemically divided into a simpler substance. A table listing both natural and artificial elements is shown in Figure 2-4. An atom is the smallest part of an element. The three principal parts of an atom are the **electron, neutron,** and **proton.** Figure 2-5 illustrates those parts of the atom.

Notice that the proton has a positive charge, the electron a negative charge, and the neutron no charge. The neutron and proton combine to form the **nucleus** of the atom. Since the neutron has no charge, the nucleus will have a net positive charge. The number of protons in the nucleus determines

Figure 2-3 Water can exist in three states, depending on temperature and pressure.

what kind of element an atom is. Oxygen, for example, contains 8 protons in its nucleus, and gold contains 79. The **atomic number** of an element is the same as the number of protons in the nucleus. The lines of force produced by the positive charge of the proton extend outward in all directions (Fig. 2-6). The nucleus may or may not contain as many neutrons as protons. For example, the nucleus of an atom of helium contains two protons and two neutrons, while an atom of copper contains 29 protons and 35 neutrons (Fig. 2-7).

The electron orbits the outside of the nucleus. Notice in Figure 2-5 that the electron is shown to be larger than the proton. Actually, an electron is about three times as large as a proton. The estimated size of a proton is 0.07 trillionth of an inch in diameter, and the estimated size of an electron is 0.22 trillionth of an inch in diameter. Although the electron is larger in size, the proton weighs about 1840 times more. Imagine comparing a soap bubble with a piece of buckshot. Compared with the electron, the proton is a massive particle. Since the electron exhibits a negative charge, the lines of force come in from all directions (Fig. 2-8).

ATOMIC NUMBER	NAME	VALENCE ELECTRONS	SYMBOL	ATOMIC NUMBER	NAME	VALENCE ELECTRONS	SYMBOL	ATOMIC NUMBER	NAME	VALENCE ELECTRONS	SYMBOL
1	HYDROGEN	1	H	37	RUBIDIUM	1	Rb	73	TANTALUM	2	Ta
2	HELIUM	2	He	38	STRONTIUM	2	Sr	74	TUNGSTEN	2	W
3	LITHIUM	1	Li	39	YTTRIUM	2	Y	75	RHENIUM	2	Re
4	BERYLLIUM	2	Be	40	ZIRCONIUM	2	Zr	76	OSMIUM	2	Os
5	BORON	3	B	41	NIOBIUM	1	Nb	77	IRIDIUM	2	Ir
6	CARBON	4	C	42	MOLYBDENUM	1	Mo	78	PLATINUM	1	Pt
7	NITROGEN	5	N	43	TECHNETIUM	2	Tc	79	GOLD	1	Au
8	OXYGEN	6	O	44	RUTHENIUM	1	Ru	80	MERCURY	2	Hg
9	FLUORINE	7	F	45	RHODIUM	1	Rh	81	THALLIUM	3	Tl
10	NEON	8	Ne	46	PALLADIUM	–	Pd	82	LEAD	4	Pb
11	SODIUM	1	Na	47	SILVER	1	Ag	83	BISMUTH	5	Bl
12	MAGNESIUM	2	Ma	48	CADMIUM	2	Cd	84	POLONIUM	6	Po
13	ALUMINUM	3	Al	49	INDIUM	3	In	85	ASTATINE	7	At
14	SILICON	4	Si	50	TIN	4	Sn	86	RADON	8	Rd
15	PHOSPHORUS	5	P	51	ANTIMONY	5	Sb	87	FRANCIUM	1	Fr
16	SULFUR	6	S	52	TELLURIUM	6	Te	88	RADIUM	2	Ra
17	CHLORINE	7	Cl	53	IODINE	7	I	89	ACTINIUM	2	Ac
18	ARGON	8	A	54	XENON	8	Xe	90	THORIUM	2	Th
19	POTASSIUM	1	K	55	CESIUM	1	Cs	91	PROTACTINIUM	2	Pa
20	CALCIUM	2	Ca	56	BARIUM	2	Ba	92	URANIUM	2	U
21	SCANDIUM	2	Sc	57	LANTHANUM	2	La				
22	TITANIUM	2	Ti	58	CERIUM	2	Ce		ARTIFICAL ELEMENTS		
23	VANADIUM	2	V	59	PRASEODYMIUM	2	Pr				
24	CHROMIUM	1	Cr	60	NEODYMIUM	2	Nd	93	NEPTUNIUM	2	Np
25	MANGANESE	2	Mn	61	PROMETHIUM	2	Pm	94	PLUTONIUM	2	Pu
26	IRON	2	Fe	62	SAMARIUM	2	Sm	95	AMERICIUM	2	Am
27	COBALT	2	Co	63	EUROPIUM	2	Eu	96	CURIUM	2	Cm
28	NICKEL	2	Ni	64	GADOLINIUM	2	Gd	97	BERKELIUM	2	Bk
29	COPPER	1	Cu	65	TERBIUM	2	Tb	98	CALIFORNIUM	2	Cf
30	ZINC	2	Zn	66	DYSPROSIUM	2	Dy	99	EINSTEINIUM	2	E
31	GALLIUM	3	Ga	67	HOLMIUM	2	Ho	100	FERMIUM	2	Fm
32	GERMANIUM	4	Ge	68	ERBIUM	2	Er	101	MENDELEVIUM	2	Mv
33	ARSENIC	5	As	69	THULIUM	2	Tm	102	NOBELIUM	2	No
34	SELENIUM	6	Se	70	YTTERBIUM	2	Yb	103	LAWRENCIUM	2	Lw
35	BROMINE	7	Br	71	LUTETIUM	2	Lu				
36	KRYPTON	8	Kr	72	HAFNIUM	2	Hf				

Figure 2-4 Table of elements.

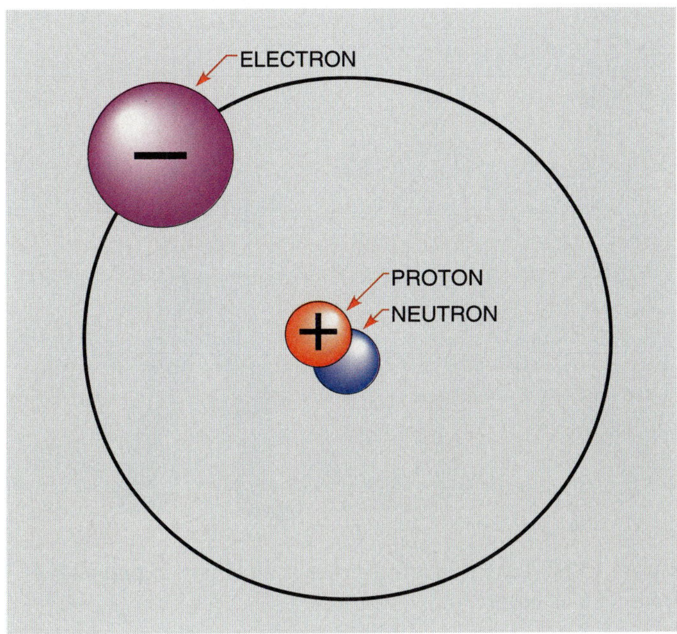

Figure 2-5 **The three principal parts of an atom.**

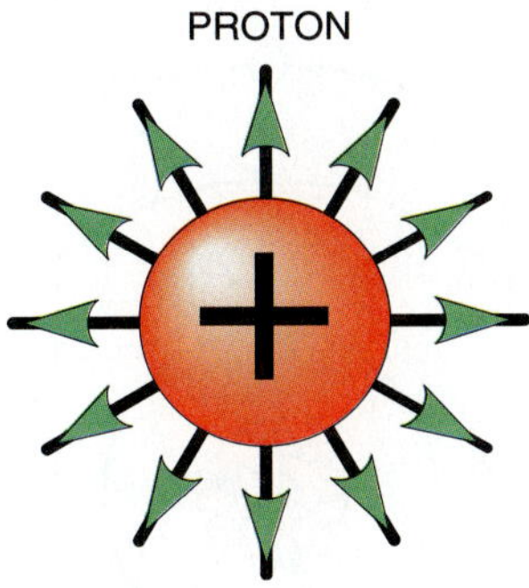

Figure 2-6 **The lines of force extend outward.**

The Law of Charges

To understand atoms, it is necessary first to understand two basic laws of physics. One of these is the law of charges, which states that **opposite charges attract and like charges repel.** In Figure 2-9, which illustrates this principle, charged balls are suspended from strings. Notice that the two balls that contain opposite charges are attracted to each other. The two positively charged balls and the two negatively charged balls repel each other. This is because lines of force can never cross each other. The outward-going lines of force of a positively charged object combine with the inward-going lines of force of a negatively charged object (Fig. 2-10). This combining produces an attraction between the two objects. If two objects with like charges come close to each other, the lines of force repel (Fig. 2-11). Since the nucleus has a net positive charge and the electron has a negative charge, the electron is attracted to the nucleus.

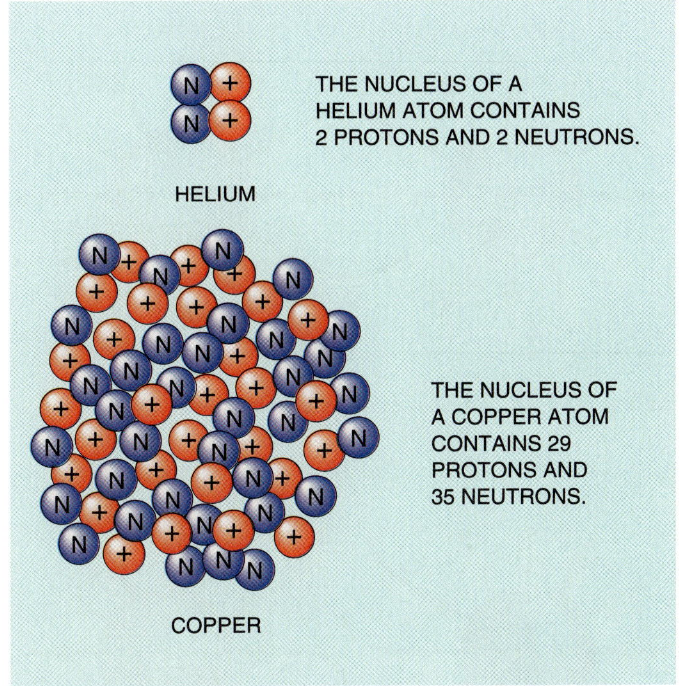

Figure 2-7 **The nucleus may or may not contain the same number of protons and neutrons.**

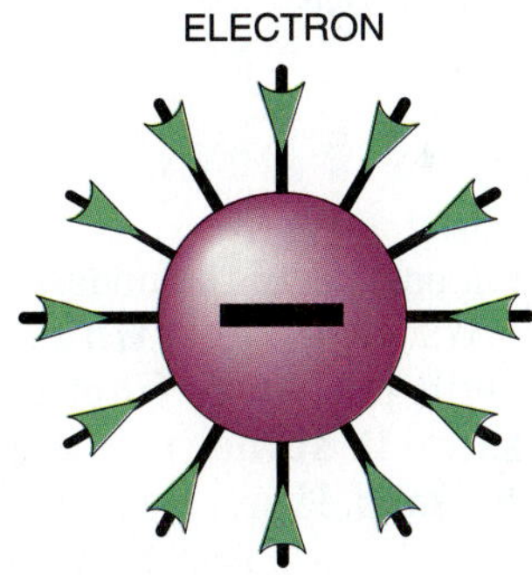

Figure 2-8 **The lines of force come inward.**

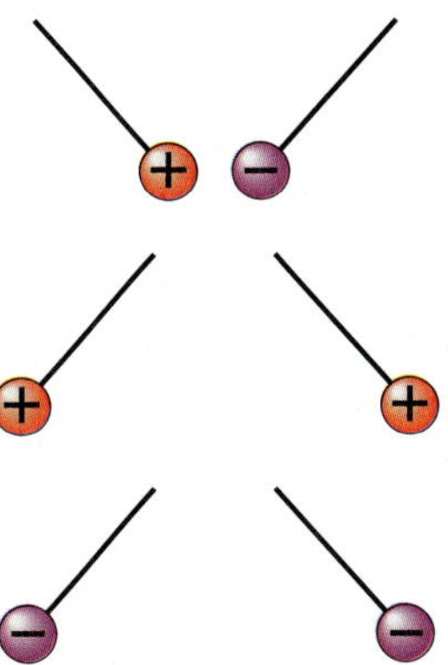

Figure 2-9 **Unlike charges attract and like charges repel.**

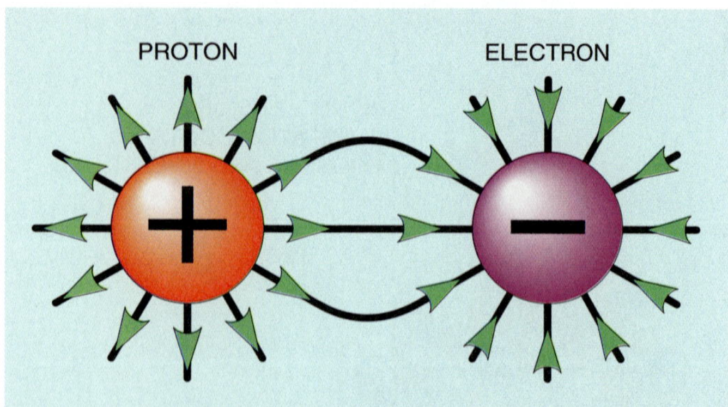

Figure 2-10 **Unlike charges attract each other.**

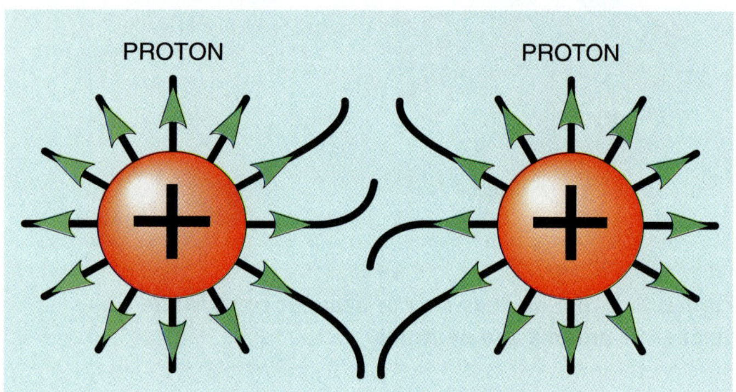

Figure 2-11 **Like charges repel each other.**

Centrifugal Force

The second law important to the understanding of atoms is **the law of centrifugal force, which states that a spinning object will pull away from its center point and that the faster it spins, the greater the centrifugal force becomes.** Figure 2-12 shows an example of this principle. If you tie an object to a string and spin it around, it will try to pull away from you. The faster the object spins, the greater the force that tries to pull the object away. Centrifugal force prevents the electron from falling into the nucleus of the atom. The faster an electron spins, the farther away from the nucleus it will be.

Figure 2-12 **Centrifugal force causes an object to pull away from its axis point.**

Electron Orbits

Each **electron orbit** of an atom contains a set number of electrons (Fig. 2-13). The number of electrons that can be contained in any one orbit, or shell, is found by using the formula ($2N^2$). The letter *N* represents the number of the orbit, or shell. For example, the first orbit can hold no more than two electrons.

$$2 \times (1)^2 \text{ or}$$
$$2 \times 1 = 2$$

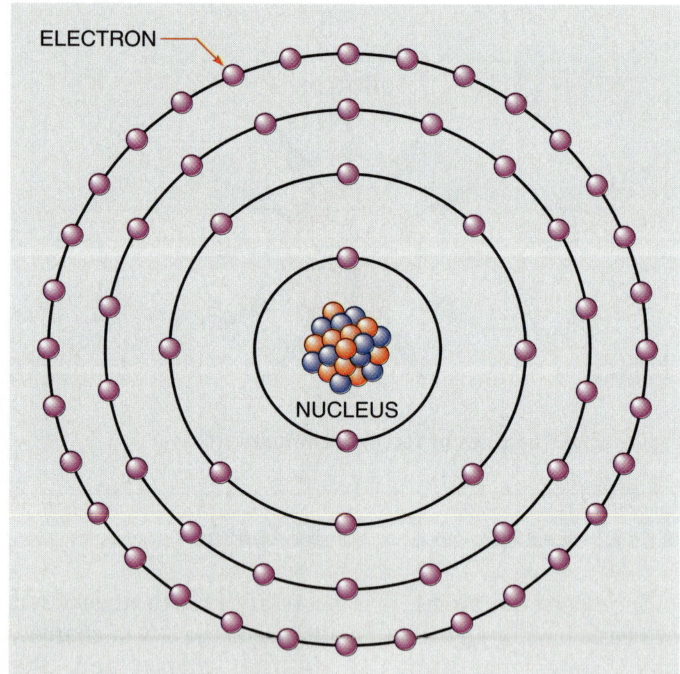

Figure 2-13 **Electron orbits.**

The second orbit can hold no more than eight electrons.

$$2 \times (2)^2 \text{ or}$$
$$2 \times 4 = 8$$

The third orbit can contain no more than 18 electrons.

$$2 \times (3)^2 \text{ or}$$
$$2 \times 9 = 18$$

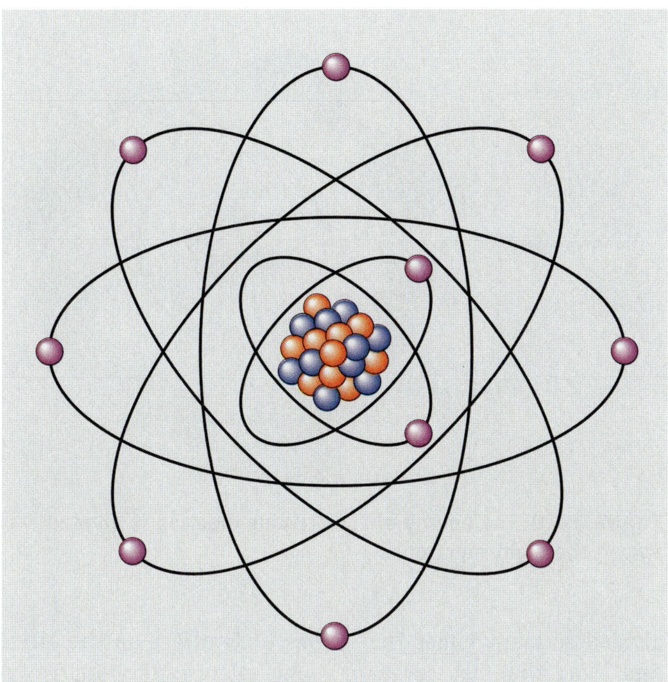

Figure 2-14 **Electrons orbit the nucleus in a circular fashion.**

The fourth and fifth orbits cannot hold more than 32 electrons. Thirty-two is the maximum number of electrons that can be contained in any orbit.

$$2 \times (4)^2 \text{ or}$$
$$2 \times 16 = 32$$

Although atoms are often drawn flat, as illustrated in Figure 2-13, electrons orbit the nucleus in a spherical fashion, as shown in Figure 2-14. Electrons travel at such a high rate of speed that they form a shell around the nucleus. For this reason, electron orbits are often referred to as *shells.*

Valence Electrons

The outer shell of an atom is known as the *valence shell.* Any electrons located in the outer shell of an atom are known as **valence electrons** (Fig. 2-15). The valence shell of an atom cannot hold more than eight electrons. The valence electrons are of primary concern in the study of electricity, because they explain much of electrical theory. A **conductor,** for instance, is generally made from a material that contains one or two valence electrons. Atoms with one or two valence electrons are unstable and can be made to give up these electrons with little effort. Conductors are materials that permit electrons to flow through them easily. When an atom has only one or two valence electrons, these electrons are loosely held by the atom and are easily given up for current flow. Silver, copper, and gold all contain one valence electron and are excellent conductors of electricity. Silver is the best natural conductor of electricity, followed by copper, gold, and aluminum. An atom of copper is shown in Figure 2-16. Although it is known that atoms containing few valence electrons are the best conductors, it is not known why some of these materials are better conductors than others. Copper, gold, platinum, and silver all contain only one

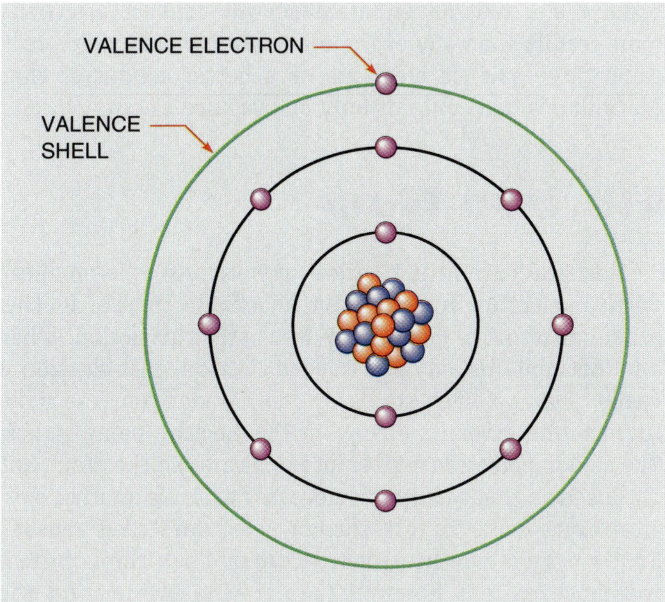

Figure 2-15 **The electrons located in the outer orbit of an atom are valence electrons.**

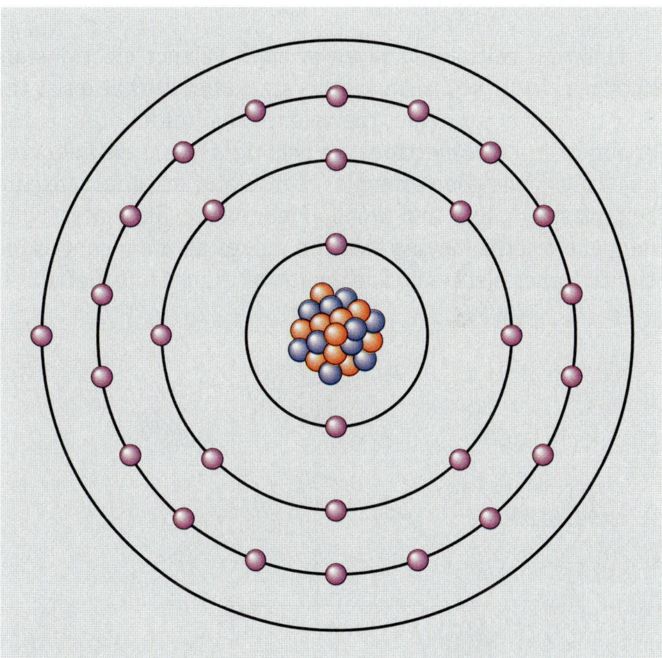

Figure 2-16 **A copper atom contains 29 electrons and has 1 valence electron.**

valence electron. Silver, however, will conduct electricity more readily than any of the others. Aluminum, which contains three valence electrons, is a better conductor than platinum, which contains only one valence electron.

Electron Flow

Electrical current is the flow of electrons. It is produced when an electron from one atom knocks electrons of another atom out of orbit. Figure 2-17 illustrates this action. When an atom contains only one valence electron, that electron is easily given up when struck by another electron. The striking electron gives its energy to the electron being struck. The striking electron settles into orbit around the atom, and the electron that was struck moves off to strike another electron. Some energy is lost when one electron strikes another. That is why a wire heats when current flows through it. If too much current flows through a wire, overheating will damage the wire and possibly become a fire hazard.

Insulators

Materials containing seven or eight valence electrons are known as **insulators.** Insulators are materials that resist the flow of electricity. When the valence shell of an atom is full or almost full, the electrons are held tightly and are not given up easily. Some good examples of insulator materials are rubber, plastic, glass, and wood. Figure 2-18 illustrates what happens when a moving electron strikes an atom containing eight valence electrons. The energy of the moving electron is divided so many times that it has little effect on the atom. Any atom that has seven or eight valence electrons is extremely stable and does not easily give up an electron.

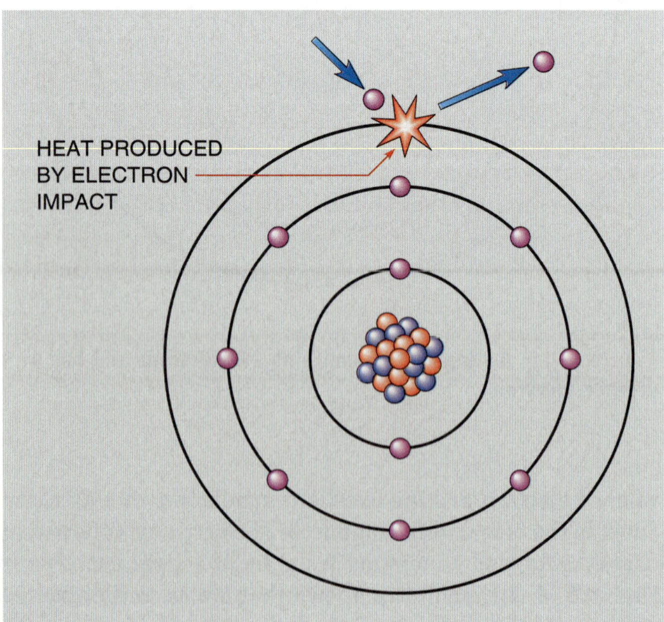

Figure 2-17 An electron of one atom knocks an electron of another atom out of orbit.

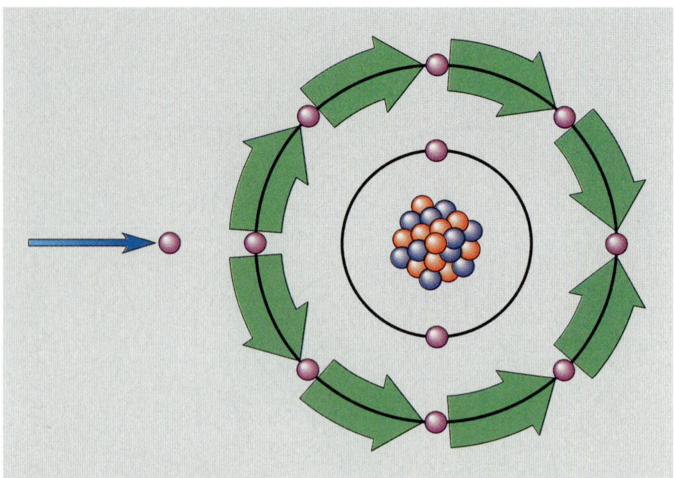

Figure 2-18 The energy of the striking electron is divided among the eight electrons.

Methods of Producing Electricity

In this unit, it has been shown that electricity is a flow of electrons. There are six basic methods for producing electricity:

1. magnetism
2. chemical
3. pressure
4. heat
5. friction
6. light

Of the six methods listed, magnetism is the most commonly used to produce electricity. Electromagnetic induction is the operating principle of all generators and alternators. These principles are covered fully later in this text.

The second most common method of producing electricity is chemical action. The chemical production of electricity involves the movement of entire ions instead of just electrons.

The production of electricity by pressure involves the striking, bending, or twisting of certain crystals. This effect is referred to as the *piezo* electric effect. The word *piezo* is derived from a Greek word meaning pressure.

Producing electricity with heat is called the *Seebeck* effect. The Seebeck effect is the operating principle of thermocouples.

Static charges are probably the best example of producing electricity by friction. A static charge occurs when certain materials are rubbed together and electrons are transferred from one object to the other.

Electricity is produced from light using particles called *photons*. In theory, photons are massless particles of pure energy that can be produced when electrons are forced to change energy levels, or are knocked out of orbit. This is the operating principle of gas-filled lights such as sodium vapor or mercury vapor. Electricity can be produced by photons when they strike a semiconductor material. The energy of the photon is given to an electron, forcing it to move out of orbit. This is the operating principle of *photovoltaic* devices called solar cells.

Summary

- The atom is the smallest part of an element.
- The three basic parts of an atom are the proton, electron, and neutron.
- Protons have a positive charge, electrons a negative charge, and neutrons no charge.
- Valence electrons are located in the outer orbit of an atom.
- Conductors are materials that provide an easy path for electron flow.
- Conductors are made from materials that contain one, two, or three valence electrons.
- Insulators are materials that do not provide an easy path for the flow of electrons.
- Insulators are generally made from materials containing seven or eight valence electrons.
- Six basic methods for producing electricity are magnetism, chemical, light, heat, pressure, and friction.
- A photon is a massless particle of pure energy.
- Photons can be produced when electrons move from one energy level to another.

Review Questions

1. **What are the three subatomic parts of an atom, and what charge does each carry?**
2. **How many times larger is an electron than a proton?**
3. **How many times more does the proton weigh than the electron?**
4. **State the law of charges.**
5. **What force keeps the electron from falling into the nucleus of the atom?**
6. **How many valence electrons are generally contained in materials used for conductors?**
7. **How many valence electrons are generally contained in materials used for insulators?**
8. **What is electricity?**

Chapter 3 Electrical Quantities and Ohm's Law

Electricity has a standard set of values. Before one can work with electricity, one must know these values and how to use them. Because the values of electrical measurement have been standardized, they are understood by everyone who uses them. For instance, carpenters use a standard system for measuring length, such as the inch, foot, meter, or centimeter. Imagine what a house would look like if it was constructed by two carpenters who used different lengths of measure for an inch or foot. The same holds true for people who work with electricity. The standards of measurement must be the same for everyone. Meters should be calibrated to indicate the same quantity of current flow, voltage, or resistance. A volt, an ampere, or an ohm is the same everywhere in the world.

OBJECTIVES

After completing this unit, the student should be able to:

- **define a coulomb.**
- **define an amp.**
- **define a volt.**
- **define an ohm.**
- **define a watt.**
- **compute different electrical values using Ohm's law formulas.**
- **discuss different types of electrical circuits.**
- **select the proper Ohm's law formula from a chart.**

Glossary of Electrical Quantity and Ohm's Law Terms

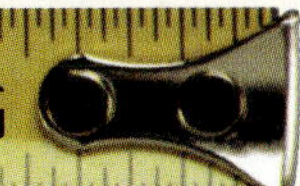

amp a measure of electric current defined as one coulomb per second

British thermal unit (BTU) the amount of heat necessary to change the temperature of one pound of water one degree Fahrenheit

complete path an electric circuit that exists from the more negative to the more positive voltage

conventional current flow theory an older theory stating that electric current flows from the more positive source to the more negative source

coulomb a quantity measure of electrons that equals $6.25^{\times 10^{18}}$ electrons

electromotive force (EMF) the force known as voltage that pushes the electrons through an electric circuit

electron theory the theory stating that electricity flows from the more negative power source to the more positive power source

grounding conductor generally the bare or green wire used to establish a low resistance path to ground

horsepower a common measure of power in which the base is established at 550 pound feet per second; in electrical quantities, 746 watts equals one horsepower

impedance the total current limiting effect in an electric circuit; impedance is measured in ohms.

joule a metric measure of power similar to the English measure watt

neutral conductor the part of an electric circuit that is generally grounded

ohm (Ω) the unit of measure used to describe the resistance to the flow of electric current

Ohm's law a set of mathematical formulas developed by the German scientist Georg S. Ohm

power the ability to do work; in an electric circuit, it is generally measured in watts

resistance the element of an electric circuit that can limit the flow of current by producing heat

volt a measure of EMF often described as electrical pressure

watt a power measurement that describes the amount of electrical energy converted to some other form

The Coulomb

A **coulomb** is a quantity measurement for electrons. One coulomb contains 6.25×10^{18}, or 6,250,000,000,000,000,000, electrons. To better understand the number of electrons contained in a coulomb, think of comparing one second with 200 billion years. Since the coulomb is a quantity measurement, it is similar to a quart, gallon, or liter. It takes a certain amount of liquid to equal a liter, just as it takes a certain amount of electrons to equal a coulomb.

The coulomb is named for a French scientist who lived in the 1700s named Charles Augustin de Coulomb. Coulomb experimented with electrostatic charges and developed a law dealing with the attraction and repulsion of these forces. The law, known as **Coulomb's law of electrostatic charges, states that the force of electrostatic attraction or repulsion is directly proportional to the product of the two charges and inversely proportional to the square of the distance between them.** The number of electrons contained in the coulomb was determined by the average charge of an electron.

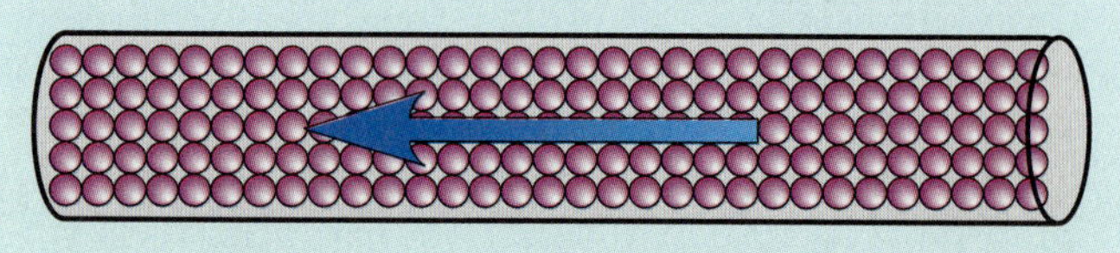

Figure 3-1 One ampere equals one coulomb per second.

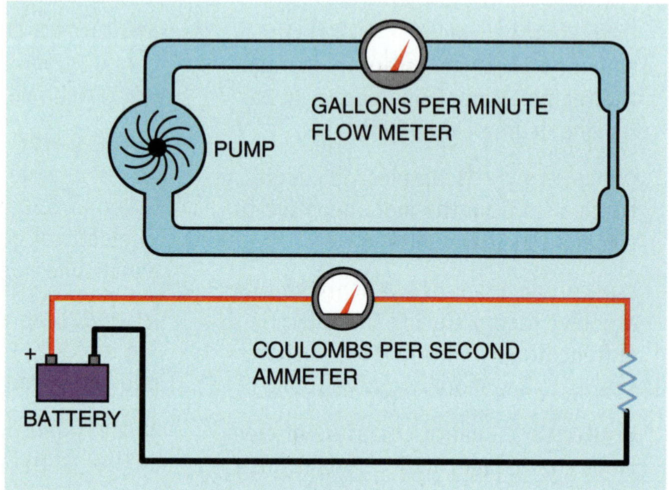

Figure 3-2 Current in an electrical circuit can be compared to flow rate in a water system.

The Amp

The **amp,** or ampere, is named for André Ampère, a scientist who lived from the late 1700s to the early 1800s. Ampère is most famous for his work dealing with electromagnetism, which will be discussed in a later chapter. The amp (A) is defined as one coulomb per second. Notice that the definition of an amp involves a quantity measurement, the coulomb, and a time measurement, the second. One amp of current flows through a wire when one coulomb flows past a point in one second (Fig. 3-1). The ampere is a measurement of the amount of electricity that is flowing through a circuit. In a water system, it would be comparable to gallons per minute or gallons per second (Fig. 3-2). The letter *I,* which stands for intensity of current, and the letter *A,* which stands for amp, are both used to represent current flow in algebraic formulas. This text will use the letter *I* in formulas to represent current.

$$F = \frac{Q_1 \times Q_3}{KD^2}$$

F = Force in dynes

Q = Strength of charge in electrostatic units

D = Separation in cm

K = Dielectric constant

The Electron Theory

There are actually two theories concerning current flow. The theory known as the **electron theory** states that since electrons are negative particles, current flows from the most negative point in the circuit to the most positive. The electron theory is the more widely accepted as being correct and is used throughout this text.

The Conventional Current Theory

The second theory, known as the **conventional current flow theory,** is older than the electron theory and states that current flows from the most positive point to the most negative. Although it has been established almost to a certainty that the electron theory is correct, the conventional current theory is still widely used for several reasons. Most electronic circuits use the negative terminal as ground or common. When the negative terminal is used as ground, the positive terminal is considered to be above ground, or hot. It is easier for most people to think of something flowing down rather than up, or from a point above ground to ground. An automobile electrical system is a good example of this type of circuit. Most people consider the positive battery terminal to be the hot terminal.

Many people who work in the electronics field prefer the conventional current flow theory because all the arrows on the semiconductor symbols point in the direction of conventional current flow. If the electron flow theory is used, it must be assumed that current flows against the arrow (Fig. 3-3). Another reason that many people prefer using the conventional current flow theory is that most electronic schematics are drawn in such a manner that it assumes current flows from the more positive to the more negative source. In Figure 3-4 the positive voltage point is shown at the top of the schematic and the negative (ground) is shown at the bottom. When tracing the flow of current through a circuit, most people find it easier to go from top to bottom than from bottom to top.

The electron theory states that current flows in the direction that electrons move; electrons always move from negative to positive.

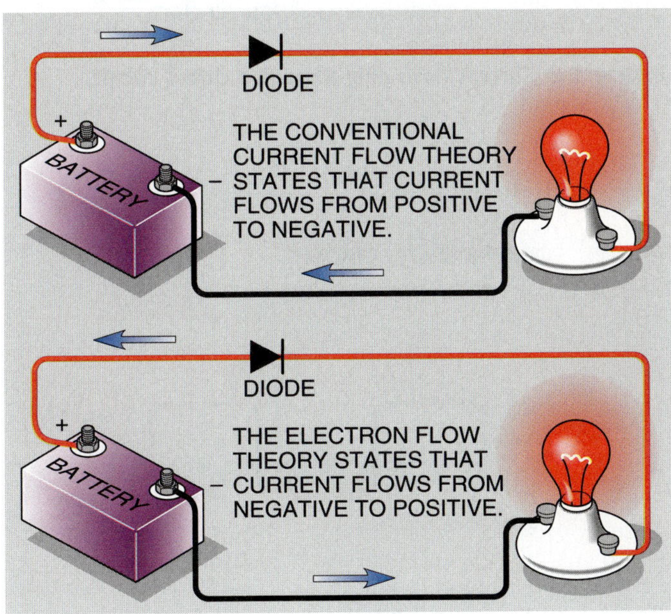

Figure 3-3 Conventional current flow theory and electron flow theory.

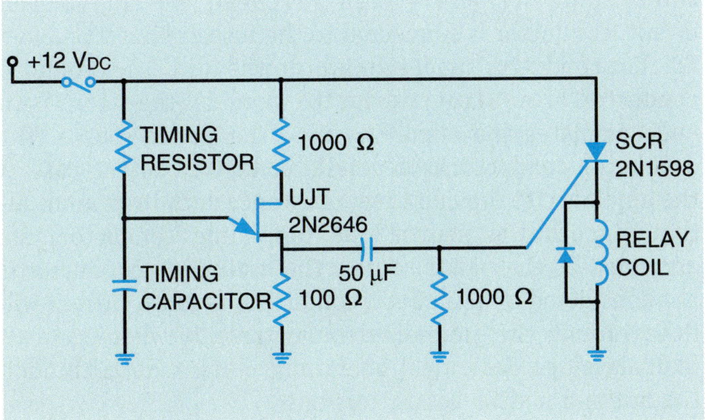

Figure 3-4 On-delay timer.

Speed of Current

To determine the speed of current flow through a wire, one must first establish exactly what is being measured. As stated previously, current is a flow of electrons through a conductive substance. Assume for a moment that it is possible to remove a single electron from a wire and identify it by painting it red. If it were possible to observe the progress of the identified electron as it moved from atom to atom, it would be seen that a single electron moves rather slowly (Fig. 3-5). It is estimated that a single electron moves at a rate of about 3 inches per hour at one ampere of current flow.

Another factor that must be considered is whether the circuit is DC, AC, or radio waves. Radio waves move at approximately the speed of light, which is 186,000 miles per second or 300,000,000 meters per second. The velocity of alternating current through a conductor is less than the speed of light because magnetic fields travel more slowly in material dielectrics than they do through free air.

In a direct current circuit the impulse of electricity can appear to be faster than the speed of light. Assume for a moment that a pipe has been filled with ping pong balls, Figure 3-6. If a ball is forced into the end of the pipe, the ball at the other end will be forced out. Each time a ball enters one end of the pipe, another ball is forced out the other end. This principle is also true for electrons in a wire. There are billions of electrons in a wire. If an electron enters one end of a wire, another electron is forced out the other end. Assume that a

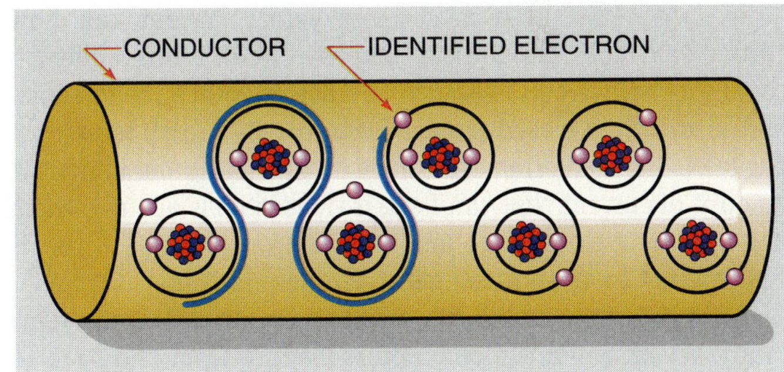

Figure 3-5 Electrons moving from atom to atom.

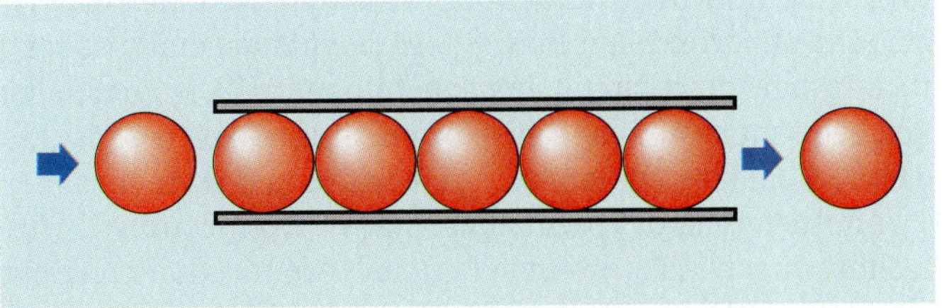

Figure 3-6 When a ball is pushed into one end, another ball is forced out the other end. This basic principle causes the instantaneous effect of electrical impulses.

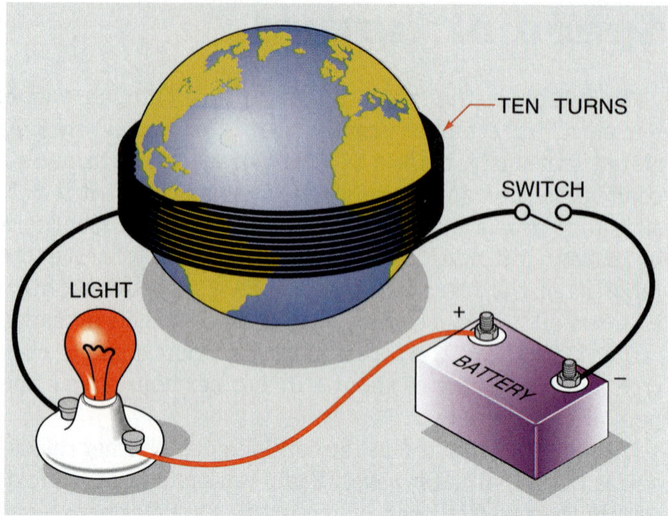

Figure 3-7 The impulse of electricity can travel faster than light.

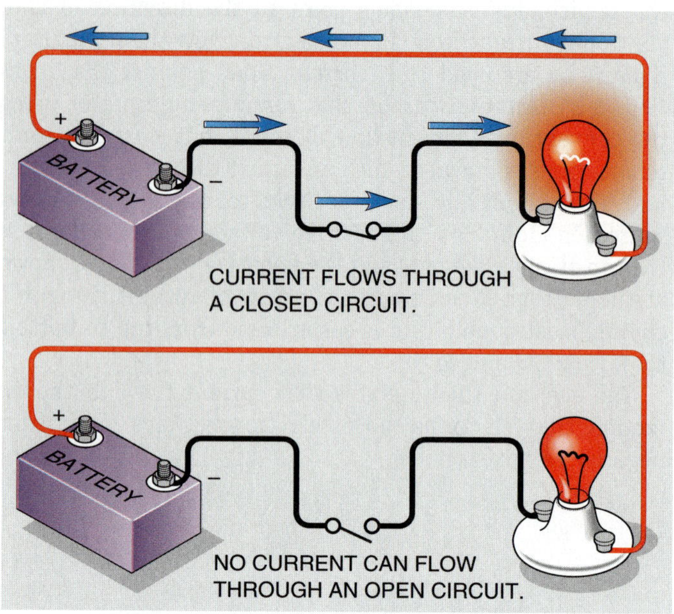

Figure 3-8 Current flows only through a closed circuit.

wire is long enough to be wound around the earth 10 times. If a power source and switch were connected at one end of the wire and a light at the other end, Figure 3-7, the light would turn on the moment the switch was closed. It would take light approximately 1.3 seconds to travel around the earth 10 times.

Basic Electrical Circuits

A **complete path** must exist before current can flow through a circuit (Fig. 3-8). A complete circuit is often referred to as a closed circuit, because the power source, conductors, and load form a closed loop. In Figure 3-8, a lamp is used as the load. The load offers resistance to the circuit and limits the amount of current that can flow. If the switch is opened, there is no longer a closed loop and no current can flow. This is often referred to as an incomplete, or open, circuit.

A short circuit, which has very little or no resistance, generally occurs when the conductors leading from and back to the power source become connected together (Fig. 3-9). In this example, a separate current path has been established that bypasses the load. Because the load is the device that limits the flow of current, when it is bypassed an excessive amount of current can flow. Short circuits generally cause a fuse to blow or a circuit breaker to open. If the circuit has not been protected by a fuse or circuit breaker, a short circuit can damage equipment, melt wires, and start fires.

Another type of circuit, one that is often confused with a short circuit, is a grounded circuit. Grounded circuits can also cause an excessive amount of current flow. They occur when a path other than the one intended is established to ground. Many circuits contain an extra conductor called the **grounding conductor.** A typical 120-volt appliance circuit is shown in Figure 3-10. In this circuit, the ungrounded, or hot, conductor is connected to the fuse or circuit breaker. The hot conductor supplies power to the load. The grounded conductor, or **neutral conductor,** provides the return path and completes the circuit back to the power source. The grounding conductor is generally connected to the case of the appliance to provide a low-resistance path to ground. Although both the neutral and grounding conductors are grounded at the power source, the grounding conductor is not considered to be a circuit conductor, because current will flow through the grounding conductor only when a circuit fault develops. In normal operation, current flows through the hot and neutral conductors only.

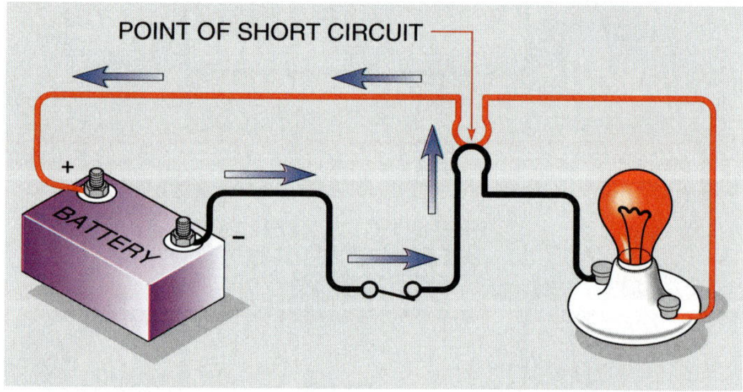

Figure 3-9 A short circuit bypasses the load and permits too much current to flow.

The grounding conductor helps prevent a shock hazard in the event that the ungrounded, or hot, conductor comes in

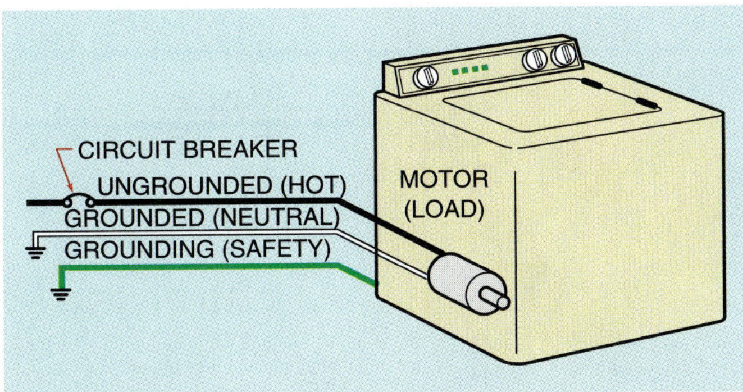

Figure 3-10 **120-V appliance circuit.**

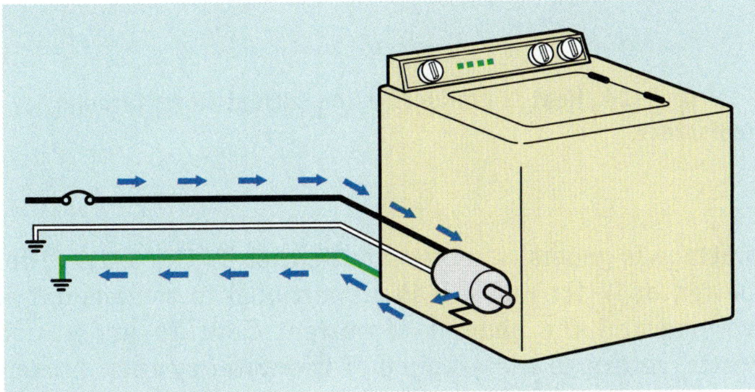

Figure 3-11 **The grounding conductor provides a low-resistance path to ground.**

contact with the case or frame of the appliance (Fig. 3-11). This condition could occur in several ways. In this example, it will be assumed that the motor winding becomes damaged and makes connection to the frame of the motor. Since the frame of the motor is connected to the frame of the appliance, the grounding conductor will provide a circuit path to ground. If enough current flows, the circuit breaker will open. Without a grounding conductor connected to the frame of the appliance, the frame would become hot (in the electrical sense) and anyone touching the case and a grounded point, such as a water line, would complete the circuit to ground. The resulting shock could be fatal. For this reason the grounding prong of a plug should never be cut off or bypassed.

The Volt

Voltage is defined as **electromotive force,** or EMF. It is the force that pushes the electrons through a wire and is often referred to as electrical pressure. A **volt** is the amount of potential necessary to cause one coulomb to produce one joule of work. Remember that voltage cannot flow. Voltage in an electrical circuit is like pressure in a water system (Fig. 3-12). To say that voltage flows through a circuit is like saying that pressure flows through a pipe. Pressure can push water through a pipe, and it is correct to say that water flows

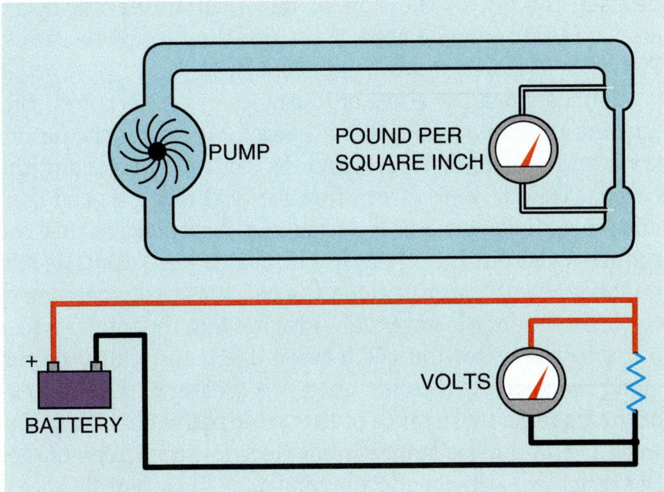

Figure 3-12 **Voltage in an electrical circuit can be compared to pressure in a water system.**

through a pipe, but it is not correct to say that pressure flows through a pipe. The same is true for voltage. Voltage pushes current through a wire, but voltage cannot flow through a wire.

Voltage is often thought of as the potential to do something. For this reason, it is frequently referred to as potential, especially in older publications and service manuals. Voltage must be present before current can flow, just as pressure must be present before water can flow. A voltage, or potential, of 120 volts is present at a common wall outlet, but there is no flow until some device is connected and a complete circuit exists. The same is true in a water system. Pressure is present, but water cannot flow until the valve is opened and a path is provided to a region of lower pressure. The letter *E*, which stands for EMF, or the letter *V*, which stands for volt, can be used to represent voltage in an algebraic formula. This text will use the letter *E* to represent voltage in an algebraic formula.

The Ohm

An **ohm** is the unit of **resistance** to current flow. It was named after the German scientist Georg S. Ohm. The symbol used to represent an ohm, or resistance, is the Greek letter omega (Ω). The letter *R*, which stands for resistance, is used to represent ohms in an algebraic formula. An ohm is the amount of resistance that allows 1 amp of current to flow when the applied voltage is 1 volt. So, one ohm equals one volt divided by one ampere. Without resistance, every electrical circuit would be a short circuit. All electrical loads, such as heating elements, lamps, motors, transformers, and so on, are measured in ohms. Just as in a water system a reducer can be used to control the flow of water, in an electrical circuit, a resistor can be used to control the flow of electrons. Figure 3-13 illustrates this concept.

To understand the effect of resistance on an electric circuit, imagine a person running along a beach. As long as the runner stays on the hard, compact sand, he can run easily along the beach. Likewise, current can flow easily through a good conductive material, such as a copper wire. Now imagine that the runner wades out into the water until it is knee deep. He will no longer be able to run along the beach as easily because of the resistance of the water. Now imagine that the runner wades out into the water until it is waist deep. His ability to run along the beach will be hindered to a greater extent because of the increased resistance of the water against his body. The same is true for resistance in an electric circuit. The higher the resistance, the greater the hindrance to current flow.

Another fact an electrician should be aware of is that any time current flows through a resistance, heat is produced (Fig. 3-14). That is why a wire becomes warm when current flows through it. The elements of an electric range become hot, and the filament of an incandescent lamp becomes extremely hot because of resistance.

Another term similar in meaning to resistance is **impedance.** Impedance is most often used in calculations of alternating current rather than direct current. Impedance will be discussed to a greater extent later in this text.

The Watt

Wattage is a measure of the amount of power that is being used in a circuit. The **watt** was named in honor of the English scientist, James Watt. In an algebraic formula, wattage is generally represented either by the letter *P*, for power, or *W*, for watts. It is proportional to the amount of voltage and the amount of current flow. To understand watts, return to the example of the water system. Assume that a water pump has a pressure of 120 pounds per square inch (PSI) and causes a flow rate of one gallon per second. Now assume that this water is used to drive a turbine, as shown in Figure 3-15. The turbine has a radius of one foot from the center shaft to the rim of the wheel. Since water weighs 8.34 pounds per gallon and it is being forced against the wheel at a pressure of 120 PSI, the turbine would develop a torque of 1000.8 foot-pounds ($120 \times 8.34 \times 1 = 1000.8$). Torque is a force that produces, or tends to produce, rotation or torsion. If the pressure is increased to 240 PSI, but the flow of water remains constant, the force against the wheel will double ($240 \times 8.34 \times 1 = 2001.6$). Notice that the amount of power developed by the turbine is determined by both the amount of pressure driving the water and the amount of flow.

The power of an electrical circuit is very similar. Figure 3-16 shows a resistor connected to a circuit with a voltage of 120 V and a current flow of 1 A. The resistor shown represents an electrical heating element. When 120 V force a current of 1 A through it, the heating element will produce 120 watts of heat ($120 \times 1 = 120$ W). If the voltage is increased to 240 V, but the current remains constant, the element will produce 240 W of heat ($240 \times 1 = 240$ W). If the voltage remains at 120 V, but the current is increased to 2 A, the heating element will again produce 240 W ($120 \times 2 = 240$). Notice that the amount of power used by the heating element is determined by the amount of current flow and the voltage driving it.

Watts is a measure of the amount of electrical energy converted into some other form.

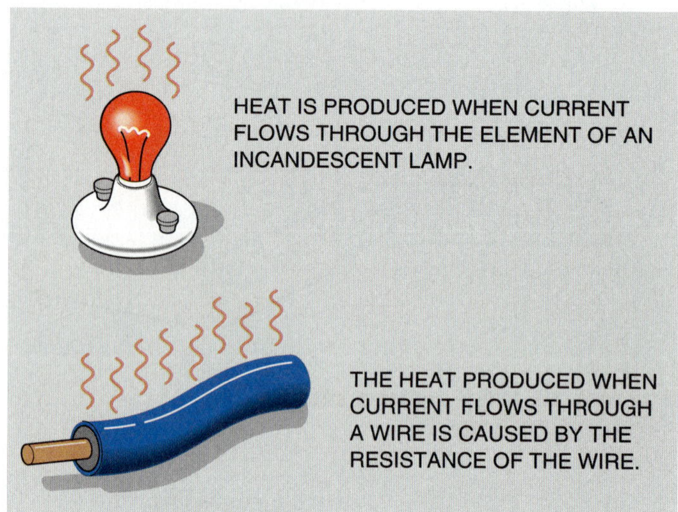

Figure 3-14 Heat is produced when current flows through resistance.

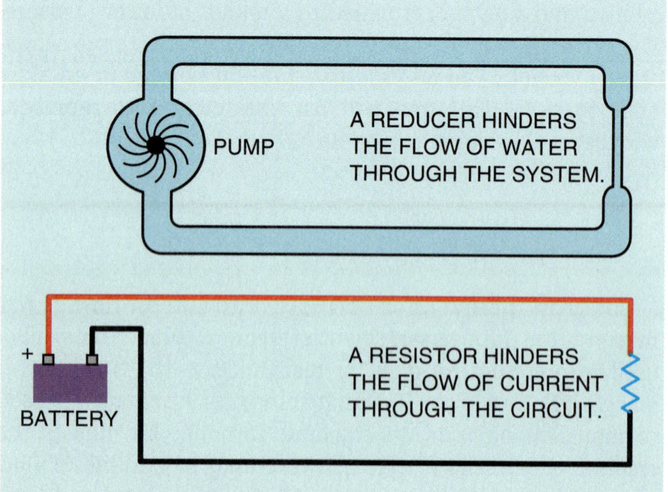

Figure 3-13 A resistor in an electrical circuit can be compared to a reducer in a water system.

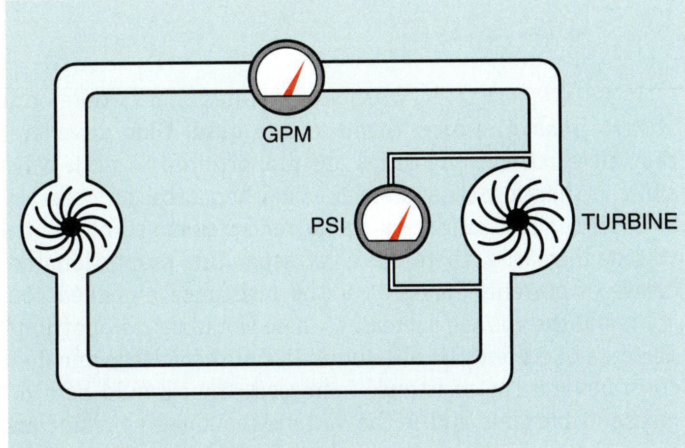

Figure 3-15 **Force equals flow rate times pressure.**

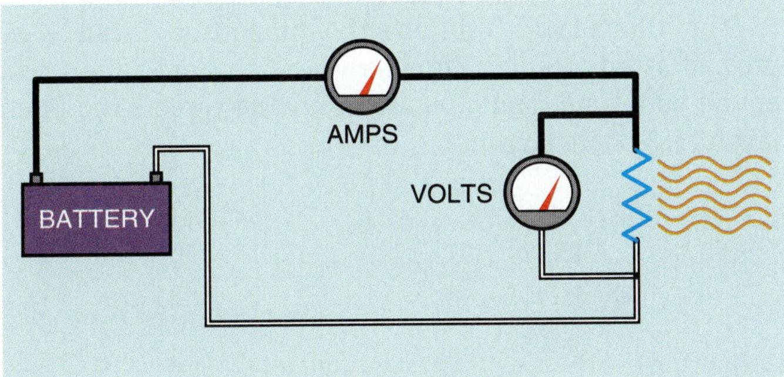

Figure 3-16 **Amps times volts equals watts.**

An important concept concerning **power** in an electrical circuit is that before true power, or watts, can exist, there must be some type of energy change or conversion. In other words, electrical energy must be changed or converted into some other form of energy before there can be power or watts. It makes no difference whether electrical energy is converted into heat energy or mechanical energy, but there must be some form of energy conversion before watts can exist.

Other Measures of Power

The watt is not the only unit of power measure. Many years ago, James Watt decided that in order to sell his steam engines, he would have to rate their power in terms that the average person could understand. He decided to compare his steam engines to the horses he hoped his engines would replace. After experimenting, Watt found that the average horse working at a steady rate could do 550 foot-pounds of work per second. A foot-pound (ft-lb) is the amount of force required to raise a one-pound weight one foot. This rate of doing work is the definition of a **horsepower** (hp).

1 hp = 550 ft-lb/s

Horsepower can also be expressed as 33,000 foot-pounds per minute (550 ft-lb × 60 s = 33,000).

1 hp = 33,000 ft-lb/min

It was later computed that the amount of electrical energy needed to produce one horsepower was 746 W.

1 hp = 746 W

Another measure of energy frequently used in the English system of measure is the **British thermal unit (BTU).** A BTU is defined as the amount of heat required to raise the temperature of one pound of water one degree Fahrenheit. In the metric system, the calorie is used instead of the BTU to measure heat. A calorie is the amount of heat needed to raise the temperature of one gram of water one degree Celsius. The **joule** is the metric equivalent of the watt. A joule is defined as one newton per meter. A newton is a force of 100,000 dynes, or about 3-1/2 ounces, and a meter is about 39 inches. The joule can also be expressed as the amount of work done by one coulomb flowing through a potential of one volt, or as the amount of work done by one watt for one second.

1 joule = 1 watt/s

The chart in Figure 3-17 gives some common conversions for different quantities of energy. These quantities can be used to calculate different values.

1 HORSEPOWER =	746 WATTS
1 HORSEPOWER =	550 FT-LB./S
1 WATT =	0.00134 HORSEPOWER
1 WATT =	3.412 BTU/HR
1 WATT/S =	1 JOULE
1 BTU/HR. × 0.293 =	WATTS
1 CAL/S =	4.19 WATTS
1 FT-LB./S =	1.36 WATTS
1 BTU =	1050 JOULES
1 JOULE =	0.2389 CAL
1 CAL =	4.186 JOULES

Figure 3-17 **Common power units.**

Example 1

An elevator must lift a load of 4000 lb to a height of 50 ft in 20 s. How much horsepower is required to operate the elevator?

SOLUTION

Find the amount of work that must be performed, and then convert that to horsepower.

$$4000\ \text{lb} \times 50\ \text{ft} = 200{,}000\ \text{ft-lb}$$

$$\frac{200{,}000\ \text{ft–lb}}{20\ s} = 10{,}000\ \text{fb–lb}/2$$

$$\frac{10{,}000\ \text{ft–lb./s}}{550\ \text{ft–lb./s}} = 18.18\text{hp}$$

Example 2

A water heater contains 40 gallons of water. Water weighs 8.34 lb per gallon. The present temperature of the water is 68°F. The water must be raised to a temperature of 160°F in 1 hour. How much power will be required to raise the water to the desired temperature?

SOLUTION

First determine the weight of the water in the tank, because a BTU is the amount of heat required to raise the temperature of one pound of water one degree Fahrenheit.

$$40\ \text{gal} \times 8.34\ \text{lb per gal} = 333.6\ \text{lb}$$

The second step is to determine how many degrees of temperature the water must be raised. This amount will be the difference between the present temperature and the desired temperature.

$$160°\text{F} - 68°\text{F} = 92°\text{F}$$

The amount of heat required in BTUs will be the product of the pounds of water and the desired increase in temperature.

$$333.6\ \text{lb} \times 92° = 30{,}691.2\ \text{BTU}$$

$$1\ \text{W} = 3.412\ \text{BTU/hr}$$

Therefore

$$\frac{30{,}691\ \text{BTU}}{3.412\ \text{BTU/hr}} = 8995.1\ \text{W/hr}$$

Ohm's Law

In its simplest form, **Ohm's law** states that **it takes one volt to push one amp through one ohm.** Ohm discovered that all electrical quantities are proportional to each other and can therefore be expressed as mathematical formulas. He found if the resistance of a circuit remained constant and the voltage increased, there was a corresponding proportional increase of current. Similarly, if the resistance remained constant and the voltage decreased, there would be a proportional decrease of current. He also found that if the voltage remained constant and the resistance increased, there would be a decrease of current, and if the voltage remained constant and the resistance decreased, there would be an increase of current. This finding led Ohm to the conclusion that **in a DC circuit, the current is directly proportional to the voltage and inversely proportional to the resistance.**

Since Ohm's law is a statement of proportion, it can be expressed as an algebraic formula when standard values such as the volt, amp, and ohm are used. The three basic Ohm's law formulas are shown.

$$E = I \times R$$

$$I = \frac{E}{R}$$

$$R = \frac{E}{I}$$

where

E = EMF, or voltage

I = intensity of current, or amperage

R = resistance

The first formula states that the voltage can be found if the current and resistance are known. Voltage is equal to amps multiplied by ohms. For example, assume a circuit has a resistance of 50 Ω and a current flow through it of 2 A. The voltage connected to this circuit is 100 V.

$$E = I \times R$$

$$E = 2 \times 50$$

$$E = 100$$

The second formula states that the current can be found if the voltage and resistance are known. In the example shown, 120 V are connected to a resistance of 30 Ω. The amount of current flow will be 4 A.

$$I = \frac{E}{R}$$

$$I = \frac{120}{30}$$

$$I = 4$$

The third formula states that if the voltage and current are known, the resistance can be found. Assume a circuit has a voltage of 240 V and a current flow of 10 A. The resistance in the circuit is 24 Ω.

$$R = \frac{E}{I}$$

$$R = \frac{240}{10}$$

$$R = 24$$

Figure 3-18 shows a simple chart that can be a great help in remembering an Ohm's law formula. To use the chart, cover the quantity that is to be found. For example, if the voltage, E, is to be found, cover the E on the chart. The chart now shows the remaining letters IR (Fig. 3-19), thus E = I × R. The same method reveals the formulas for current (I) and resistance (R).

A larger chart that shows the formulas needed to find watts as well as voltage, amperage, and resistance is shown in Figure 3-20. The letter *P* (power) is used to represent the value of watts. Notice that this chart is divided into four sections and that each section contains three different formulas. To use this chart, select the section containing the quantity to be found and then choose the proper formula from the given quantities.

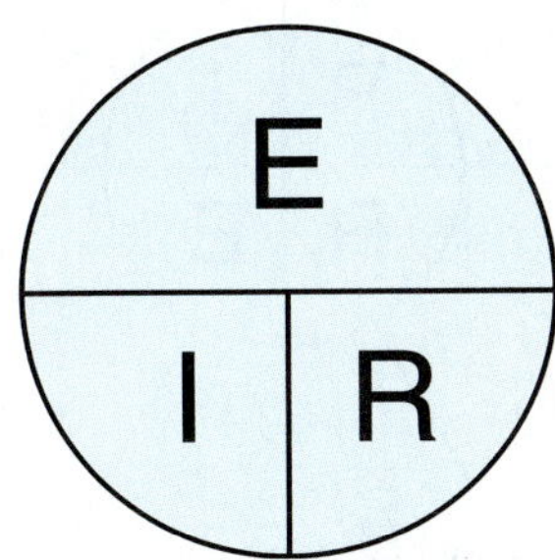

Figure 3-18 **Chart for finding values of voltage, current, and resistance.**

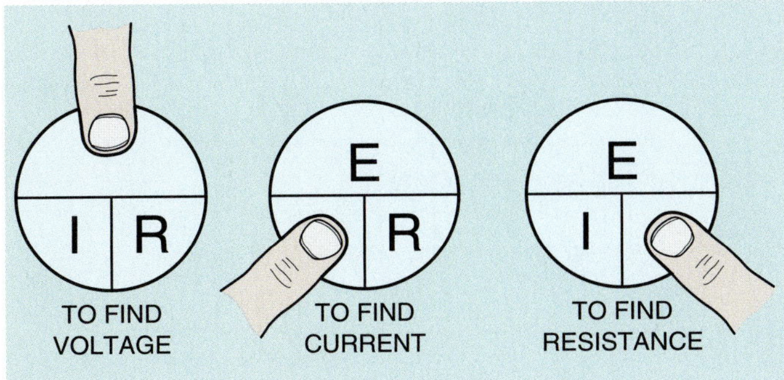

Figure 3-19 **Using the Ohm's law chart.**

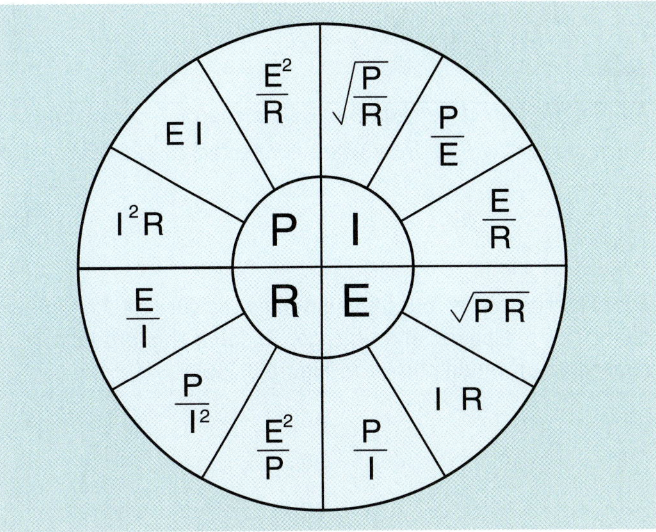

Figure 3-20 **Formula chart for finding values of voltage, current, resistance, and power.**

Example 3

An electric iron is connected to 120 V and has a current draw of 8 A. How much power is used by the iron?

SOLUTION

The quantity to be found is watts, or power. The known quantities are voltage and amperage. The proper formula to use is shown in Figure 3-21.

$$P = EI$$

$$P = 120 \times 8$$

$$P = 960 \text{ W}$$

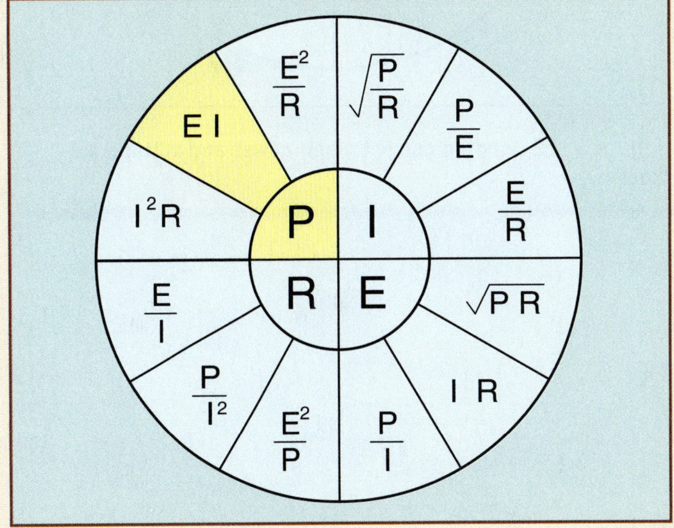

Figure 3-21 **Finding power when voltage and current are known.**

Example 4

An electric hair dryer has a power rating of 1000 W. How much current will it draw when connected to 120 V?

SOLUTION

The quantity to be found is amperage, or current. The known quantities are power and voltage. To solve this problem, choose the formula shown in Figure 3-22.

$$I = \frac{P}{E}$$

$$I = \frac{1000}{120}$$

$$I = 8.33\ \text{A}$$

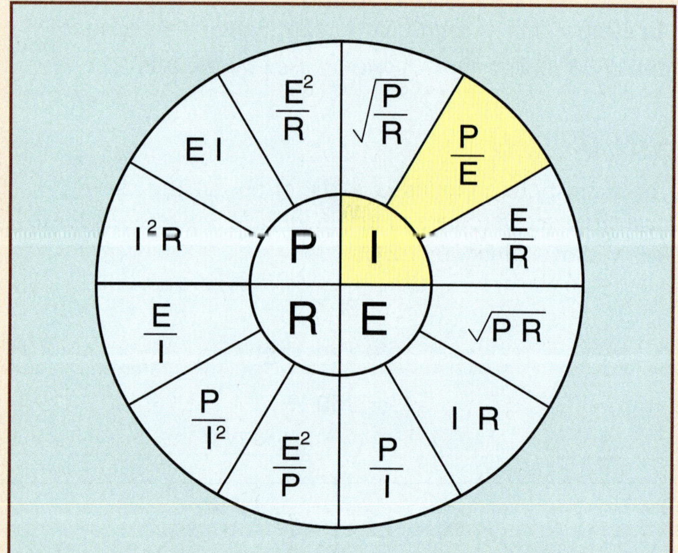

Figure 3-22 **Finding current when power and voltage are known.**

Example 5

An electric hotplate has a power rating of 1440 W and a current draw of 12 A. What is the resistance of the hotplate?

SOLUTION

The quantity to be found is resistance, and the known quantities are power and current. Use the formula shown in Figure 3-23.

$$R = \frac{P}{I^2}$$

$$R = \frac{1440}{12 \times 12}$$

$$R = \frac{1440}{144}$$

$$R = 10\ \Omega$$

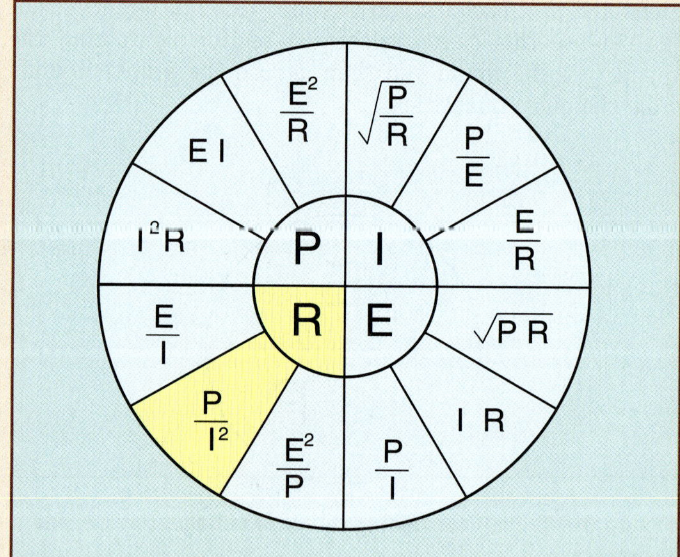

Figure 3-23 **Finding resistance when power and current are known.**

KILO	1000
HECTO	100
DEKA	10
BASE UNIT	1
DECI	1/10 OR 0.1
CENTI	1/100 OR 0.01
MILLI	1/1000 OR 0.001

Figure 3-24 **Standard units of metric measure.**

ENGINEERING UNIT	SYMBOL	MULTIPLY BY	
TERA	T	1,000,000,000,000	X 10^{12}
GIGA	G	1,000,000,000	X 10^{9}
MEGA	M	1,000,000	X 10^{6}
KILO	k	1,000	X 10^{3}
BASE UNIT		1	
MILLI	m	0.001	X 10^{-3}
MICRO	μ	0.000,001	X 10^{-6}
NANO	n	0.000,000,001	X 10^{-9}
PICO	p	0.000,000,000,001	X 10^{-12}

Figure 3-25 **Standard units of engineering notation.**

Metric Units

Metric units of measure are used in the electrical field just as they are in most scientific fields. A special type of metric notation, known as engineering notation, is used in electrical measurements. Engineering notation is the same as any other metric measure except that it is in steps of one thousand instead of ten. The chart in Figure 3-24 shows standard metric units. The first step above the base unit is deka, which means 10. The second unit is hecto, which means 100, and the third unit is kilo, which means 1000. The first unit below the base unit is deci, which means 1/10; the second unit is centi, which means 1/100; and the third is milli, which means 1/1000.

The chart in Figure 3-25 shows engineering units. The first unit above the base unit is kilo, or 1000; the second unit is mega, or 1,000,000; and the third unit is giga, or 1,000,000,000. Notice that each unit is 1000 times greater than the previous unit. The chart also shows that the first unit below the base unit is milli, or 1/1000; the second is micro, represented by the Greek mu (μ), or 1/1,000,000; and the third is nano, or 1/1,000,000,000.

Metric units are used in almost all scientific measurements for ease of notation. It is much simpler to write a value such as 10 MΩ than it is to write 10,000,000 ohms, or to write 0.5 ns than to write 0.000,000,000,5 second. Once the metric system has been learned, measurements such as 47 kilohms (kΩ) or 50 milliamps (mA) become commonplace to the technician.

Summary

- A coulomb is a quantity measurement of electrons.
- An amp (A) is one coulomb per second.
- Either the letter ***I***, which stands for intensity of current flow, or the letter ***A***, which stands for amps, can be used in Ohm's law formulas.
- Voltage is referred to as electrical pressure, potential difference, or electromotive force. An ***E*** or a ***V*** can be used to represent voltage in Ohm's law formulas.
- An ohm (Ω) is a measurement of resistance (R) in an electrical circuit.
- The watt (W) is a measurement of power in an electrical circuit. It is represented by either a ***W*** or a ***P*** (power) in Ohm's law formulas.
- Electrical measurements are generally expressed in engineering notation.
- Engineering notation differs from the standard metric system in that it uses steps of 1000 instead of steps of 10.
- Before current can flow, there must be a complete circuit.
- A short circuit has little or no resistance.

Review Questions

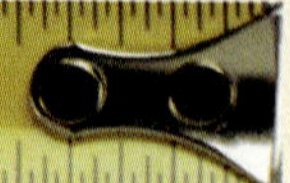

1. What is a coulomb?
2. What is an amp?
3. Define voltage.
4. Define ohm.
5. Define watt.
6. An electric heating element has a resistance of 16 Ω and is connected to a voltage of 120 V. How much current will flow in this circuit?
7. How many watts of heat are being produced by the heating element in question 6?
8. A 240-V circuit has a current flow of 20 A. How much resistance is connected in the circuit?
9. An electric motor has an apparent resistance of 15 Ω. If 8 A of current are flowing through the motor, what is the connected voltage?
10. A 240-V air conditioning compressor has an apparent resistance of 8 Ω. How much current will flow in the circuit?
11. How much power is being used by the motor in question 8?
12. A 5-kW electric heating unit is connected to a 240-V line. What is the current flow in the circuit?
13. If the voltage in question 12 is reduced to 120 V, how much current would be needed to produce the same amount of power?
14. Is it less expensive to operate the electric heating unit in question 12 on 240 V or 120 V?

Practice Problems

Ohm's Law

Fill in the missing values.

Volts (E)	Amps (I)	Ohms (R)	Watts (P)
153	0.056		
	0.65	470 Ω	
24			124
	0.00975		0.035
		6.8 kΩ	0.86
460		72 Ω	
48	1.2		
	154	0.8 Ω	
277			760
	0.0043		0.0625
		130 kΩ	0.0225
96		2.2 kΩ	

Chapter 4 Resistors

The resistor is one of the most common components found in electrical circuits. The unit of measure for resistance (R) is the ohm, *which was named for a German scientist, Georg S. Ohm. The symbol used to represent resistance is the Greek letter omega* (Ω)*. Resistors come in various sizes, types, and ratings to accommodate the needs of almost any circuit application.*

OBJECTIVES

After studying this unit, you should be able to:

- **list the major types of fixed resistors.**
- **determine the resistance of a resistor using the color code.**
- **determine if a resistor is operating within its power rating.**
- **connect a variable resistor for use as a potentiometer.**

Glossary of Resistor Terms

carbon film resistor a resistor made by coating a nonconducting material with a film of carbon

color code a method of marking the resistance and tolerance of a resistor with bands of color

composition carbon resistor a resistor that is made by combining carbon with a resin material

metal film resistor a resistor made by coating a nonconducting material with a film of metal

metal glaze resistor a resistor made by combining glass and metal

multiturn variable resistors resistors constructed so that their ohmic value can be adjusted by turning a shaft from three to ten turns

pot slang for potentiometer

potentiometer a variable resistor employed in such a manner that it can provide a variable voltage

rheostat a variable resistor that contains only two terminals

short circuit a circuit that contains almost no resistance to limit the flow of current

tolerance the range over which a resistor is considered to be within its marked value

variable resistor a resistor whose resistance value can be changed

voltage divider a circuit generally constructed of series connected resistors to provide different amounts of voltage

wire wound resistor a resistor made with a high-resistance metal wire such as Nichrome

Uses of Resistors

Resistors are commonly used to perform two functions in a circuit. One is to limit the flow of current through the circuit. In Figure 4-1 a 30-Ω resistor is connected to a 15-V battery. The current in this circuit is limited to a value of 0.5 A.

$$I = \frac{E}{R}$$

$$I = \frac{15}{30}$$

$$I = 0.5\ \text{A}$$

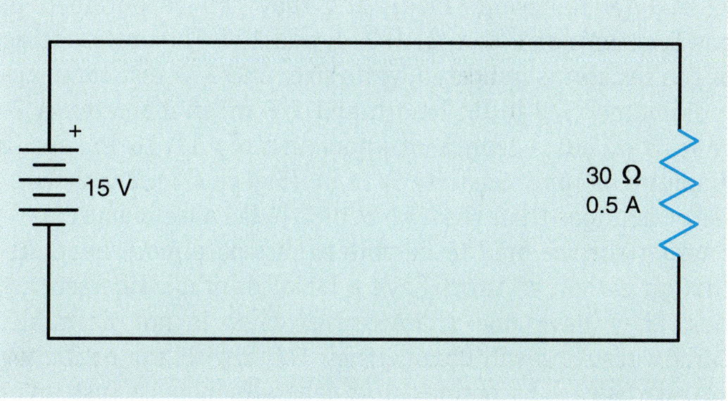

Figure 4-1 Resistor used to limit the flow of current.

If this resistor were not present, the circuit current would be limited only by the resistance of the conductor, which would be very low, and a large amount of current would flow. Assume for example that the wire has a resistance of 0.0001 Ω. When the wire is connected across the 15-V power source, a current of 150,000 A would try to flow through the circuit (15/0.0001 = 150,000). This is commonly known as a **short circuit.**

The second principal function of resistors is to produce a **voltage divider.** The three resistors shown in Figure 4-2 are connected in series with a 17.5-V battery. If the leads of a voltmeter were connected between different points in the circuit, it would indicate the following voltages:

A to B, 1.5 V
A to C, 7.5 V
A to D, 17.5 V
B to C, 6 V
B to D, 16 V
C to D, 10 V

By connecting resistors of the proper value, almost any voltage desired can be obtained. Voltage dividers were used to a large extent in vacuum tube circuits many years ago. Voltage divider circuits are still used today in applications involving field effect transistors (FETs) and in multirange voltmeter circuits.

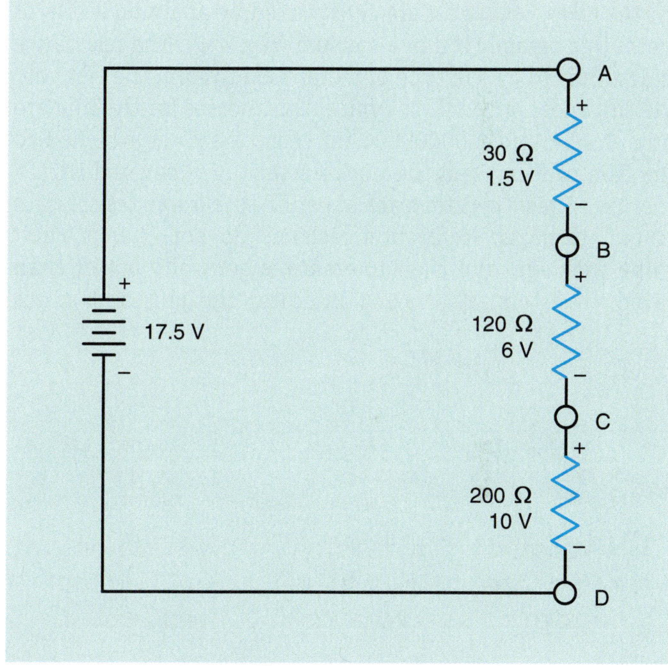

Figure 4-2 Resistors used as a voltage divider.

Fixed Resistors

Fixed resistors have only one ohmic value, which cannot be changed or adjusted. There are several different types of fixed resistors. One of the most common types of fixed resistors is the **composition carbon resistor.** Carbon resistors are made from a compound of carbon graphite and a resin bonding material. The proportions of carbon and resin material determine the value of resistance. This compound is enclosed in a case of nonconductive material with connecting leads (Fig. 4-3).

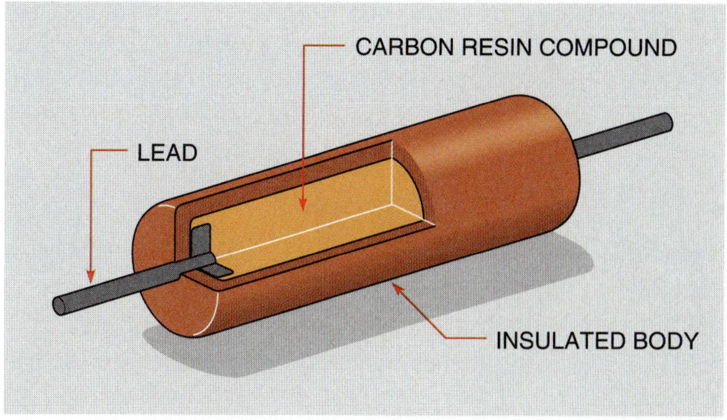

Figure 4-3 Composition carbon resistor.

Carbon resistors are very popular for most applications because they are inexpensive and readily available. They are made in standard values that range from about 1 Ω to about

22 MΩ (M represents meg), and they can be obtained in power ratings of 1/8, 1/4, 1/2, 1, and 2 W. The power rating of the resistor is indicated by its size. A 1/2-W resistor is approximately 3/8 in. in length and 1/8 in. in diameter. A 2-W resistor has a length of approximately 11/16 in. and a diameter of approximately 5/16 in. (Figure 4-4). The 2-W resistor is larger than the 1/2-W or 1-W because it must have a larger surface area to be able to dissipate more heat. Although carbon resistors have a lot of desirable characteristics, they have one characteristic that is not desirable. Carbon resistors will change their value with age or if they are overheated. Carbon resistors generally increase instead of decrease in value.

Metal Film Resistors

Another type of fixed resistor is the metal film resistor. **Metal film resistors** are constructed by applying a film of metal to a ceramic rod in a vacuum (Fig. 4-5). The resistance is determined by the type of metal used to form the film and the thickness of the film. Typical thicknesses for the film are from 0.00001 to 0.00000001 in. Leads are then attached to the film coating, and the entire assembly is covered with a coating. These resistors are superior to carbon resistors in several respects. Metal film resistors do not change their value with age, and their tolerance is generally better than carbon resistors. **Tolerance** indicates the plus and minus limits of a resistor's ohmic value. Carbon resistors commonly have a tolerance range of 20%, 10%, or 5%. Metal film resistors generally range in tolerance from 2% to 0.1%. The disadvantage of the metal film resistor is that it costs more.

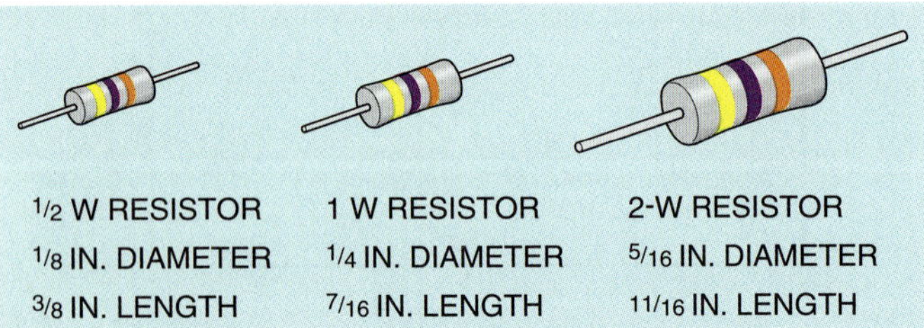

Figure 4-4 Power rating is indicated by size.

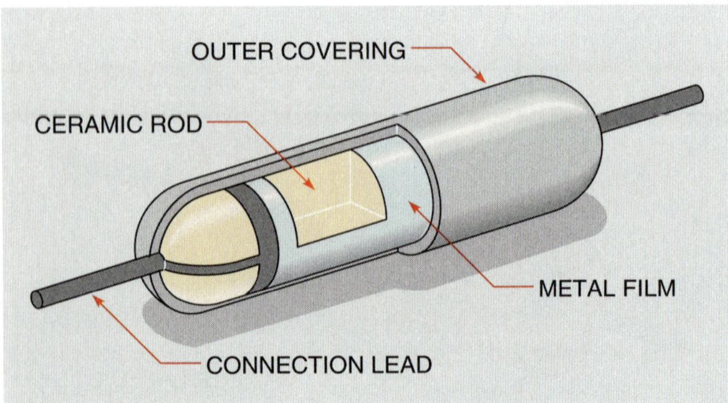

Figure 4-5 Metal film resistor.

Carbon Film Resistors

Another type of fixed resistor that is constructed in a similar manner is the **carbon film resistor.** This resistor is made by coating a ceramic rod with a film of carbon instead of metal. Carbon film resistors are less expensive to manufacture than metal film resistors and can have a higher tolerance rating than composition carbon resistors.

Metal Glaze Resistors

The **metal glaze resistor** is also a fixed resistor, similar to the metal film resistor. This resistor is made by combining metal with glass. The compound is then applied to a ceramic base as a thick film. The resistance is determined by the amount of metal used in the compound. Tolerance ratings of 2% and 1% are common.

Wire Wound Resistors

Wire wound resistors are fixed resistors that are made by winding a piece of resistive wire around a ceramic core (Fig. 4-6). The resistance of a wire wound resistor is determined by three factors:

1. the type of material used to make the resistive wire,
2. the diameter of the wire, and
3. the length of the wire.

Wire wound resistors can be found in various case styles and sizes. These resistors are generally used when a high power rating is needed. Wire wound resistors can operate at higher temperatures than any other type of resistor. A wire wound resistor with a hollow center is shown in Figure 4-7. This type of resistor should be mounted vertically and not horizontally. The center of the resistor is hollow for a very good reason. When the resistor is mounted vertically, the heat from the resistor produces a chimney effect and causes air to circulate through the center (Fig. 4-8). This increase

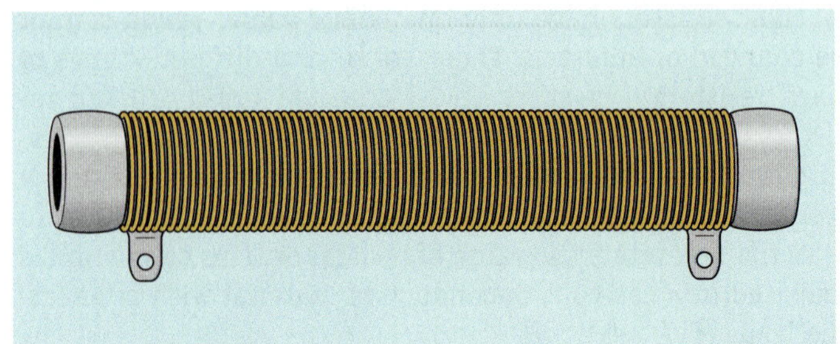

Figure 4-6 Wire wound resistor.

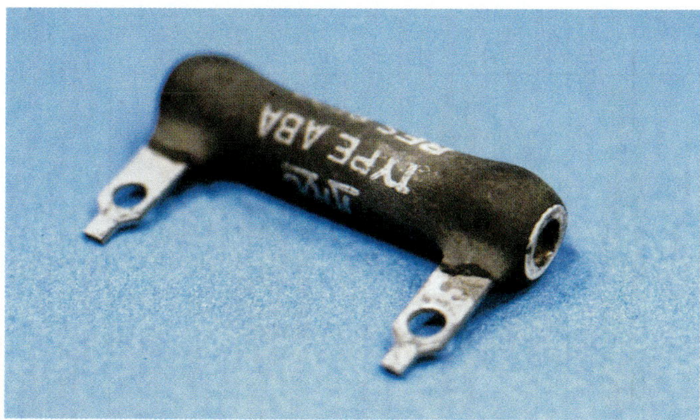

Figure 4-7 **Wire wound resistor with hollow core.**

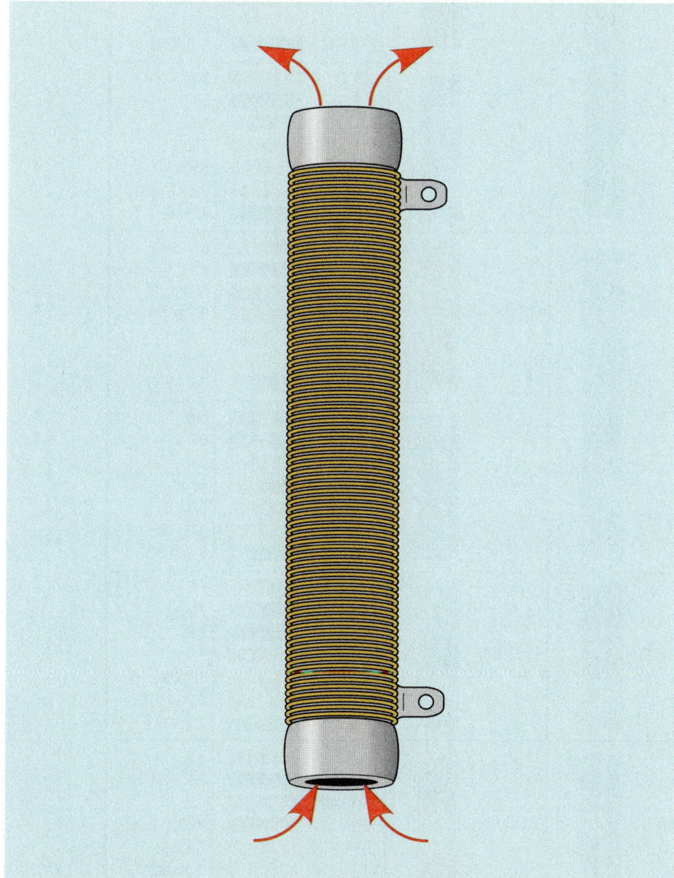

Figure 4-8 **Air flow helps cool the resistor.**

of air flow dissipates heat at a faster rate to help keep the resistor from overheating. The disadvantage of wire wound resistors is they are expensive and generally require a large amount of space for mounting. They can also exhibit an amount of inductance in circuits that operate at high frequencies. This added inductance can cause problems to the rest of the circuit. Inductance will be covered in later units.

Color Code

The values of a resistor can often be determined by the **color code.** Many resistors have bands of color that are used to determine the resistance value, tolerance, and in some cases reliability. The color bands represent numbers. Each color represents a different numerical value, as shown in Table 4-1. Figure 4-9 lists the colors and the number value assigned to each color. The resistor shown beside the color chart illustrates how to determine the value of a resistor. Resistors can have from three to five bands of color. Resistors that have a tolerance of ±20% have only three color bands. Most resistors contain four bands of color. For resistors with tolerances that range from ±10% to ±2%, the first two color bands represent number values. The third color band is called the multiplier. The first two numbers are combined and multiplied by the power of 10 indicated by the value of the third band. The fourth band indicates the tolerance. For example, assume a resistor has color bands of brown, green, red, and silver (Fig. 4-10). The first two bands represent the numbers 1 and 5 (brown is 1 and green is 5). The third band is red, which has a number value of 2. The number 15 should be multiplied by 10^2, or 100. The value of the resistor is 1500 Ω. Another method, which is simpler to understand, is to add the number of zeros indicated by the multiplier band to the combined first two numbers. The multiplier band in this example is red, which has a numeric value of 2. Add two zeros to the first two numbers. The number 15 becomes 1500.

The fourth band is the tolerance band. The tolerance band in this example is silver, which means ±10%. This resistor should be 1500 Ω plus or minus 10%. To determine the value limits of this resistor, find 10% of 1500.

$$1500 \times 0.10 = 150$$

The value can range from 1500 + 10%, or 1500 + 150 = 1650 Ω, to 1500 − 10% or 1500 − 150 = 1350 Ω.

Resistors that have a tolerance of ±1%, as well as some military resistors, contain five bands of color.

Color	Value	Tolerance	
Black	0	No color	± 20%
Brown	1	Silver	± 10%
Red	2	Gold	± 5%
Orange	3	Red	± 2%
Yellow	4	Brown	± 1%
Green	5		
Blue	6		
Violet	7		
Gray	8		
White	9		

Table 4-1 **Colors and numeric values.**

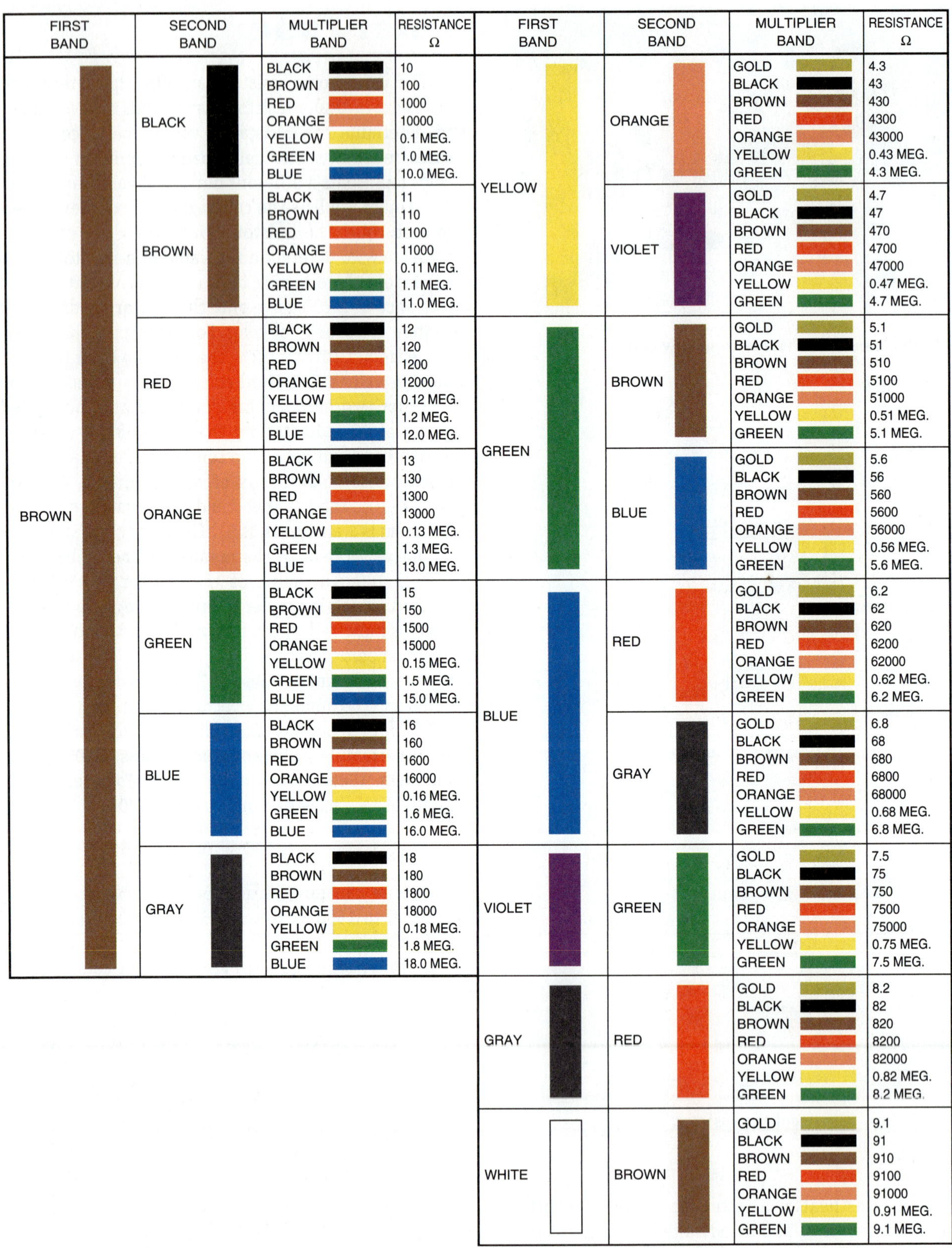

FIRST BAND	SECOND BAND	MULTIPLIER BAND	RESISTANCE Ω
BROWN	BLACK	BLACK	10
		BROWN	100
		RED	1000
		ORANGE	10000
		YELLOW	0.1 MEG.
		GREEN	1.0 MEG.
		BLUE	10.0 MEG.
	BROWN	BLACK	11
		BROWN	110
		RED	1100
		ORANGE	11000
		YELLOW	0.11 MEG.
		GREEN	1.1 MEG.
		BLUE	11.0 MEG.
	RED	BLACK	12
		BROWN	120
		RED	1200
		ORANGE	12000
		YELLOW	0.12 MEG.
		GREEN	1.2 MEG.
		BLUE	12.0 MEG.
	ORANGE	BLACK	13
		BROWN	130
		RED	1300
		ORANGE	13000
		YELLOW	0.13 MEG.
		GREEN	1.3 MEG.
		BLUE	13.0 MEG.
	GREEN	BLACK	15
		BROWN	150
		RED	1500
		ORANGE	15000
		YELLOW	0.15 MEG.
		GREEN	1.5 MEG.
		BLUE	15.0 MEG.
	BLUE	BLACK	16
		BROWN	160
		RED	1600
		ORANGE	16000
		YELLOW	0.16 MEG.
		GREEN	1.6 MEG.
		BLUE	16.0 MEG.
	GRAY	BLACK	18
		BROWN	180
		RED	1800
		ORANGE	18000
		YELLOW	0.18 MEG.
		GREEN	1.8 MEG.
		BLUE	18.0 MEG.

FIRST BAND	SECOND BAND	MULTIPLIER BAND	RESISTANCE Ω
YELLOW	ORANGE	GOLD	4.3
		BLACK	43
		BROWN	430
		RED	4300
		ORANGE	43000
		YELLOW	0.43 MEG.
		GREEN	4.3 MEG.
	VIOLET	GOLD	4.7
		BLACK	47
		BROWN	470
		RED	4700
		ORANGE	47000
		YELLOW	0.47 MEG.
		GREEN	4.7 MEG.
GREEN	BROWN	GOLD	5.1
		BLACK	51
		BROWN	510
		RED	5100
		ORANGE	51000
		YELLOW	0.51 MEG.
		GREEN	5.1 MEG.
	BLUE	GOLD	5.6
		BLACK	56
		BROWN	560
		RED	5600
		ORANGE	56000
		YELLOW	0.56 MEG.
		GREEN	5.6 MEG.
BLUE	RED	GOLD	6.2
		BLACK	62
		BROWN	620
		RED	6200
		ORANGE	62000
		YELLOW	0.62 MEG.
		GREEN	6.2 MEG.
	GRAY	GOLD	6.8
		BLACK	68
		BROWN	680
		RED	6800
		ORANGE	68000
		YELLOW	0.68 MEG.
		GREEN	6.8 MEG.
VIOLET	GREEN	GOLD	7.5
		BLACK	75
		BROWN	750
		RED	7500
		ORANGE	75000
		YELLOW	0.75 MEG.
		GREEN	7.5 MEG.
GRAY	RED	GOLD	8.2
		BLACK	82
		BROWN	820
		RED	8200
		ORANGE	82000
		YELLOW	0.82 MEG.
		GREEN	8.2 MEG.
WHITE	BROWN	GOLD	9.1
		BLACK	91
		BROWN	910
		RED	9100
		ORANGE	91000
		YELLOW	0.91 MEG.
		GREEN	9.1 MEG.

Figure 4-9 Resistor color code chart.

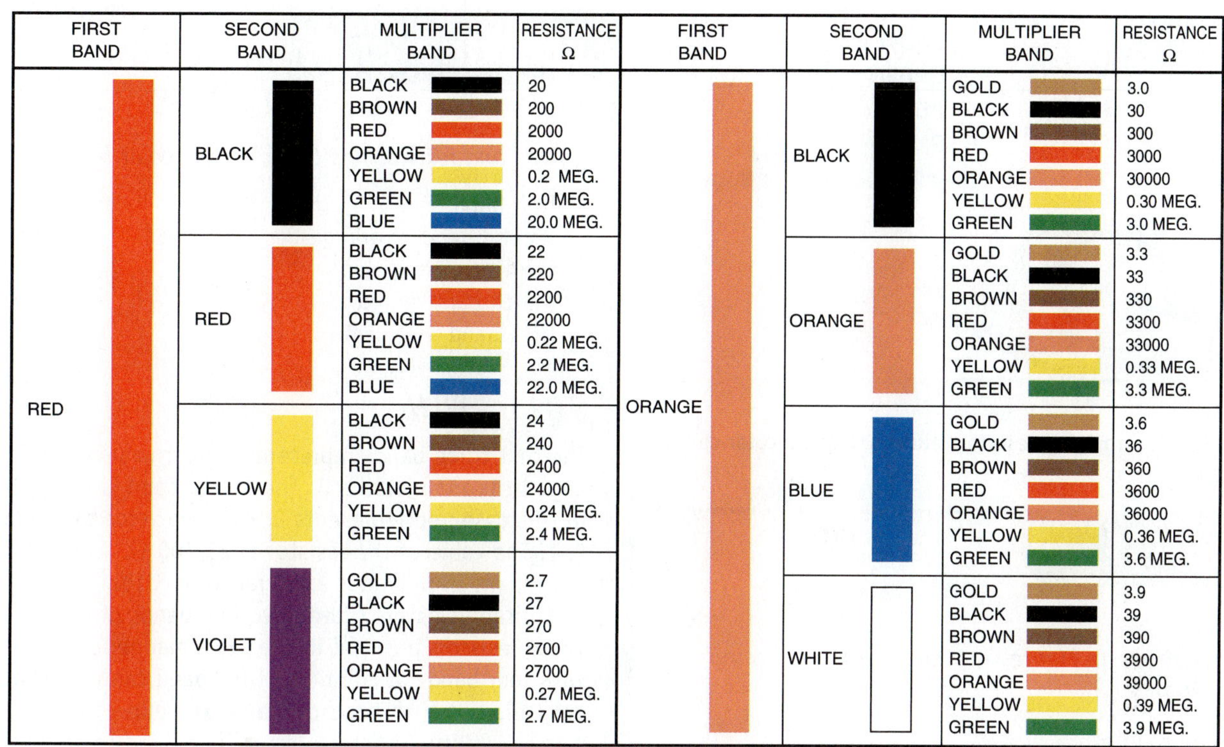

FIRST BAND	SECOND BAND	MULTIPLIER BAND	RESISTANCE Ω
RED	BLACK	BLACK	20
		BROWN	200
		RED	2000
		ORANGE	20000
		YELLOW	0.2 MEG.
		GREEN	2.0 MEG.
		BLUE	20.0 MEG.
	RED	BLACK	22
		BROWN	220
		RED	2200
		ORANGE	22000
		YELLOW	0.22 MEG.
		GREEN	2.2 MEG.
		BLUE	22.0 MEG.
	YELLOW	BLACK	24
		BROWN	240
		RED	2400
		ORANGE	24000
		YELLOW	0.24 MEG.
		GREEN	2.4 MEG.
	VIOLET	GOLD	2.7
		BLACK	27
		BROWN	270
		RED	2700
		ORANGE	27000
		YELLOW	0.27 MEG.
		GREEN	2.7 MEG.
ORANGE	BLACK	GOLD	3.0
		BLACK	30
		BROWN	300
		RED	3000
		ORANGE	30000
		YELLOW	0.30 MEG.
		GREEN	3.0 MEG.
	ORANGE	GOLD	3.3
		BLACK	33
		BROWN	330
		RED	3300
		ORANGE	33000
		YELLOW	0.33 MEG.
		GREEN	3.3 MEG.
	BLUE	GOLD	3.6
		BLACK	36
		BROWN	360
		RED	3600
		ORANGE	36000
		YELLOW	0.36 MEG.
		GREEN	3.6 MEG.
	WHITE	GOLD	3.9
		BLACK	39
		BROWN	390
		RED	3900
		ORANGE	39000
		YELLOW	0.39 MEG.
		GREEN	3.9 MEG.

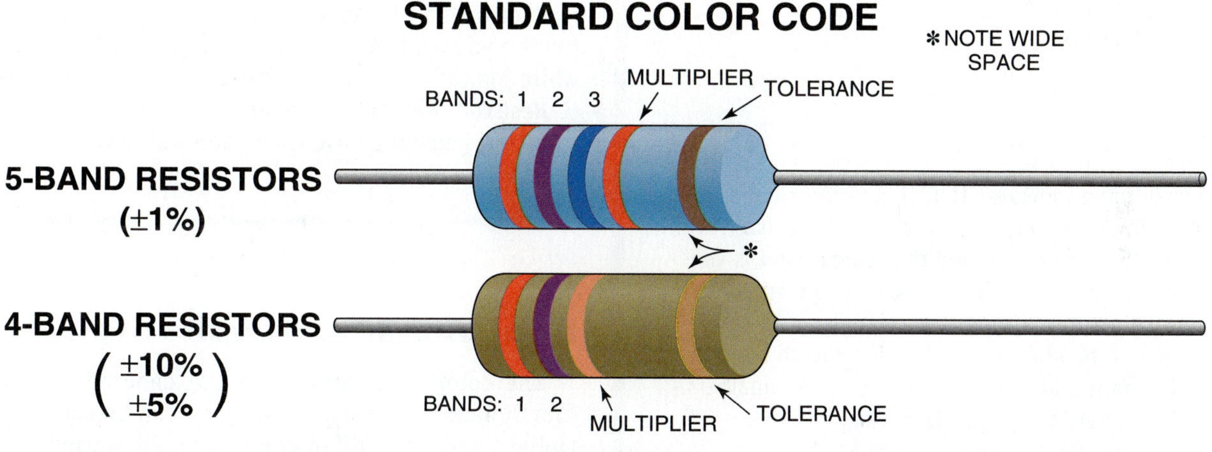

BAND 1 1ST DIGIT	
COLOR	DIGIT
BLACK	0
BROWN	1
RED	2
ORANGE	3
YELLOW	4
GREEN	5
BLUE	6
VIOLET	7
GRAY	8
WHITE	9

BAND 2 2ND DIGIT	
COLOR	DIGIT
BLACK	0
BROWN	1
RED	2
ORANGE	3
YELLOW	4
GREEN	5
BLUE	6
VIOLET	7
GRAY	8
WHITE	9

BAND 3 (IF USED) 3RD DIGIT	
COLOR	DIGIT
BLACK	0
BROWN	1
RED	2
ORANGE	3
YELLOW	4
GREEN	5
BLUE	6
VIOLET	7
GRAY	8
WHITE	9

MULTIPLIER	
COLOR	MULTIPLIER
BLACK	1
BROWN	10
RED	100
ORANGE	1,000
YELLOW	10,000
GREEN	100,000
BLUE	1,000,000
SILVER	0.01
GOLD	0.1

RESISTANCE TOLERANCE	
COLOR	TOLERANCE
SILVER	±10%
GOLD	± 5%
BROWN	± 1%

Figure 4-9 **Continued.**

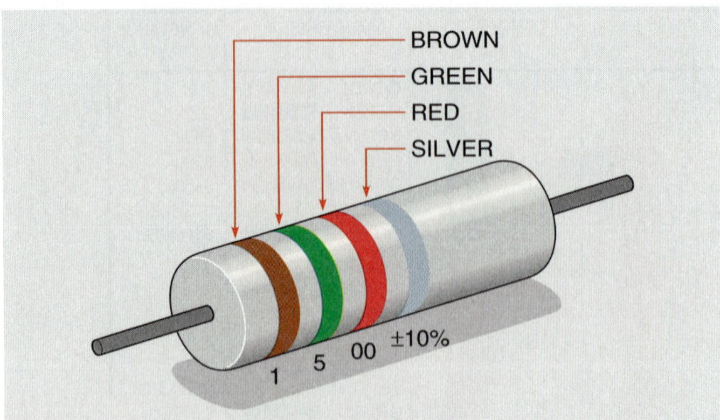

Figure 4-10 Determining resistor values using the color code.

Example 1

The resistor shown in Figure 4-11 contains the following bands of color:

first band 5 brown
second band 5 black
third band 5 black
fourth band 5 brown
fifth band 5 brown

SOLUTION

The brown fifth band indicates that this resistor has a tolerance of ±1%. To determine the value of a 1% resistor, the first three bands are numbers, and the fourth band is the multiplier. In this example, the first band is brown, which has a number value of 1. The next two bands are black, which represents a number value of 0. The fourth band is brown, which means add one 0 to the first three numbers. The value of this resistor is 1000 Ω ±1%.

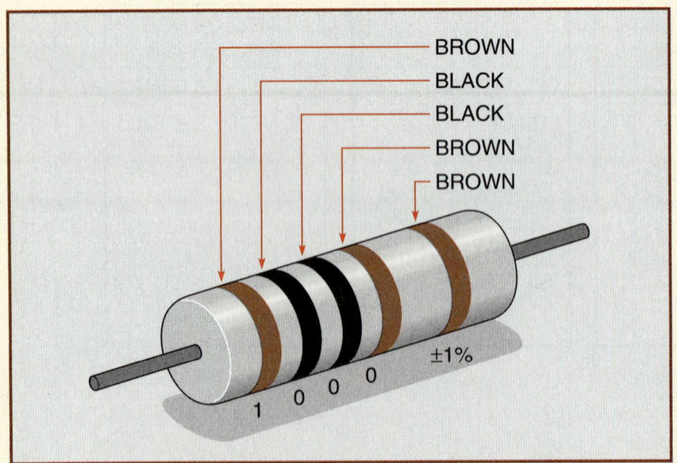

Figure 4-11 Determining the value of a ±1% resistor.

Example 2

A five-band resistor has the following color bands:

first band = red
second band = orange
third band = violet
fourth band = red
fifth band = brown

SOLUTION

The first three bands represent number values. Red is 2, orange is 3, and violet is 7. The fourth band is the multiplier; in this case, red represents 2. Add two zeros to the number 237. The value of the resistor is 23,700 Ω. The fifth band is brown, which indicates a tolerance of ±1%.

Military resistors often have five bands of color also. These resistors are read in the same manner as a resistor with four bands of color. The fifth band can represent different things. A fifth band of orange or yellow is used to indicate reliability. Resistors with a fifth band of orange have a reliability good enough to be used in missile systems, and a resistor with a fifth band of yellow can be used in space flight equipment. A military resistor with a fifth band of white indicates the resistor has solderable leads.

Resistors with tolerance ratings ranging from 0.5% to 0.1% will generally have their values printed directly on the resistor.

Gold and Silver as Multipliers

The colors gold and silver are generally found in the fourth band of a resistor, but they can be used in the multiplier band also. When the color gold is used as the multiplier band, it means to divide the combined first two numbers by 10. If silver is used as the multiplier band, it means to divide the combined first two numbers by 100. For example, assume a resistor has color bands of orange, white, gold, gold. The value of this resistor is 3.9 Ω with a tolerance of ±5% (orange = 3; white = 9; gold means to divide 39 by 10 = 3.9; and gold in the fourth band means ±5% tolerance).

Standard Resistance Values of Fixed Resistors

Fixed resistors are generally produced in standard values. The higher the tolerance value, the fewer resistance values available. Standard resistor values are listed in the chart

STANDARD RESISTANCE VALUES (Ω)									
0.1% 0.25% 0.5%	1%	0.1% 0.25% 0.5%	1%	0.1% 0.25% 0.5%	1%	0.1% 0.25% 0.5%	1%	0.1% 0.25% 0.5%	1%
10.0	10.0	17.2	–	29.4	29.4	50.5	–	86.6	86.6
10.1	–	17.4	17.4	29.8	–	51.1	51.1	87.6	–
10.2	10.2	17.6	–	30.1	30.1	51.7	–	88.7	88.7
10.4	–	17.8	17.8	30.5	–	52.3	52.3	89.8	–
10.5	10.5	18.0	–	30.9	30.9	53.0	–	90.9	90.9
10.6	–	18.2	18.2	31.2	–	53.6	53.6	92.0	–
10.7	10.7	18.4	–	31.6	31.6	54.2	–	93.1	93.1
10.9	–	18.7	18.7	32.0	–	54.9	54.9	94.2	–
11.0	11.0	18.9	–	32.4	32.4	55.6	–	95.3	95.3
11.1	–	19.1	19.1	32.8	–	56.2	56.2	96.5	–
11.3	11.3	19.3	–	33.2	33.2	56.9	–	97.6	97.6
11.4	–	19.6	19.6	33.6	–	57.6	57.6	98.8	–
11.5	11.5	19.8	–	34.0	34.0	58.3	–		
11.7	–	20.0	20.0	34.4	–	59.0	59.0		
11.8	11.8	20.3	–	34.8	34.8	59.7	–		
12.0	–	20.5	20.5	35.2	–	60.4	60.4		
12.1	12.1	20.8	–	35.7	35.7	61.2	–		
12.3	–	21.0	21.0	36.1	–	61.9	61.9		
12.4	12.4	21.3	–	36.5	36.5	62.6	–		
12.6	–	21.5	21.5	37.0	–	63.4	63.4		
12.7	12.7	21.8	–	37.4	37.4	64.2	–	2%,5%	10%
12.9	–	22.1	22.1	37.9	–	64.9	64.9	10	10
13.0	13.0	22.3	–	38.3	38.3	65.7	–	11	–
13.2	–	22.6	22.6	38.8	–	66.5	66.5	12	12
13.3	13.3	22.9	–	39.2	39.2	67.3	–	13	–
13.5	–	23.2	23.2	39.7	–	68.1	68.1	15	15
13.7	13.7	23.4	–	40.2	40.2	69.0	–	16	–
13.8	–	23.7	23.7	40.7	–	69.8	69.8	18	18
14.0	14.0	24.0	–	41.2	41.2	70.6	–	20	–
14.2	–	24.3	24.3	41.7	–	71.5	71.5	22	22
14.3	14.3	24.6	–	42.2	42.2	72.3	–	24	–
14.5	–	24.9	24.9	42.7	–	73.2	73.2	27	27
14.7	14.7	25.2	–	43.2	43.2	74.1	–	30	–
14.9	–	25.5	25.5	43.7	–	75.0	75.0	33	33
15.0	15.0	25.8	–	44.2	44.2	75.9	–	36	–
15.2	–	26.1	26.1	44.8	–	76.8	76.8	39	39
15.4	15.4	26.4	–	45.3	45.3	77.7	–	43	–
15.6	–	26.7	26.7	45.9	–	78.7	78.7	47	47
15.8	15.8	27.1	–	46.4	46.4	79.6	–	51	–
16.0	–	27.4	27.4	47.0	–	80.6	80.6	56	56
16.2	16.2	27.7	–	47.5	47.5	81.6	–	62	–
16.4	–	28.0	28.0	48.1	–	82.5	82.5	68	68
16.5	16.5	28.4	–	48.7	48.7	83.5	–	75	–
16.7	–	28.7	28.7	49.3	–	84.5	84.5	82	82
16.9	16.9	29.1	–	49.9	49.9	85.6	–	91	–

Figure 4-12 **Standard resistance values.**

shown in Figure 4-12. In the column under 10% only 12 values of resistors are listed. These standard values, however, can be multiplied by factors of 10. Notice that one of the standard values listed is 33 Ω. There are also standard values in 10% resistors of 0.33, 3.3, 330, 3300, 33,000, 330,000, and 3,300,000 Ω. The 2% and 5% column shows 24 resistor values, and the 1% column lists 96 values. All of the values listed in the chart can be multiplied by factors of 10 to obtain other resistance values.

Power Ratings

Resistors also have a power rating in watts that should not be exceeded or the resistor will be damaged. The amount of heat that must be dissipated (given off to the surrounding air) by the resistor can be determined using one of the following formulas.

$$P = I^2R$$

$$P = \frac{E^2}{R}$$

$$P = EI$$

Example 3

The resistor shown in Figure 4-13 has a value of 100 Ω and a power rating of 1/2 W. If the resistor is connected to a 10-V power supply, will it be damaged?

SOLUTION

Using the formula $P = \frac{E^2}{R}$ determine the amount of heat that will be dissipated by the resistor.

$$P = \frac{E^2}{R}$$

$$P = \frac{10 \times 10}{100}$$

$$P = \frac{100}{100}$$

$$P = 1\ W$$

Since the resistor has a power rating of 1/2 W, and the amount of heat that will be dissipated is 1 W, the resistor will be damaged.

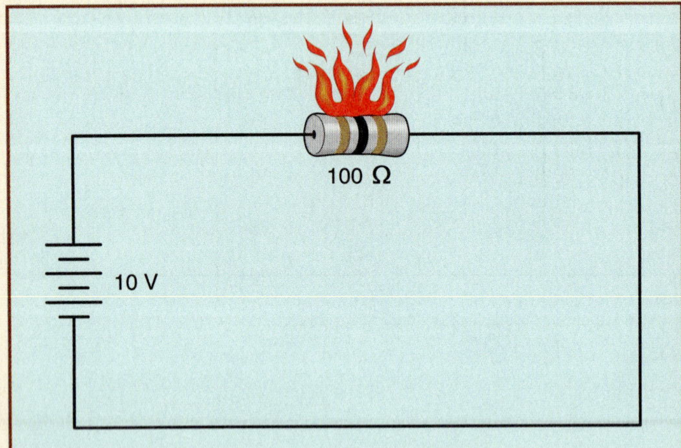

Figure 4-13 Exceeding the power rating causes damage to the resistor.

Variable Resistors

A **variable resistor** is a resistor whose values can be changed or varied over a range. Variable resistors can be obtained in different case styles and power ratings. Figure 4-14 illustrates how a variable resistor is constructed. In this example, a resistive wire is wound in a circular pattern, and a sliding tap makes contact with the wire. The value of resistance can be adjusted between one end of the resistive wire and the sliding tap. If the resistive wire has a total value of 100 Ω, the resistor can be set between the values of 0 and 100 Ω.

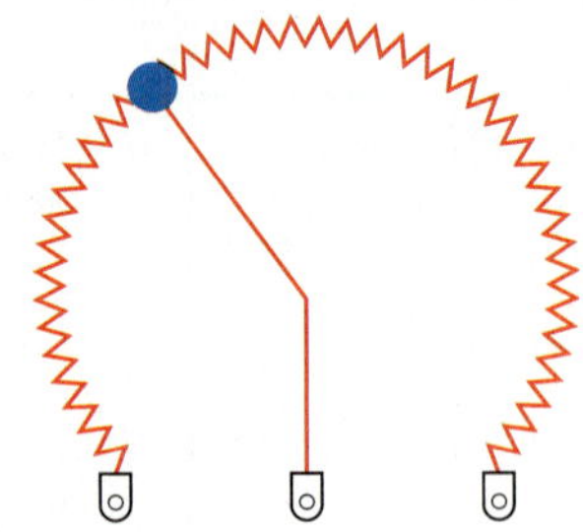

Figure 4-14 Variable resistor.

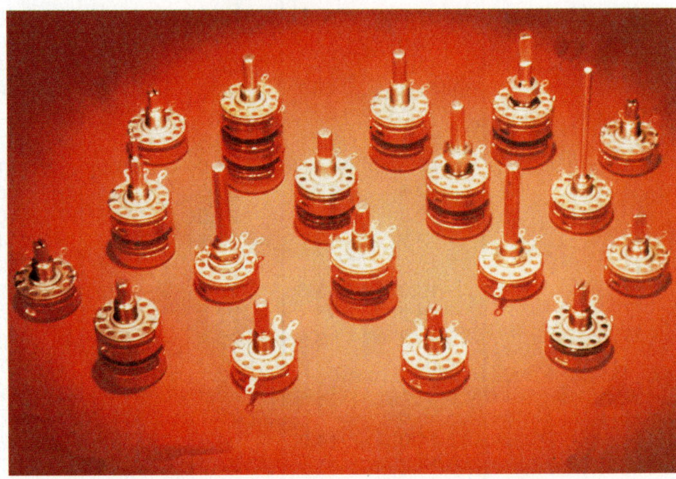

Figure 4-15 Variable resistors with three terminals. *Courtesy of Allen Bradley Co. Inc., a Rockwell International company.*

A variable resistor with three terminals is shown in Figure 4-15. This type of resistor has a wiper arm inside the case that makes contact with the resistive element. The full resistance value is between the two outside terminals, and the wiper arm is connected to the center terminal. The resistance between the center terminal and either of the two outside terminals can be adjusted by turning the shaft and changing the position of the wiper arm. Wire wound variable resistors of this type can be obtained also (Fig. 4-16). The advantage of the wire wound type is a higher power rating.

The resistor shown in Figure 4-15 can be adjusted from its minimum to maximum value by turning the control approximately three-quarters of a turn. In some types of electrical equipment, this range of adjustment may be too coarse to allow for sensitive adjustments. When this becomes a problem, a multiturn resistor (Fig. 4-17) can be used. **Multiturn variable resistors** operate by moving the wiper arm with a screw of some number of turns. They generally range from 3 turns to 10 turns. If a 10-turn variable resistor is used, it

Figure 4-16 **Wire wound variable resistor.**

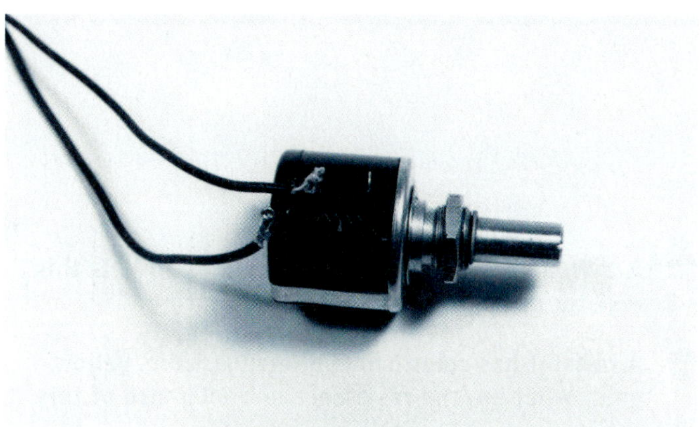

Figure 4-17 **Multiturn variable resistor.**

will require 10 turns of the control knob to move the wiper from one end of the resistor to the other end instead of three-quarters of a turn.

Variable Resistor Terminology

Variable resistors are known by several common names. The most popular name is **pot,** which is shortened from the word *potentiometer.* Another common name is **rheostat.** A rheostat is actually a variable resistor that has two terminals. They are used to adjust the current in a circuit to a certain value. A potentiometer is a variable resistor that has three terminals. Potentiometers can be used as rheostats, by only using two of their three terminals. A **potentiometer** describes how a variable resistor is used rather than some specific type of resistor. The word *potentiometer* comes from the word *potential,* or voltage. A potentiometer is a variable resistor used to provide a variable voltage, as shown in Figure 4-18. In this example, one end of a variable resistor is connected to + 12 V, and the other end is connected to ground. The middle terminal, or wiper, is connected to the positive terminal of a voltmeter and the negative lead is connected to ground. If the wiper is moved to the upper end of the resistor, the voltmeter will indicate a potential of 12 V. If the wiper is moved to the bottom, the voltmeter will indicate a value of 0 V. The wiper can be adjusted to provide any value of voltage between 12 and 0 V.

Schematic Symbols

Electrical schematics use symbols to represent the use of a resistor. Unfortunately, the symbol used to represent a resistor is not standard. Figure 4-19 illustrates several schematic symbols used to represent both fixed and variable resistors.

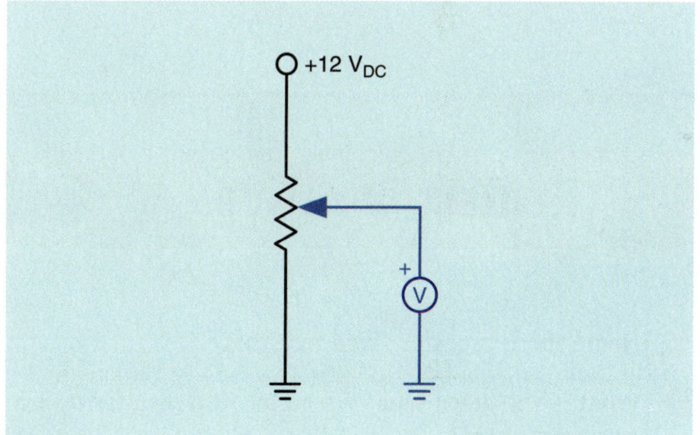

Figure 4-18 **Variable resistor used as a potentiometer.**

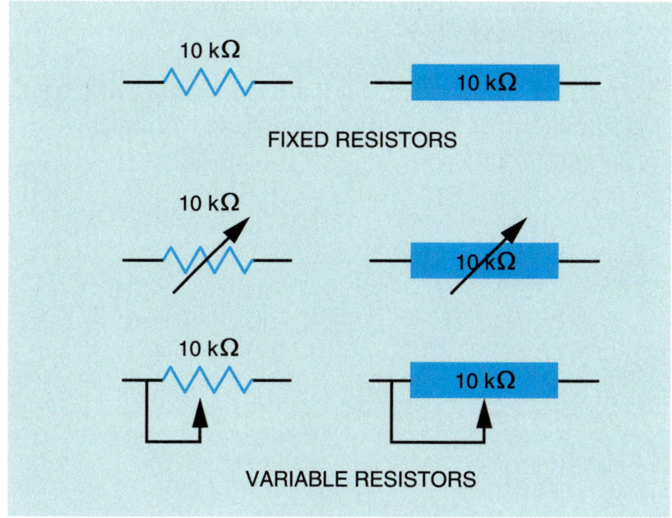

Figure 4-19 **Schematic symbols used to represent resistors.**

Summary

- Resistors are used in two main applications: as voltage dividers and to limit the flow of current in a circuit.
- The value of fixed resistors cannot be changed.
- There are several types of fixed resistors such as composition carbon, metal film, and wire wound.
- Carbon resistors change their resistance with age or if overheated.
- Metal film resistors never change their value, but are more expensive than carbon resistors.
- The advantage of wire wound resistors is their high power ratings.
- Resistors often have bands of color to indicate their resistance value and tolerance.
- Resistors are produced in standard values. The number of values between 0 and 100 Ω is determined by the tolerance.
- Variable resistors can change their value within the limit of their full value.
- A potentiometer is a variable resistor used as a voltage divider.

Review Questions

1. **Name three types of fixed resistors.**
2. **What is the advantage of a metal film resistor over a carbon resistor?**
3. **What is the advantage of a wire wound resistor?**
4. **How should tubular wire wound resistors be mounted and why?**
5. **A 0.5-W, 2000-Ω resistor has a current flow of 0.01 A through it. Is this resistor operating within its power rating?**
6. **A 1-W, 350-Ω resistor is connected to 24 V. Is this resistor operating within its power rating?**
7. **A resistor has color bands of orange, blue, yellow, gold. What are the resistance and tolerance of this resistor?**
8. **A 10,000-Ω resistor has a tolerance of 5%. What are the minimum and maximum ratings of this resistor?**
9. **Is 51,000 Ω a standard value for a 5% resistor?**
10. **What is a potentiometer?**

Practice Problems

Resistors

Fill in the missing values.

1ST Band	2ND Band	3RD Band	4TH Band	Value	%Tol
Red	Yellow	Brown	Silver		
				6800 Ω	5
Orange	Orange	Orange	Gold		
				12 Ω	2
Brown	Green	Silver	Silver		
				1.8 MΩ	10
Brown	Black	Yellow	None		
				10 kΩ	5
Violet	Green	Black	Red		
				4.7 kΩ	20
Gray	Red	Green	Red		
				5.6 Ω	2

Chapter 5 Series Circuits

Electrical circuits can be divided into three major types: series, parallel, and combination. Combination circuits contain both series and parallel paths. The first type discussed is the series circuit.

OBJECTIVES

After studying this unit, you should be able to:

- **discuss the properties of series circuits.**
- **list three rules for solving electrical values of series circuits.**
- **compute values of voltage, current, resistance, and power for series circuits.**
- **compute the values of voltage drop in a series circuit using the voltage divider formula.**

Glossary of Series Circuit Terms

chassis ground a conductor connected to the metal case of a piece of equipment for the purpose of placing the case at ground potential

circuit breakers devices that disconnect power to a circuit in the event of an excessive amount of current

earth ground point at which a metal rod is driven into the ground to provide a low-resistance path to the earth

fuses devices that disconnect power to a circuit in the event of an excessive amount of current

general voltage divider formula a formula employed to determine the value of resistance needed to produce a certain voltage drop

ground point point at which a circuit or device is grounded

resistance adds one of the major rules concerning series circuits

series circuit a circuit that contains only one path for current flow

series resistance resistance connected in series with other circuit components

voltage divider a circuit used to provide different values of voltage

voltage drop the amount of voltage existing across any component that limits the flow of current; voltage drop indicates the amount of voltage necessary to push current through a limiting element

voltage polarity the polarity, positive or negative, of the voltage at a certain point in a circuit

Series Circuits

A **series circuit is a circuit that has only one path for current flow** (Fig. 5-1). Because there is only one path for current flow, the current is the same at any point in the circuit. Imagine that an electron leaves the negative terminal of the battery. This electron must flow through each resistor before it can complete the circuit to the positive battery terminal.

One of the best examples of a series-connected device is a fuse or circuit breaker (Fig. 5-2). Since **fuses** and **circuit breakers** are connected in series with the rest of the circuit, all the circuit current must flow through them. If the current becomes excessive, the fuse or circuit breaker will open and disconnect the rest of the circuit from the power source.

Voltage Drops in a Series Circuit

Voltage is the force that pushes the electrons through a resistance. The amount of voltage required is determined by the amount of current flow and resistance. If a voltmeter is connected across a resistor (Fig. 5-3), the amount of voltage necessary to push the current through that resistor will be indicated by the meter. This amount is known as **voltage drop.** It is similar to pressure drop in a water system. **In a series circuit, the sum of all the voltage drops across all the resistors must equal the voltage applied to the circuit.** The amount of voltage drop across each resistor will be proportional to its resistance and the circuit current.

In the circuit shown in Figure 5-4 four resistors are connected in series. It is assumed that all four resistors have the same value. The circuit is connected to a 24-V battery. Since all the resistors have the same value, the voltage drop across each will be 6 V (24 V/4 resistors = 6 V). Note that all four

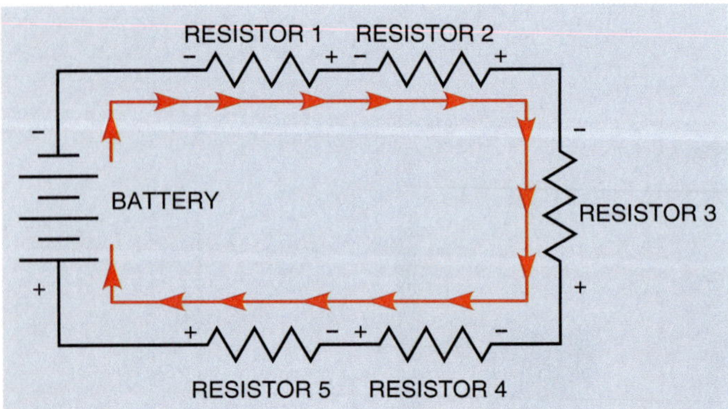

Figure 5-1 In a series circuit there is only one path for current flow.

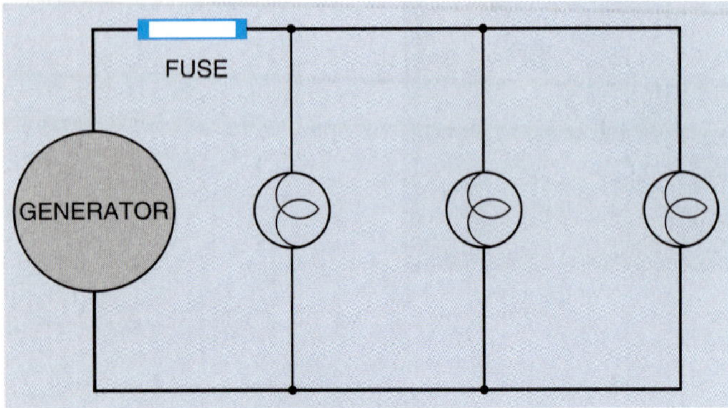

Figure 5-2 All the current must flow through the fuse.

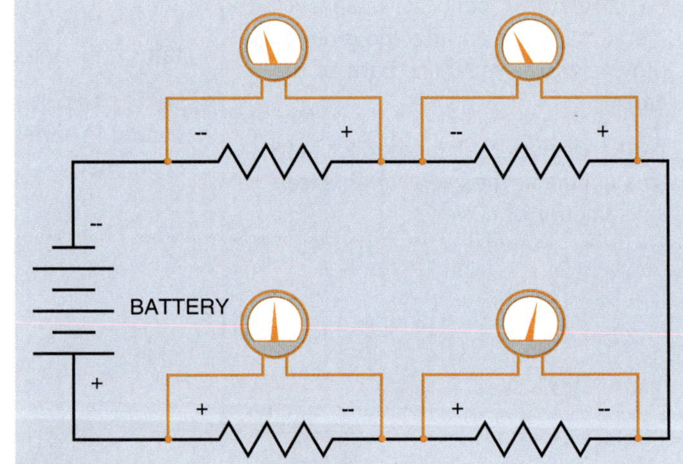

Figure 5-3 The voltage drops in a series circuit must equal the applied voltage.

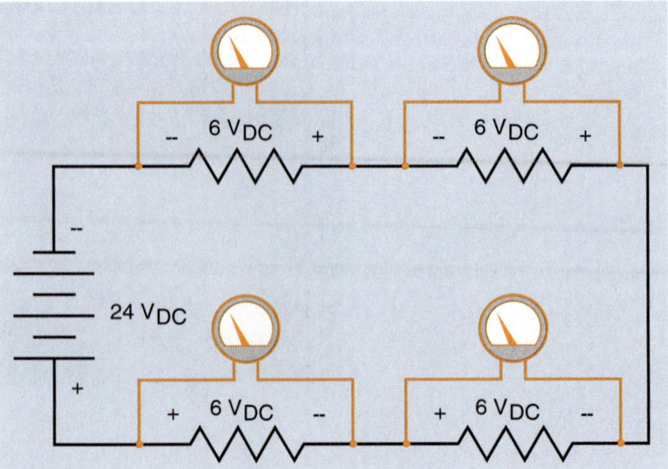

Figure 5-4 The voltage drop across each resistor is proportional to its resistance.

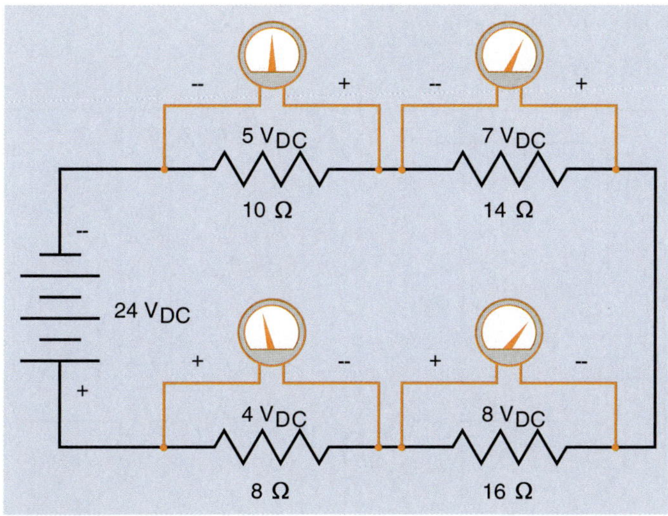

Figure 5-5 Series circuit with four resistors having different voltage drops.

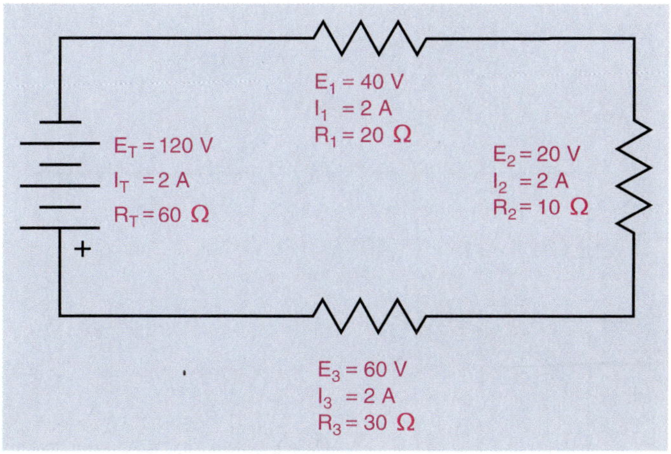

Figure 5-6 Series circuit values.

resistors will have the same voltage drop only if they all have the same value. The circuit shown in Figure 5-5 illustrates a series circuit comprising resistors of different values. Notice that the voltage drop across each resistor is proportional to its resistance. Also notice that the sum of the voltage drops is equal to the applied voltage of 24 V.

Resistance in a Series Circuit

Because there is only one path for the current to flow through a series circuit, it must flow through each resistor in the circuit (Fig. 5-1). Each resistor limits or impedes the flow of current in the circuit. Therefore, the total amount of **series resistance** to current flow in a series circuit is equal to the sum of the resistances in that circuit.

Calculating Series Circuit Values

Three rules can be used with Ohm's law for finding values of voltage, current, resistance, and power in any series circuit.

1. **The current is the same at any point in the circuit.**
2. **The total resistance is the sum of the individual resistors.**
3. **The applied voltage is equal to the sum of the voltage drops across all the resistors.**

The circuit shown in Figure 5-6 shows the values of current flow, voltage drop, and resistance for each of the resistors. The total resistance (R_T) of the circuit can be found by adding the values of the three resistors (**resistance adds**).

$$R_T = R_1 + R_2 + R_3$$

$$R_T = 20\ \Omega + 10\ \Omega + 30\ \Omega$$

$$R_T = 60\ \Omega$$

The amount of current flow in the circuit can be found using Ohm's law.

$$I = \frac{E}{R}$$

$$I = \frac{120}{60}$$

$$I = 2\ A$$

A current of 2 A flows through each resistor in the circuit.

$$I_T = I_1 = I_2 = I_3$$

Since the amount of current flowing through resistor R_1 is known, the voltage drop across the resistor can be found using Ohm's law.

$$E_1 = I_1 \times R_1$$

$$E_1 = 2\ A \times 20\ \Omega$$

$$E_1 = 40\ V$$

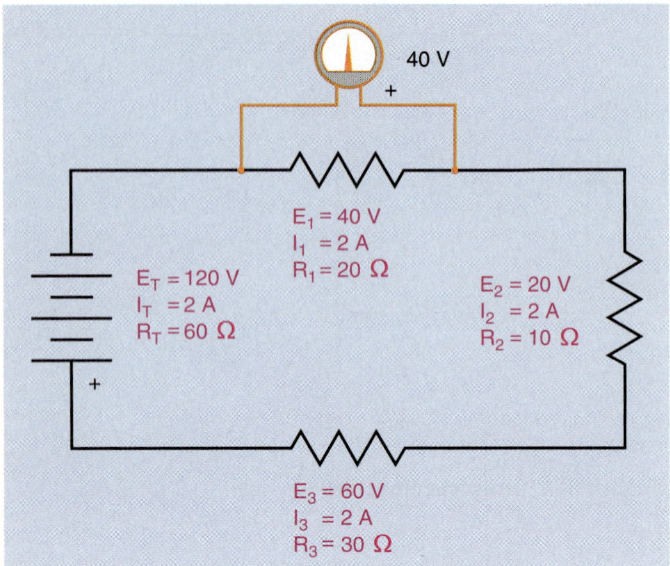

Figure 5-7 The voltmeter indicates a voltage drop of 40 V.

In other words, it takes 40 V to push 2 A of current through 20 Ω of resistance. If a voltmeter were connected across resistor R_1, it would indicate a value of 40 V (Fig. 5-7). The voltage drop across resistors R_2 and R_3 can be found in the same way.

$$E_2 = I_2 \times R_2$$
$$E_2 = 2\text{ A} \times 10\ \Omega$$
$$E_2 = 20\text{ V}$$

$$E_3 = I_3 \times R_3$$
$$E_3 = 2\text{ A} \times 30\ \Omega$$
$$E_3 = 60\text{ V}$$

If the voltage drop across all the resistors is added, it equals the total applied voltage (E_T).

$$E_T = E_1 + E_2 + E_3$$
$$E_T = 40\text{ V} + 20\text{ V} + 60\text{ V}$$
$$E_T = 120\text{ V}$$

Solving Circuits

In the following problems, circuits that have missing values are shown. The missing values can be found using the rules for series circuits and Ohm's law.

Example 1

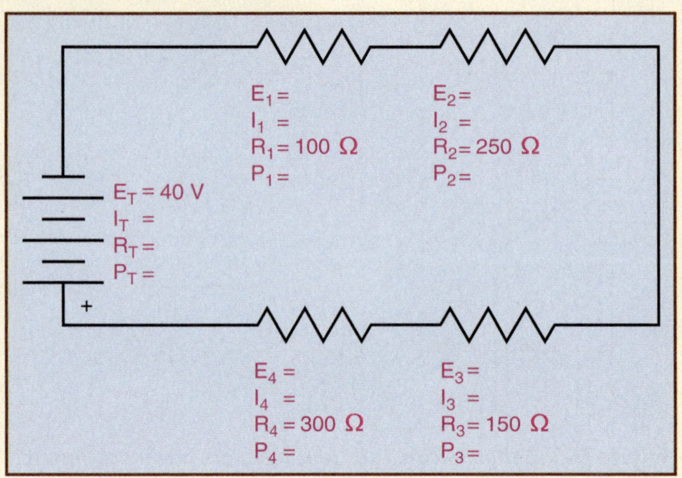

Figure 5-8 Series circuit, Example 1.

The first step in finding the missing values in the circuit shown in Figure 5-8 is to find the total resistance (R_T). This can be done using the second rule of series circuits, which states that resistances add to equal the total resistance of the circuit.

$$R_T = R_1 + R_2 + R_3 + R_4$$
$$R_T = 100\ \Omega + 250\ \Omega + 150\ \Omega + 300\ \Omega$$
$$R_T = 800\ \Omega$$

Now that the total voltage and total resistance are known, the current flow through the circuit can be found using Ohm's law.

$$I = \frac{E}{R}$$
$$I = \frac{40}{800}$$
$$I = 0.050\text{ A}$$

The first rule of series circuits states that current remains the same at any point in the circuit. Therefore, 0.050 A flows through each resistor in the circuit (Fig. 5-9). The voltage drop across each resistor can now be found using Ohm's law (Fig. 5-10).

$$E_1 = I_1 \times R_1$$
$$E_1 = 0.050 \times 100$$
$$E_1 = 5\text{ V}$$

Example 1 Continued

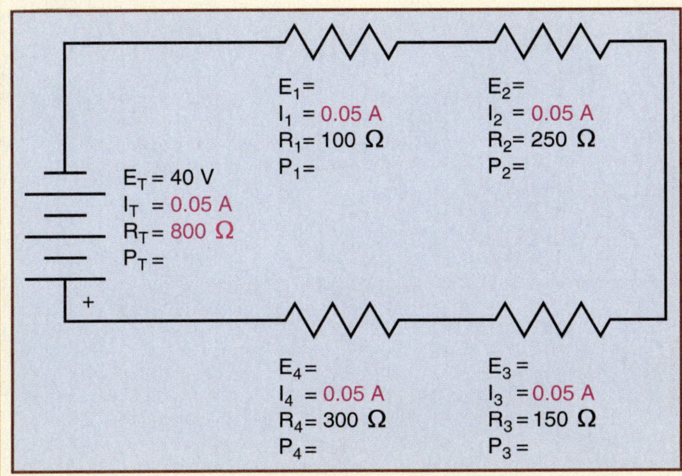

Figure 5-9 The current is the same at any point in a series circuit.

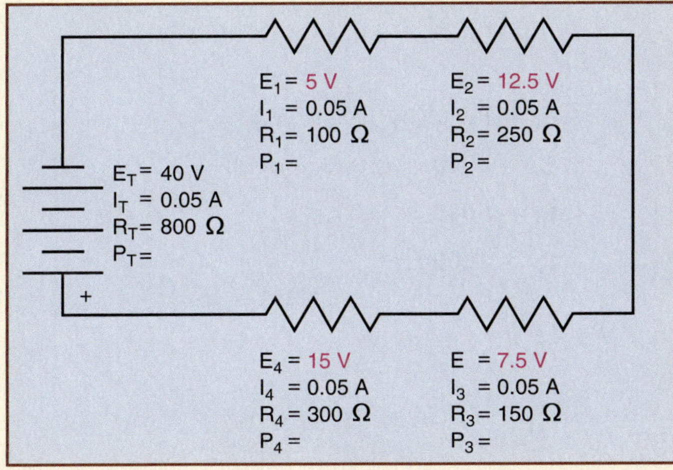

Figure 5-10 The voltage drop across each resistor can be found using Ohm's law.

$$E_2 = I_2 \times R_2$$
$$E_2 = 0.050 \times 250$$
$$E_2 = 12.5\ V$$

$$E_3 = I_3 \times R_3$$
$$E_3 = 0.050 \times 150$$
$$E_3 = 7.5\ V$$

$$E_4 = I_4 \times R_4$$
$$E_4 = 0.050 \times 300$$
$$E_4 = 15\ V$$

Several formulas can be used to determine the amount of power dissipated (converted into heat) by each resistor. The power dissipation of resistor R_1 will be found using the formula

$$P_1 = E_1 \times I_1$$
$$P_1 = 5 \times 0.05$$
$$P_1 = 0.25\ W$$

The amount of power dissipation for resistor R_2 will be computed using the formula

$$P_2 = \frac{E_2^2}{R_2}$$
$$P_2 = \frac{156.25}{250}$$
$$P_2 = 0.625\ W$$

The amount of power dissipation for resistor R_3 will be computed using the formula

$$P_3 = I_3^2 \times R_3$$
$$P_3 = 0.0025 \times 150$$
$$P_3 = 0.375\ W$$

The amount of power dissipation for resistor R_4 will be found using the formula

$$P_4 = E_4 \times I_4$$
$$P_4 = 15 \times 0.05$$
$$P_4 = 0.75\ W$$

A good rule to remember when calculating values of electrical circuits is that **the total power used in a circuit is equal to the sum of the power used by all parts.** That is, the total power can be found in any kind of a circuit—series, parallel, or combination—by adding the power dissipation of all the parts. The total power for this circuit can be found using the formula

$$P_T = P_1 + P_2 + P_3 + P_4$$
$$P_T = 0.25 + 0.625 + 0.375 + 0.75$$
$$P_T = 2\ W$$

Example 1 Continued

Now that all the missing values have been found (Fig. 5-11), the circuit can be checked using the third rule of series circuits, which states that voltage drops add to equal the applied voltage.

$$E_T = E_1 + E_2 + E_3 + E_4$$

$$E_T = 5 + 12.5 + 7.5 + 15$$

$$E_T = 40\text{ V}$$

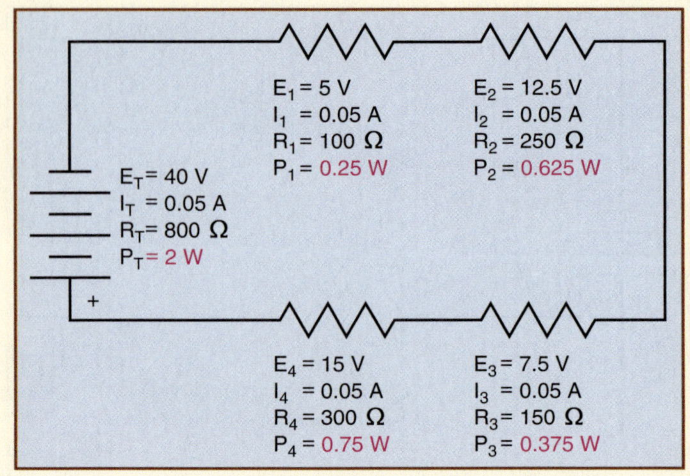

Figure 5-11 The final values for the circuit in Example 1.

Example 2

The second circuit to be solved is shown in Figure 5-12. In this circuit the total resistance is known, but the value of resistor R_2 is not. The second rule of series circuits states that resistances add to equal the total resistance of the circuit. Since the total resistance is known, the missing resistance of R_2 can be found by adding the values of the other resistors and subtracting their sum from the total resistance of the circuit (Fig. 5-13).

$$R_2 = R_T - (R_1 + R_3 + R_4)$$

$$R_2 = 6000 - (1000 + 2000 + 1200)$$

$$R_2 = 6000 - 4200$$

$$R_2 = 1800\ \Omega$$

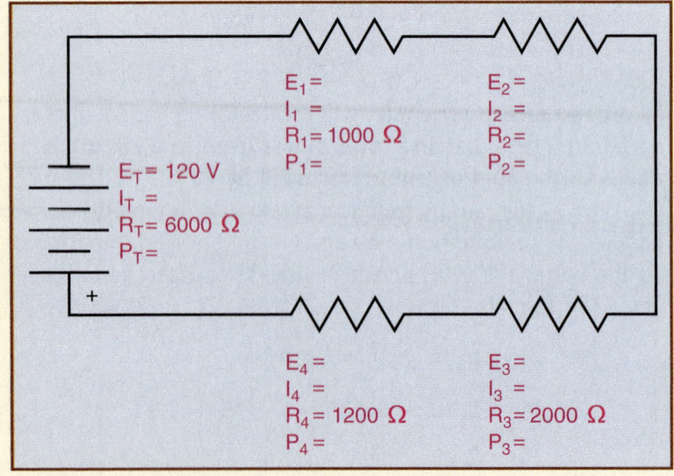

Figure 5-12 Series circuit, Example 2.

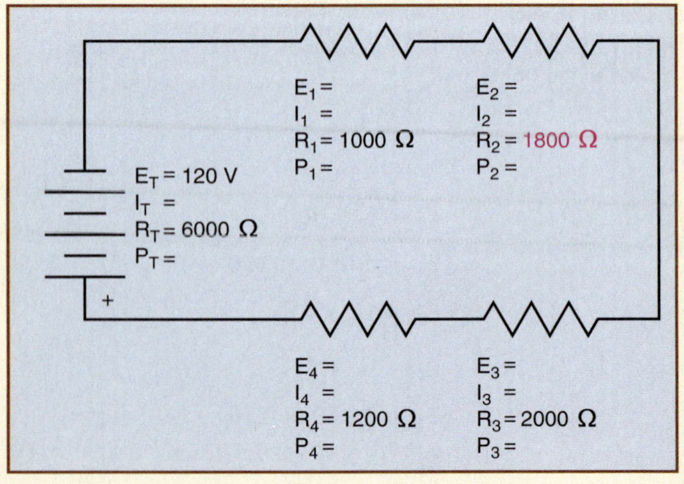

Figure 5-13 The missing resistor value.

Example 2 Continued

The amount of current flow in the circuit can be found using Ohm's law.

$$I = \frac{E}{R}$$

$$I = \frac{120}{6000}$$

$$I = 0.020\ \text{A}$$

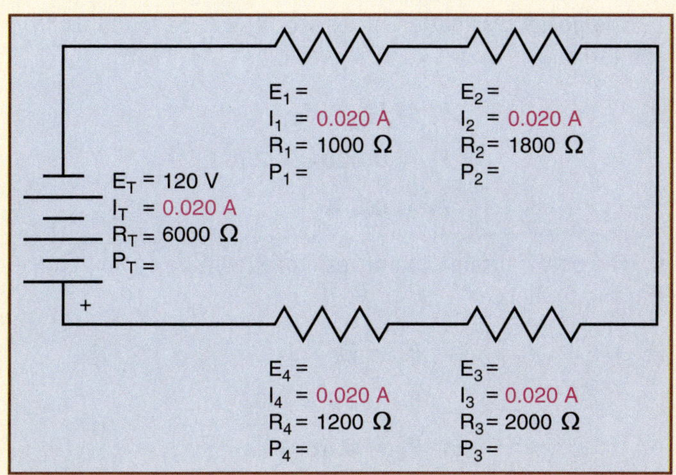

Figure 5-14 The current is the same through each circuit element.

Since the amount of current flow is the same through all elements of a series circuit (Fig. 5-14), the voltage drop across each resistor can be found using Ohm's law (Fig. 5-15).

$$E_1 = I_1 \times R_1$$

$$E_1 = 0.020 \times 1000$$

$$E_1 = 20\ \text{V}$$

$$E_2 = I_2 \times R_2$$

$$E_2 = 0.020 \times 1800$$

$$E_2 = 36\ \text{V}$$

$$E_3 = I_3 \times R_3$$

$$E_3 = 0.020 \times 2000$$

$$E_3 = 40\ \text{V}$$

$$E_4 = I_4 \times R_4$$

$$E_4 = 0.020 \times 1200$$

$$E_4 = 24\ \text{V}$$

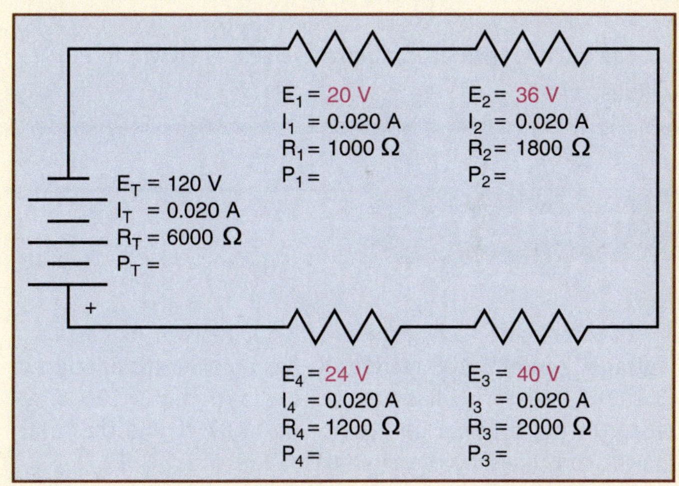

Figure 5-15 The voltage drops across each resistor.

The third rule of series circuits can be used to check the answers.

$$E_T = E_1 + E_2 + E_3 + E_4$$

$$E_T = 20 + 36 + 40 + 24$$

$$E_T = 120\ \text{V}$$

The amount of power dissipation for each resistor in the circuit can be computed using the same method used to solve the circuit in Example 1. The power dissipated by resistor R_1 will be computed using the formula

$$P_1 = E_1 \times I_1$$

$$P_1 = 20 \times 0.02$$

$$P_1 = 0.4\ \text{W}$$

The amount of power dissipation for resistor R_2 will be found using the formula

$$P_2 = \frac{E_2{}^2}{R_2}$$

$$P_2 = \frac{1296}{1800}$$

$$P_2 = 0.72\ \text{W}$$

Example 2 Continued

The power dissipation of resistor R_3 will be found using the formula

$$P_3 = I_3^2 \times R_3$$

$$P_3 = 0.0004 \times 2000$$

$$P_3 = 0.8 \text{ W}$$

The power dissipation of resistor R_4 will be computed using the formula

$$P_4 = E_4 \times I_4$$

$$P_4 = 24 \times 0.02$$

$$P_4 = 0.48 \text{ W}$$

The total power will be computed using the formula

$$P_T = E_T \times I_T$$

$$P_T = 120 \times 0.02$$

$$P_T = 2.4 \text{ W}$$

The circuit, with all computed values, is shown in Figure 5-16.

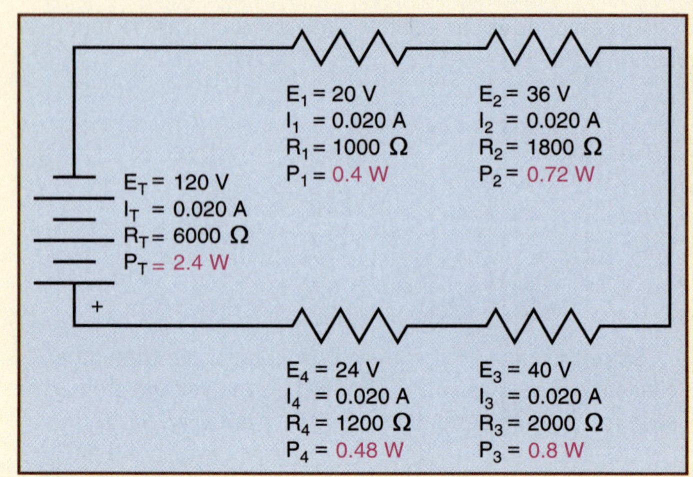

Figure 5-16 The remaining unknown values for the circuit in Example 2.

Example 3

In the circuit shown in Figure 5-17, resistor R_1 has a voltage drop of 6.4 V, resistor R_2 has a power dissipation of 0.102 W, resistor R_3 has a power dissipation of 0.154 W, resistor R_4 has a power dissipation of 0.307 W, and the total power consumed by the circuit is 0.768 W.

The only value that can be found with the given quantities is the amount of power dissipated by resistor R_1. Since the total power is known and the power dissipated by the three other resistors is known, the power dissipated by resistor R_1 can be found by subtracting the power dissipated by resistors R_2, R_3, and R_4 from the total power used in the circuit.

$$P_1 = P_T - (P_2 + P_3 + P_4)$$

or

$$P_1 = P_T - P_2 - P_3 - P_4$$

$$P_1 = 0.768 - 0.102 - 0.154 - 0.307$$

$$P_1 = 0.205 \text{ W}$$

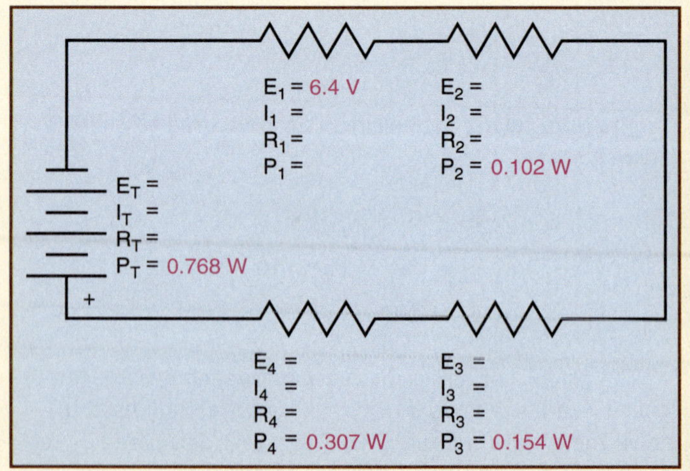

Figure 5-17 Series circuit, Example 3.

Example 3 Continued

Now that the amount of power dissipated by resistor R_1 and the voltage drop across R_1 are known, the current flow through resistor R_1 can be found using the formula

$$I = \frac{P}{E}$$

$$I = \frac{0.205}{6.4}$$

$$I = 0.032 \text{ A}$$

Since the current in a series circuit must be the same at any point in the circuit, it must be the same through all circuit components (Fig. 5-18).

Now that the power dissipation of each resistor and the amount of current flowing through each resistor are known, the voltage drop of each resistor can be computed (Fig. 5-19).

$$E_2 = \frac{P_2}{I_2}$$

$$E_2 = \frac{0.102}{0.032}$$

$$E_2 = 3.2 \text{ V}$$

$$E_3 = \frac{P_3}{I_3}$$

$$E_3 = \frac{0.154}{0.032}$$

$$E_3 = 4.8 \text{ V}$$

$$E_4 = \frac{P_4}{I_4}$$

$$E_4 = \frac{0.307}{0.032}$$

$$E_4 = 9.6 \text{ V}$$

Ohm's law can now be used to find the Ohmic value of each resistor in the circuit (Fig. 5-20).

$$R_1 = \frac{E_1}{I_1}$$

$$R_1 = \frac{6.4}{0.032}$$

$$R_1 = 200\ \Omega$$

$$R_2 = \frac{E_2}{I_2}$$

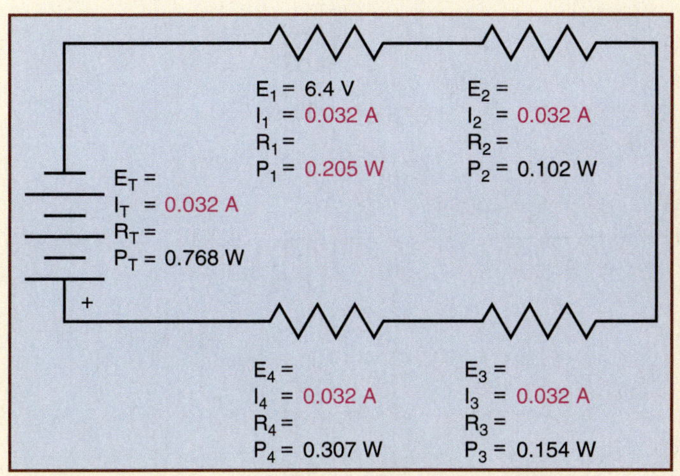

Figure 5-18 The current flow in the circuit in Example 3.

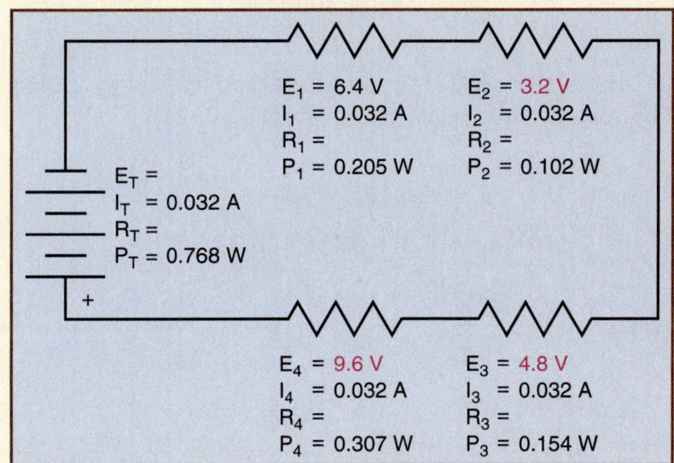

Figure 5-19 The voltage drops across each resistor.

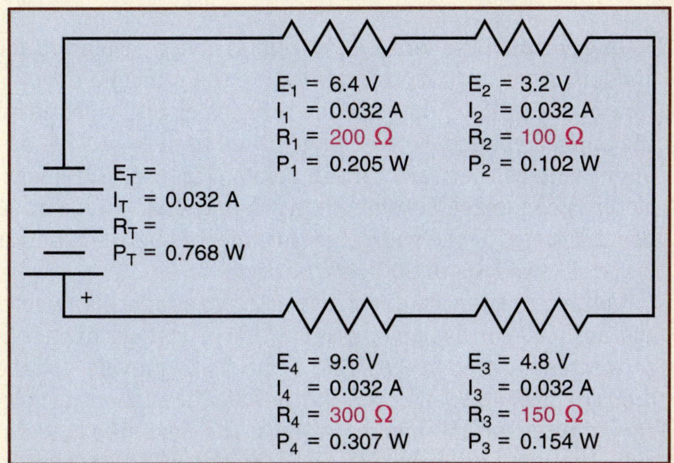

Figure 5-20 The Ohmic value of each resistor.

Example 3 Continued

$$R_2 = \frac{3.2}{0.032}$$

$$R_2 = 100\ \Omega$$

$$R_3 = \frac{E_3}{I_3}$$

$$R_3 = \frac{4.8}{0.032}$$

$$R_3 = 150\ \Omega$$

$$R_4 = \frac{E_4}{I_4}$$

$$R_4 = \frac{9.6}{0.032}$$

$$R_4 = 300\ \Omega$$

The voltage applied to the circuit can be found by adding the voltage drops across the resistor (Fig. 5-21).

$$E_T = E_1 + E_2 + E_3 + E_4$$

$$E_T = 6.4 + 3.2 + 4.8 + 9.6$$

$$E_T = 24\ V$$

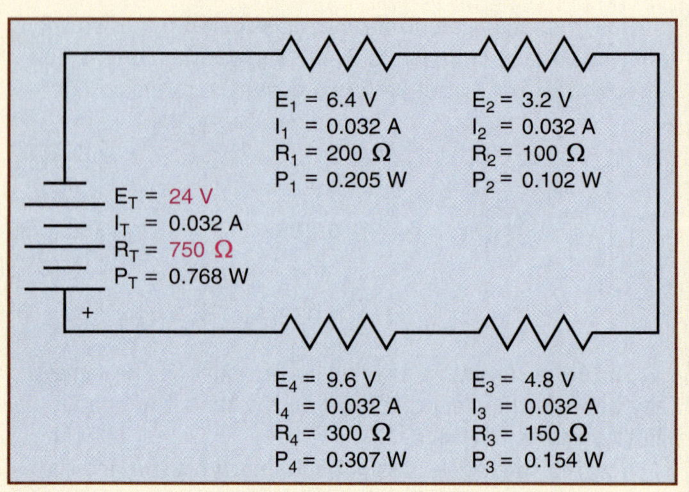

Figure 5-21 The applied voltage and the total resistance.

The total resistance of the circuit can be found in a similar manner (Fig. 5-21). The total resistance is equal to the sum of all the resistive elements in the circuit.

$$R_T = R_1 + R_2 + R_3 + R_4$$

$$R_T = 200 + 100 + 150 + 300$$

$$R_T = 750\ \Omega$$

Voltage Dividers

One common use for series circuits is constructing voltage dividers. A **voltage divider** works on the principle that the sum of the voltage drops across a series circuit must equal the applied voltage. Voltage dividers are used to provide different voltages between certain points (Fig. 5-22). If a voltmeter is connected between points A and B, a voltage of 20 V will be seen. If the voltmeter is connected between points B and D, a voltage of 80 V will be seen.

Voltage dividers can be constructed to produce any voltage desired. For example, assume that a voltage divider is connected to a source of 120 V and is to provide voltage drops of 36 V, 18 V, and 66 V. Notice that the sum of the voltage drops equals the applied voltage. The next step is to decide how much current is to flow through the circuit. Because there is only one path for current flow, the current will be the same through all the resistors. In this circuit, a

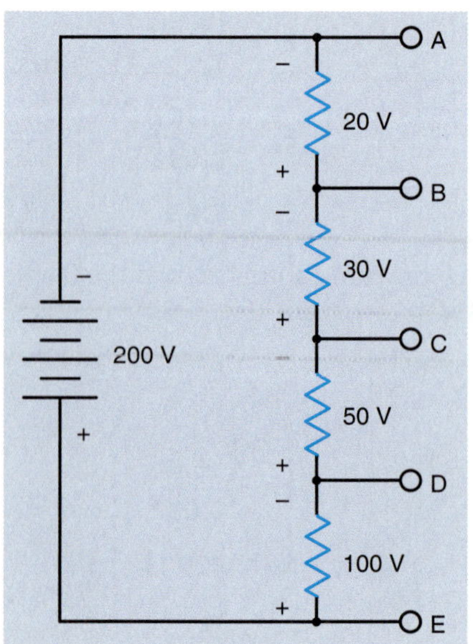

Figure 5-22 Series circuit used as a voltage divider.

current flow of 15 mA (0.015 A) will be used. The resistance value of each resistor can now be determined.

$$R = \frac{E}{I}$$

$$R_1 = \frac{36}{0.015}$$

$$R_1 = 2.4\ k\Omega\ (2400\ \Omega)$$

$$R_2 = \frac{18}{0.015}$$

$$R_2 = 1.2\ k\Omega\ (1200\ \Omega)$$

$$R_3 = \frac{66}{0.015}$$

$$R_3 = 4.4\ k\Omega\ (4400\ \Omega)$$

The General Voltage Divider Formula

Another method of determining the voltage drop across series elements is the **general voltage divider formula.** Since the current flow through a series circuit is the same at all points in the circuit, the voltage drop across any particular resistance is equal to the total circuit current times the value of that resistor.

$$E_X = I_T \times R_X$$

The total circuit current is proportional to the source voltage (E_T) and the total resistance of the circuit.

$$I_T = \frac{E_T}{R_T}$$

If the value of I_T is substituted for E_T/R_T in the previous formula, the expression now becomes

$$E_X = \left(\frac{E_T}{R_T}\right)R_X$$

If the formula is rearranged, it becomes what is known as the general voltage divider formula.

$$E_X = \left(\frac{R_X}{R_T}\right)E_T$$

The voltage drop across any series component (E_X) can be computed by substituting the value of R_X for the resistance value of that component when the source voltage and total resistance are known.

Example 4

Three resistors are connected in series to a 24 volt source. Resistor R_1 has a resistance of 200 Ω, resistor R_2 has a value of 300 Ω, and resistor R_3 has a value of 160 Ω. What is the voltage drop across each resistor?

SOLUTION

Find the total resistance of the circuit.

$$R_T = R_1 + R_2 + R_3$$

$$R_T = 200 + 300 + 160$$

$$R_T = 660\ \Omega$$

Now use the voltage divider formula to compute the voltage drop across each resistor.

$$E_1 = \left(\frac{R_1}{R_T}\right)E_T$$

$$E_1 = \times \left(\frac{200}{660}\right)24$$

$$E_1 = 7.273 \text{ volts}$$

$$E_2 = \left(\frac{R_2}{R_T}\right)E_T$$

$$E_2 = \left(\frac{300}{660}\right)24$$

$$E_2 = 10.91 \text{ volts}$$

$$E_3 = \left(\frac{R_3}{R_T}\right)E_T$$

$$E_3 = \left(\frac{160}{660}\right)24$$

$$E_3 = 5.818 \text{ volts}$$

Voltage Polarity

It is often necessary to know the polarity of the voltage developed across a resistor. A specific point is not negative or positive itself, it is negative or positive with respect to some other point. This principle holds true with voltage as well. A specific point has no voltage itself, however, the difference of potential between that point and another point, is in fact a voltage. **Voltage polarity** can be determined by observing the direction of current flow through the circuit. In the circuit shown in Figure 5-22 it will be assumed that the current flows from the negative terminal of the battery to the positive terminal. Point A is connected to the negative battery terminal, and point E is connected to the positive terminal. If a voltmeter is connected across terminals A and B, terminal B will be positive with respect to A. If a voltmeter is connected across terminals B and C, however, terminal B will be negative with respect to terminal C. Notice that terminal B is closer to the negative terminal of the battery than terminal C is. Consequently, electrons flow through the resistor in a direction that makes terminal B more negative than C. Terminal C would be negative with respect to terminal D for the same reason.

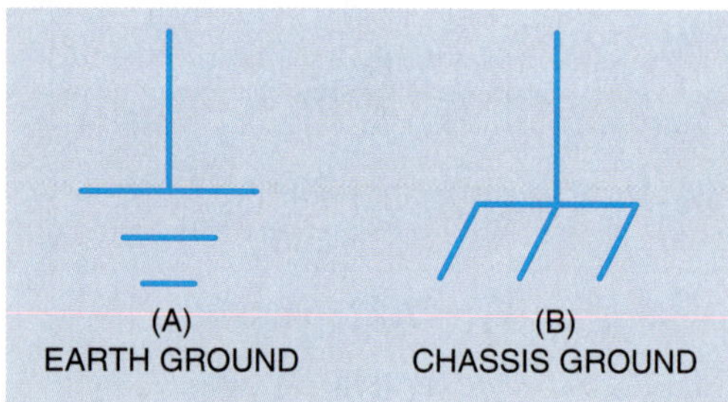

Figure 5-23 Ground symbols.

Using Ground as a Reference

Two symbols are used to represent ground (Fig. 5-23). The symbol shown in Figure 5-23A is an **earth ground** symbol. It symbolizes a **ground point** that is made by physically driving an object such as a rod or a pipe into the ground. The symbol shown in Figure 5-23B symbolizes a **chassis ground.** This point is used as a common connection for other parts of a circuit, but it is not actually driven into the ground. Although the symbol shown in Figure 5-23B is the accepted symbol for a chassis ground, the symbol shown in Figure 5-23A is often used to represent a chassis ground also.

An excellent example of using a chassis ground as a common connection can be found in the electrical system of an automobile. The negative terminal of the battery is grounded to the frame or chassis of the vehicle. The frame of the automobile is not connected directly to earth ground; it is insulated from the ground by rubber tires. In the case of an automobile electrical system, the chassis of the vehicle is the negative side of the circuit. An electrical circuit using ground as a common connection point is shown in Figure 5-24. This circuit is an electronic burglar alarm. Notice the numerous ground points in the schematic. In practice, when the circuit is connected, all the ground points will be connected together.

In voltage divider circuits, ground is often used to provide a common reference point to produce voltages that are above and below ground (Fig. 5-25). An above ground voltage is

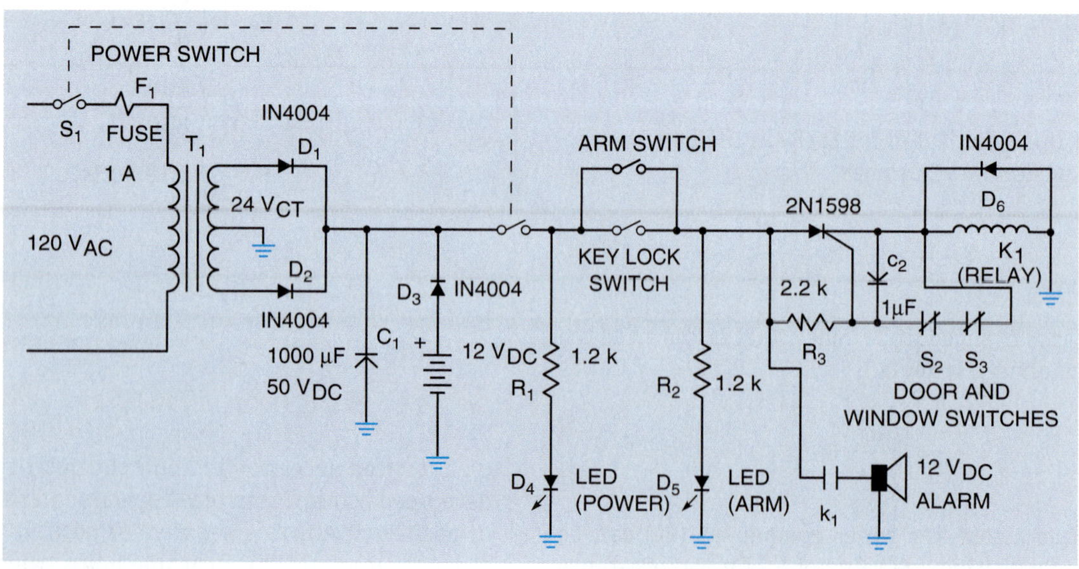

Figure 5-24 Burglar alarm with battery back-up.

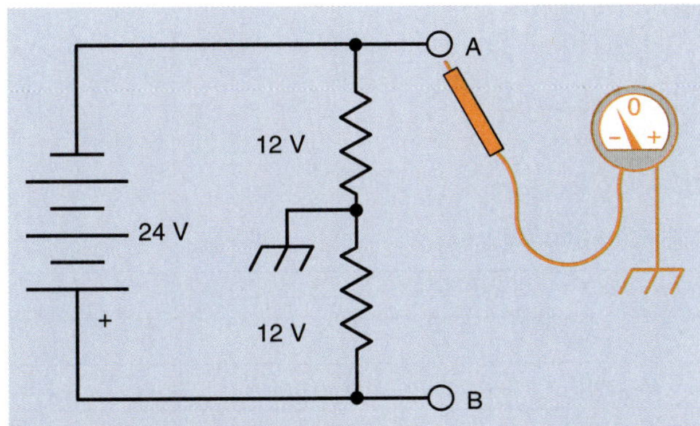

Figure 5-25 A common ground used to produce above and below ground voltage.

positive with respect to ground. A below ground voltage is negative with respect to ground. In Figure 5-25, one terminal of a zero-center voltmeter is connected to ground. If the probe is connected to point A, the pointer of the voltmeter will give a negative indication for voltage. If the probe is connected to point B, the pointer will indicate a positive voltage.

Summary

- Series circuits have only one path for current flow.
- The individual voltage drops in a series circuit can be added to equal the applied voltage.
- The current is the same at any point in a series circuit.
- The individual resistors can be added to equal the total resistance of the circuit.
- Fuses and circuit breakers are connected in series with the devices they are intended to protect.
- The total power in any circuit is equal to the sum of the power dissipated by all parts of the circuit.
- When the source voltage and total resistance are known, the voltage drop across each element can be computed using the general voltage divider formula.

Review Questions

1 A series circuit has individual resistor values of 200 Ω, 86 Ω, 91 Ω, 180 Ω, and 150 Ω. What is the total resistance of the circuit?

2 A series circuit contains four resistors. The total resistance of the circuit is 360 Ω. Three of the resistors have values of 56 Ω, 110 Ω, and 75 Ω. What is the value of the fourth resistor?

3 A series circuit contains five resistors. The total voltage applied to the circuit is 120 V. Four resistors have voltage drops of 35 V, 28 V, 22 V, and 15 V. What is the voltage drop of the fifth resistor?

4 A circuit has three resistors connected in series. Resistor R_2 has a resistance of 220 Ω and a voltage drop of 44 V. What is the current flow through resistor R_3?

5 A circuit has four resistors connected in series. If each resistor has a voltage drop of 60 V, what is the voltage applied to the circuit?

6 Define series circuit.

7 State the three rules for series circuits.

8 A series circuit has resistance values of 160 Ω, 100 Ω, 82 Ω, and 120 Ω. What is the total resistance of this circuit?

9 If a voltage of 24 V is applied to the circuit in question 8, what will be the total amount of current flow in the circuit?

10 What will be the voltage drop across each of the resistors?

160 Ω, _____ V

100 Ω, _____ V

82 Ω, _____ V

120 Ω, _____ V

11 A series circuit contains the following values of resistors: $R_1 = 510\ \Omega$; $R_2 = 680\ \Omega$; $R_3 = 390\ \Omega$; and $R_4 = 750\ \Omega$. Assume a source voltage of 48 V. Use the general voltage divider formula to compute the voltage drop across each of the resistors.

E_1 = _____ V E_2 = _____ V E_3 = _____ V E_4 = _____ V

Practice Problems

Series Circuits

1. Using the three rules for series circuits and Ohm's law, solve for the missing values.

E_T 120	E_1	E_2	E_3	E_4	E_5
I_T	I_1	I_2	I_3	I_4	I_5
R_T	R_1 430 Ω	R_2 360 Ω	R_3 750 Ω	R_4 1000 Ω	R_5 620 Ω
P_T	P_1	P_2	P_3	P_4	P_5

E_T	E_1	E_2	E_3 11	E_4	E_5
I_T	I_1	I_2	I_3	I_4	I_5
R_T	R_1	R_2	R_3	R_4	R_5
P_T 0.25	P_1 0.03	P_2 0.0825	P_3	P_4 0.045	P_5 0.0375

E_T 340	E_1 44	E_2 94	E_3 60	E_4 40	E_5
I_T	I_1	I_2	I_3	I_4	I_5
R_T	R_1	R_2	R_3	R_4	R_5
P_T	P_1	P_2	P_3	P_4	P_5 0.204

2. Use the general voltage divider formula to compute the values of voltage drop for the following series connected resistors. Assume a source voltage of 120 V.

R_1 = 1K Ω; R_2 = 2.2K Ω; R_3 = 1.8K Ω; R_4 = 1.5K Ω

E_1 = _____ V E_2 = _____ V E_3 = _____ V E_4 = _____ V

Chapter 6 Parallel Circuits

Parallel circuits are probably the type of circuit with which most people are familiar. Most devices such as lights and receptacles in homes and office buildings are connected in parallel. Imagine if the lights in your home were wired in series. All the lights in the home would have to be turned on in order for any light to operate, and if one were to burn out, all the lights would go out. The same is true for receptacles. If receptacles were connected in series, some device would have to be connected into each receptacle before power could be supplied to any other device.

OBJECTIVES

After studying this unit, you should be able to:

- discuss the characteristics of parallel circuits.
- state three rules for solving electrical values of parallel circuits.
- solve the missing values in a parallel circuit using the three rules and Ohm's law.
- discuss the operation of a current divider circuit.
- calculate current values using the current divider formula.

Glossary of Parallel Circuit Terms

circuit branch (a) in the case of parallel circuits, a circuit derived from a main set of circuit conductors that supply power to multiple other circuits; (b) normally the circuit that supplies power to electrical equipment from the last protective device, such as a fuse or circuit breaker

current adds one of the three rules used to solve values of voltage, current, resistance, and power in a parallel circuit

current divider formula a formula used to calculate the amount of current flow in any branch when the total circuit current and resistance are known

load the amount of power a circuit is supplying to a device

parallel circuits circuits with more than one path for current flow

product over sum formula

$$R_T = \frac{R_1 \times R_2}{R_1 \times R_2}$$

one of three formulas used to determine the total resistance of resistors connected in parallel

reciprocal formula

$$R_T = \frac{1}{\frac{1}{R_1} + \frac{1}{R_2} + \frac{1}{R_3} + \frac{1}{R_N}}$$

one of three formulas used to determine the total resistance of resistors connected in parallel

Parallel Circuit Values

Total Current

Parallel circuits are circuits that have more than one path for current flow (Fig. 6-1). If it is assumed that current leaves terminal A and returns to terminal B, it can be seen that the electrons can take three separate paths. In Figure 6-1, 3 A of current leave terminal A. One amp flows through resistor R_1 and 2 A flow to resistors R_2 and R_3. At the junction of resistors R_2 and R_3, 1 A flows through resistor R_2, and 1 A flows to resistor R_3. Notice that the power supply, terminals A and B, must furnish all the current that flows through each individual resistor, or **circuit branch.** One of the rules for parallel circuits states that **the total current flow in the circuit is equal to the sum of the currents through all of the branches; this is known as current adds.** ($I_{TOTAL} = I_1 + I_2 + \ldots I_n$) Notice that the amount of current leaving the source **must** return to the source.

Voltage Drop

Figure 6-2 shows another parallel circuit and gives the values of voltage, current, and resistance for each individual resistor or branch. Notice that the voltage drop across all three resistors is the same. If the circuit is traced, it can be seen that each resistor is connected directly to the power source. A second rule for parallel circuits states that **the voltage drop across any branch of a parallel circuit is the same as the applied voltage.**

For this reason, most electrical circuits in homes are connected in parallel. Each lamp and receptacle is supplied with 120 V (Fig. 6-3).

Total Resistance

In the circuit shown in Figure 6-4, three separate resistors have values of 15 Ω, 10 Ω, and 30 Ω. The total resistance of the circuit, however, is 5 Ω. **The total resistance of a parallel circuit is always less than the resistance of the lowest-value resistor, or branch, in the circuit.** Each time another element is connected in parallel, there is less opposition to the flow of current through the entire circuit. Imagine a water system consisting of a holding tank, a pump, and return lines to the tank (Fig. 6-5). Although large return pipes have less resistance to the flow of water than small pipes, the small pipes do provide a return path to the holding tank. Each time another return path is added, regardless of size, there is less overall resistance to flow, and the rate of flow increases.

That concept often causes confusion concerning the definition of **load** among students of electricity. Students often think that an increase of resistance constitutes an increase of load. An increase of current, not resistance, results in an increase of load. In laboratory exercises, students often see the circuit current increase each time a resistive element is connected to the circuit, and they conclude that an increase of resistance must, therefore, cause an increase of current. That conclusion is, of course, completely contrary to Ohm's

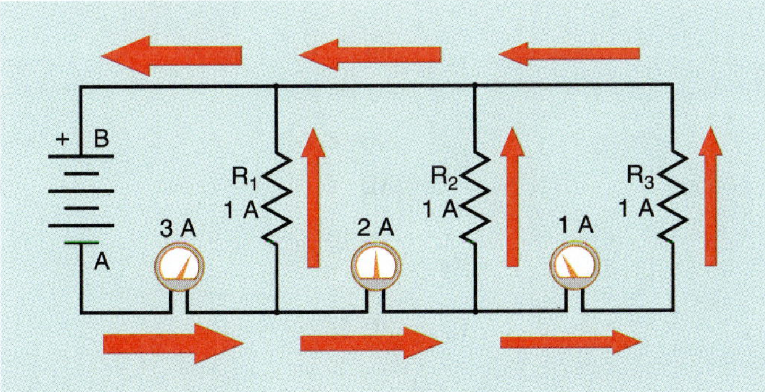

Figure 6-1 Parallel circuits provide more than one path for current flow.

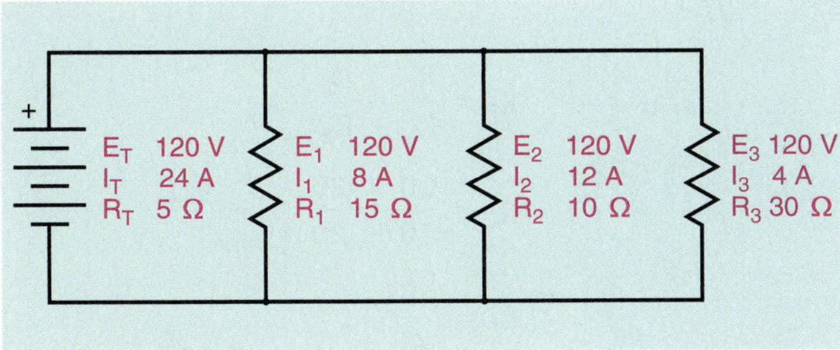

Figure 6-2 Parallel circuit values.

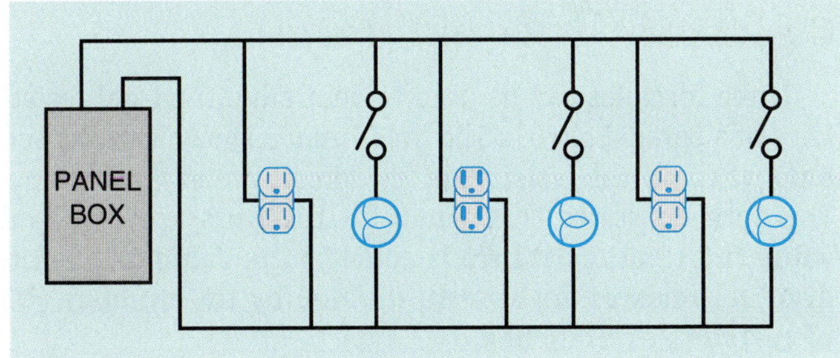

Figure 6-3 Lights and receptacles are connected in parallel.

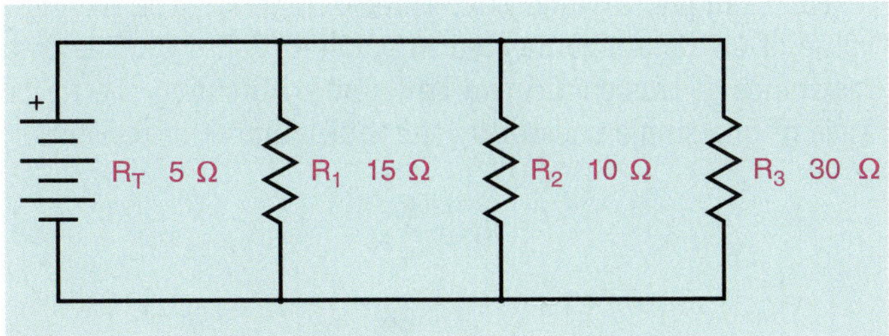

Figure 6-4 Total resistance is always less than the resistance of any single branch.

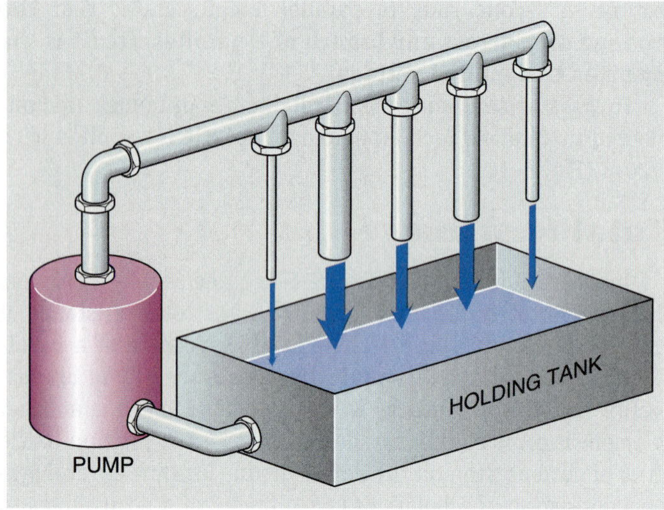

Figure 6-5 Each new path reduces the total resistance to the flow of water.

law, which states that an increase of resistance must cause a proportional decrease of current. The false concept that an increase of resistance causes an increase of current can be overcome once the student understands that if the resistive elements are being connected in parallel, the circuit resistance is actually being decreased and not increased.

Parallel Resistance Formulas

Resistors of Equal Value

Three formulas can be used to determine the total resistance of a parallel circuit. The first formula shown can be used only when all the resistors in the circuit are of equal value. This formula states that **when all resistors are of equal value the total resistance is equal to the value of one individual resistor, or branch, divided by the number (N) of resistors or branches.**

$$R_T = \frac{R}{N}$$

For example, assume that three resistors, each having a value of 24 Ω, are connected in parallel (Fig. 6-6). The total resistance of this circuit can be found by dividing the resistance of one single resistor by the total number of resistors.

$$R_T = \frac{R}{N}$$

$$R_T = \frac{24}{3}$$

$$R_T = 8\ \Omega$$

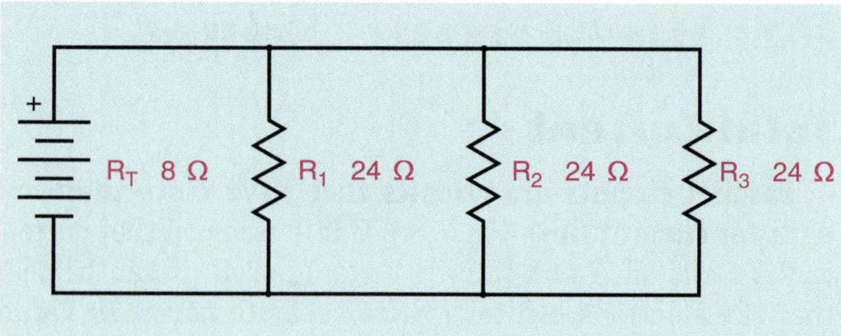

Figure 6-6 Finding the total resistance when all resistors have the same value.

Product Over Sum

The second formula used to determine the total resistance in a parallel circuit divides the product of pairs of resistors by their sum sequentially until only one pair is left. This is commonly referred to as the **product over sum formula** for finding total resistance.

$$R_T = \frac{R_1 \times R_2}{R_1 + R_2}$$

In the circuit shown in Figure 6-7, three branches having resistors with values of 20 Ω, 30 Ω, and 60 Ω are connected in parallel. To find the total resistance of the circuit using the product over sum method, find the total resistance of any two branches in the circuit (Fig. 6-8).

$$R_T = \frac{R_2 + R_3}{R_2 + R_3}$$

$$R_T = \frac{30 \times 60}{30 + 60}$$

$$R_T = \frac{1800}{90}$$

$$R_T = 20\ \Omega$$

The total resistance of the last two resistors in the circuit is 20 Ω. This 20 Ω, however, is connected in parallel with a 20 Ω resistor. The total resistance of the last two resistors is now substituted for the value of R_1 in the formula, and the value of the first resistor is substituted for the value of R_2 (Fig. 6-9).

$$R_T = \frac{R_1 \times R_2}{R_1 + R_2}$$

$$R_T = \frac{20 \times 20}{20 + 20}$$

$$R_T = \frac{400}{40}$$

$$R_T = 10\ \Omega$$

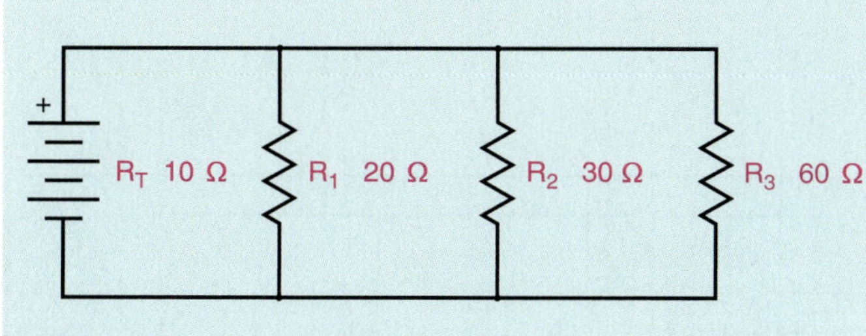

Figure 6-7 Finding the total resistance of a parallel circuit by dividing the product of two resistors by their sum.

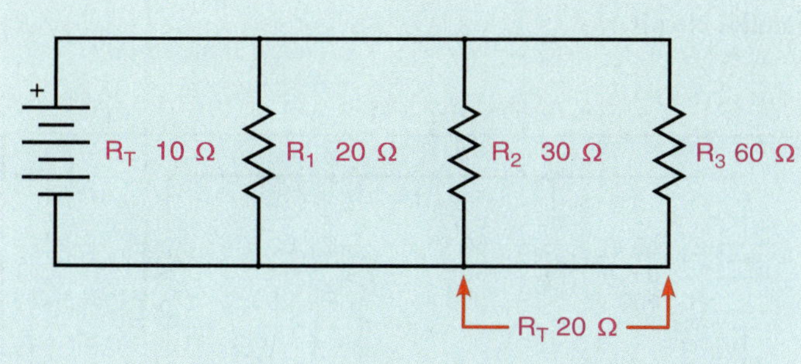

Figure 6-8 The total resistance of the last two branches.

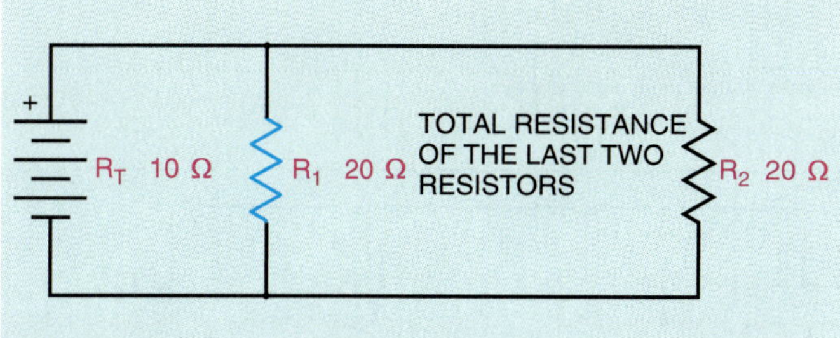

Figure 6-9 The total value of the first two resistors is used as resistor 2.

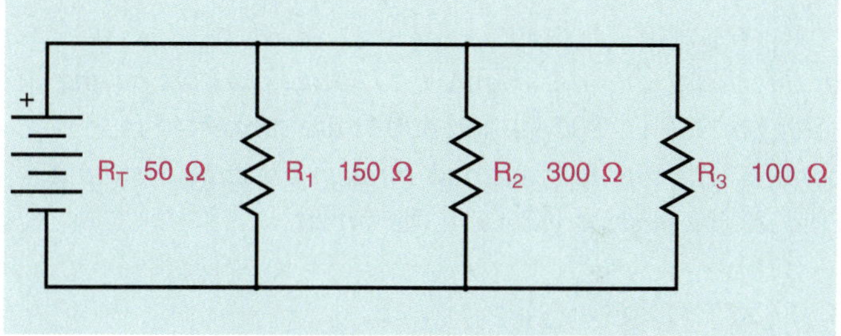

Figure 6-10 Finding the total resistance using the reciprocal method.

Reciprocal Formula

The third formula used to find the total resistance of a parallel circuit is

$$\frac{1}{R_T} = \frac{1}{R_1} + \frac{1}{R_2} + \frac{1}{R_3} + \frac{1}{R_N}$$

Notice that this formula actually finds the reciprocal of the total resistance, instead of the total resistance. To make the formula equal to the total resistance it can be rewritten as follows:

$$R_T = \frac{1}{\frac{1}{R_1} + \frac{1}{R_2} + \frac{1}{R_3} + \frac{1}{R_N}}$$

The value R_N stands for the number of resistors in the circuit. If the circuit has 25 resistors connected in parallel, for example, the last resistor in the formula would be R_{25}.

This is known as the **reciprocal formula.** The reciprocal of any number is that number divided into 1. The reciprocal of 4, for example, is 0.25 because 1/4 = 0.25. Another rule of parallel circuits is **the total resistance of a parallel circuit is the reciprocal of the sum of the reciprocals of the individual branches.** A modified version of this formula is used in several different applications to find values other than resistance. Some of those other formulas will be covered later.

Before the invention of hand-held calculators, the slide rule was often employed to help with the mathematical calculations in electrical work. At that time, the product over sum method of finding total resistance was the most popular. Since the invention of calculators, however, the reciprocal formula has become the most popular, because scientific calculators have a reciprocal key (1/X), which makes computing total resistance using the reciprocal method very easy.

In Figure 6-10, three resistors having values of 150 Ω, 300 Ω, and 100 Ω are connected in parallel. The total resistance can be found using the reciprocal formula.

$$R_T = \frac{1}{\frac{1}{R_1} + \frac{1}{R_2} + \frac{1}{R_3}}$$

$$R_T = \frac{1}{\frac{1}{150} + \frac{1}{300} + \frac{1}{100}}$$

$$R_T = \frac{1}{0.006667 + 0.003333 + 0.01}$$

$$R_T = \frac{1}{0.02}$$

$$R_T = 50\ \Omega$$

Example 1

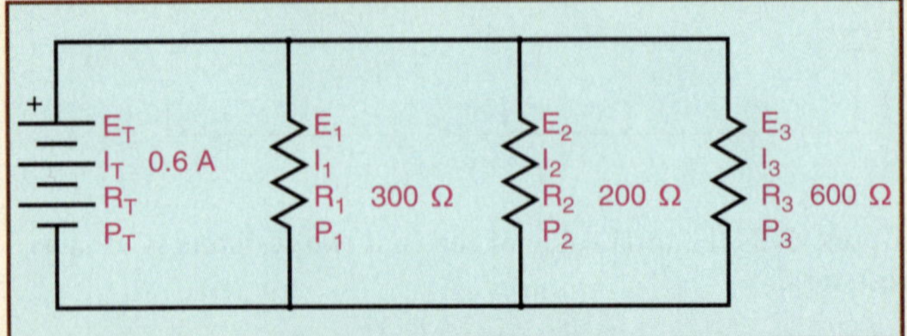

Figure 6-11 Parallel circuit, Example 1.

In the circuit shown in Figure 6-11, three resistors having values of 300 Ω, 200 Ω, and 600 Ω are connected in parallel. The total current flow through the circuit is 0.6 A. Find all the missing values in the circuit.

SOLUTION

The first step is to find the total resistance of the circuit. The reciprocal formula will be used.

$$R_T = \frac{1}{\frac{1}{R_1} + \frac{1}{R_2} + \frac{1}{R_3}}$$

$$R_T = \frac{1}{\frac{1}{300} + \frac{1}{200} + \frac{1}{600}}$$

$$R_T = \frac{1}{0.0033 + 0.0050 + 0.0017}$$

$$R_T = \frac{1}{0.01}$$

$$R_T = 100\ \Omega$$

Now that the total resistance of the circuit is known, the voltage applied to the circuit can be found using the total current value and Ohm's law.

$$E_T = I_T \times R_T$$

$$E_T = 0.6 \times 100$$

$$E_T = 60\ \text{V}$$

One of the rules for parallel circuits states that the voltage drops across all the parts of a parallel circuit are the same as the total voltage. Therefore, the voltage drop across each resistor is 60 V (Fig. 6-12).

$$E_T = E_1 = E_2 = E_3$$

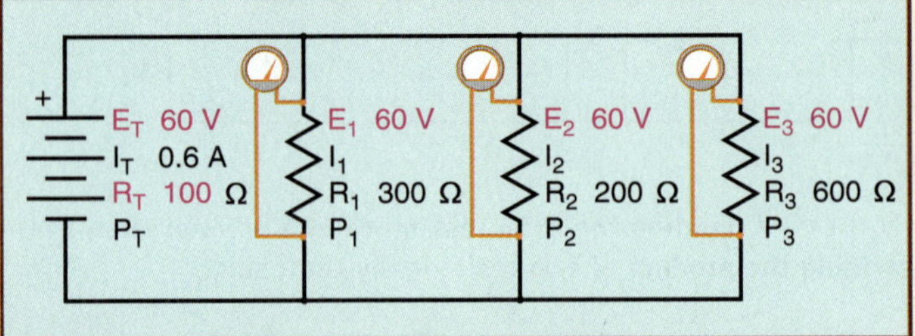

Figure 6-12 The voltage is the same across all branches of a parallel circuit.

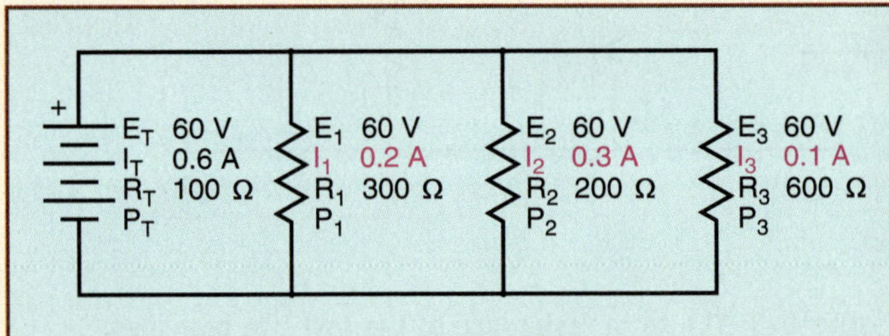

Figure 6-13 Ohm's law is used to compute the amount of current through each branch.

Since the voltage drop and resistance of each resistor are known, Ohm's law can be used to determine the amount of current flow through each resistor (Fig. 6-13).

$$I_1 = \frac{E_1}{R_1}$$

$$I_1 = \frac{60}{300}$$

$$I_1 = 0.2\ \text{A}$$

$$I_2 = \frac{E_2}{R_2}$$

$$I_2 = \frac{60}{200}$$

$$I_2 = 0.3\ \text{A}$$

$$I_3 = \frac{E_3}{R_3}$$

$$I_3 = \frac{60}{600}$$

$$I_3 = 0.1\ \text{A}$$

Example 1 Continued

The amount of power (watts) used by each resistor can be found using Ohm's law. A different formula will be used to find the amount of electrical energy converted into heat by each of the resistors.

$$P_1 = \frac{E_1{}^2}{R_1}$$

$$P_1 = \frac{60 \times 60}{300}$$

$$P_1 = 12 \text{ W}$$

$$P_2 = I_2{}^2 \times R_2$$

$$P_2 = 0.3 \times 0.3 \times 200$$

$$P_2 = 18 \text{ W}$$

$$P_3 = E_2 \times I_3$$

$$P_3 = 60 \times 0.1$$

$$P_3 = 6 \text{ W}$$

In Chapter 5, it was stated that the total amount of power in a circuit is equal to the sum of the power used by all the parts. This is true for any type of circuit. Therefore, the total amount of power used by this circuit can be found by taking the sum of the power used by all the resistors (Fig. 6-14).

$$P_T = P_1 + P_2 + P_3$$

$$P_T = 12 + 18 + 6$$

$$P_T = 36 \text{ W}$$

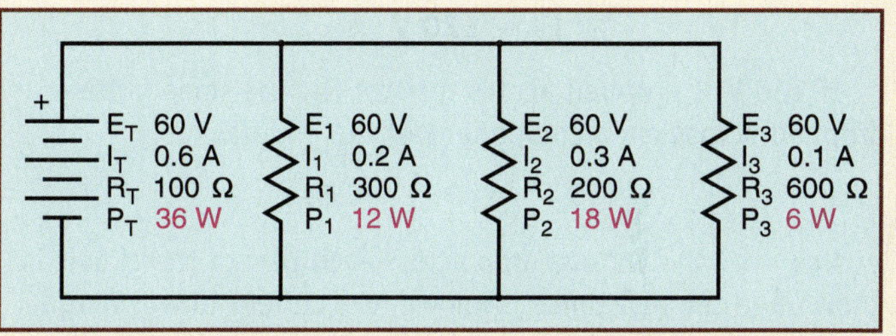

Figure 6-14 **The amount of power used by the circuit.**

Example 2

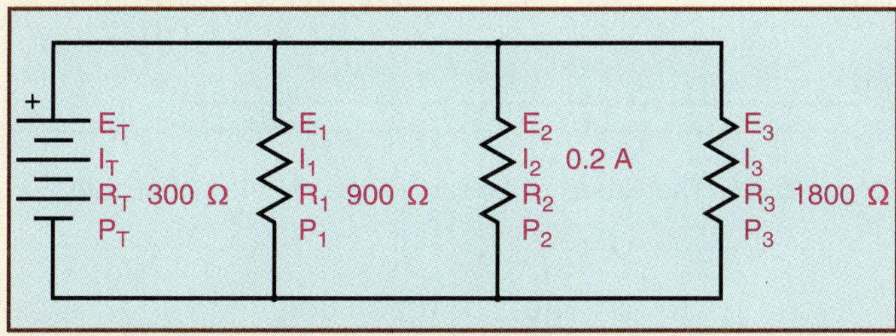

Figure 6-15 **Parallel circuit, Example 2.**

In the circuit shown in Figure 6-15, three resistors are connected in parallel. Two of the resistors have a value of 900 Ω and 1800 Ω. The value of resistor R_2 is unknown. The total resistance of the circuit is 300 Ω. Resistor R_2 has a current flow of 0.2 A. Find the missing circuit values.

SOLUTION

The first step in solving this problem is to find the missing resistor value. This can be done by changing the reciprocal formula as shown:

$$\frac{1}{R_2} = \frac{1}{R_T} - \frac{1}{R_1} - \frac{1}{R_3}$$

or

$$R_2 = \frac{1}{\frac{1}{R_T} - \frac{1}{R_1} - \frac{1}{R_3}}$$

One of the rules for parallel circuits states that the total resistance is equal to the reciprocal of the sum of the reciprocals of the individual resistors. Therefore, the reciprocal of any individual resistor is equal to the reciprocal of the difference between the reciprocal of the total resistance and the sum of the reciprocals of the other resistors in the circuit.

$$R_2 = \frac{1}{\frac{1}{R_T} - \frac{1}{R_1} - \frac{1}{R_3}}$$

$$R_2 = \frac{1}{\frac{1}{300} - \frac{1}{900} - \frac{1}{1800}}$$

$$R_2 = \frac{1}{0.003333 - 0.001111 - 0.0005556}$$

Example 2 Continued

$$R_2 = \frac{1}{0.001666}$$

$$R_2 = 600\ \Omega$$

Now that the resistance of resistor R_2 has been found, the voltage drop across resistor R_2 can be determined using the current flow through the resistor and Ohm's law (Fig. 6-16).

$$E_2 = I_2 \times R_2$$

$$E_2 = 0.2 \times 600$$

$$E_2 = 120\ V$$

If 120 V is dropped across resistor R_2, the same voltage is dropped across each component of the circuit.

$$E_2 = E_T = E_1 = E_3$$

Now that the voltage drop across each part of the circuit is known and the resistance is known, the current flow through each branch can be determined using Ohm's law (Fig. 6-17).

$$I_T = \frac{E_T}{R_T}$$

$$I_T = \frac{120}{300}$$

$$I_T = 0.4\ A$$

$$I_1 = \frac{E_1}{R_1}$$

$$I_1 = \frac{120}{900}$$

$$I_1 = 0.1333\ A$$

$$I_3 = \frac{E_3}{R_3}$$

$$I_3 = \frac{120}{1800}$$

$$I_3\ 0.0666\ A$$

The amount of power used by each resistor can be found using Ohm's law (Fig. 6-18).

$$P_1 = \frac{E_1^2}{R_1}$$

$$P_1 = \frac{120 \times 120}{900}$$

$$P_1 = 16\ W$$

$$P_2 = I_2^2 \times R_2$$

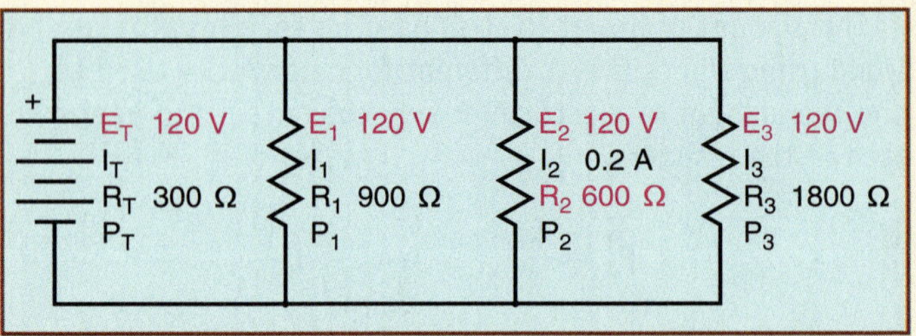

Figure 6-16 The missing resistor and voltage values.

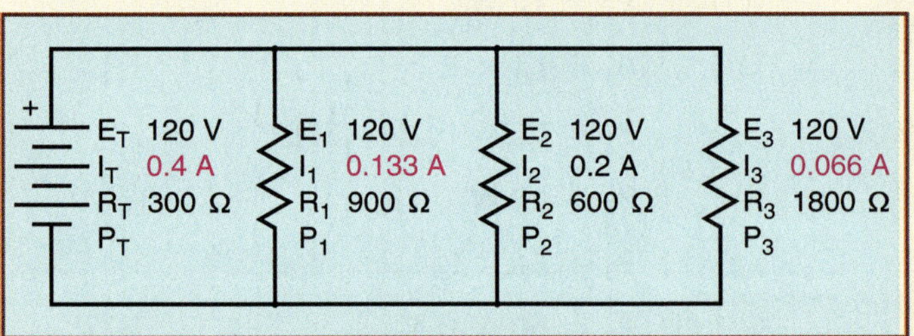

Figure 6-17 Determining the current using Ohm's law.

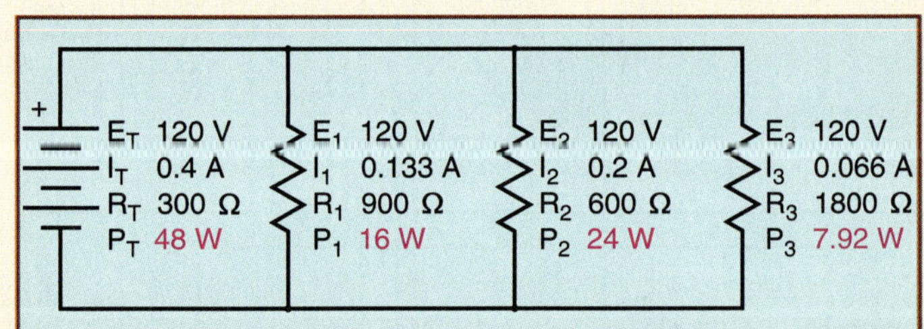

Figure 6-18 The values of power for the circuit in Example 2.

$$P_2 = 0.2 \times 0.2 \times 600$$

$$P_2 = 24\ W$$

$$P_3 = E_3 \times I_3$$

$$P_3 = 120 \times 0.066$$

$$P_3 = 7.92\ W$$

$$P_T = E_T \times I_T$$

$$P_T = 120 \times 0.4$$

$$P_T = 48\ W$$

If the wattage values of the three resistors are added to compute total power for the circuit, it will be seen that their total is 47.92 W instead of the computed 48 W. The small difference in answers is caused by the rounding off of other values. In this instance, the current of resistor R_3 was rounded from 0.066666666 to 0.066.

Example 3

In the circuit shown in Figure 6-19, three resistors are connected in parallel. Resistor R_1 is producing 0.075 W of heat, R_2 is producing 0.45 W of heat, and R_3 is producing 0.225 W of heat. The circuit has a total current of 0.05 A.

SOLUTION

Since the amount of power dissipated by each resistor is known, the total power for the circuit can be determined by finding the sum of the power used by the components.

$$P_T = P_1 + P_2 + P_3$$

$$P_T = 0.075 + 0.45 + 0.225$$

$$P_T = 0.75 \text{ W}$$

Now that the amount of total current and total power for the circuit are known, the applied voltage can be found using Ohm's law (Fig. 6-20).

$$E_T = \frac{P_T}{I_T}$$

$$E_T = \frac{0.75}{0.05}$$

$$E_T = 15 \text{ V}$$

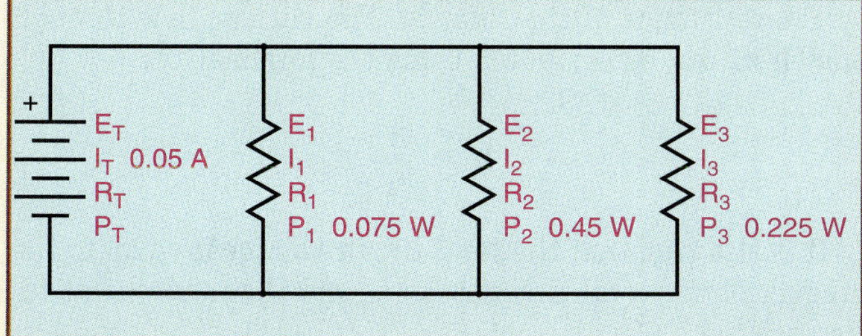

Figure 6-19 Parallel circuit, Example 3.

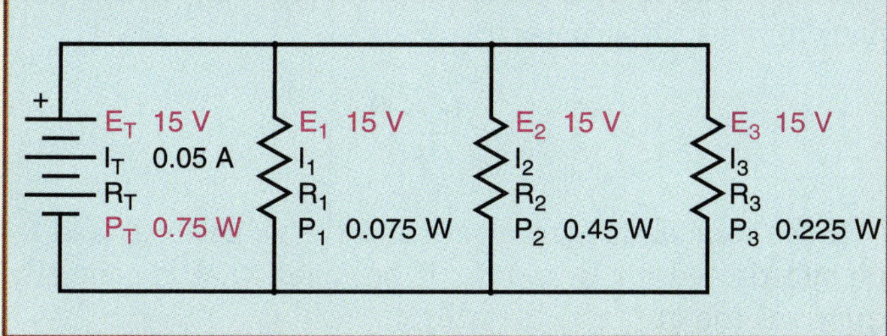

Figure 6-20 The applied voltage for the circuit.

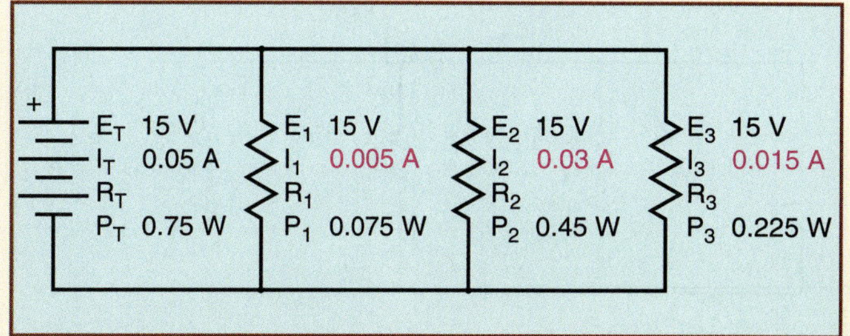

Figure 6-21 The current through each branch.

The amount of current flow through each resistor can now be found using Ohm's law (Fig. 6-21).

$$I_1 = \frac{P_1}{E_1}$$

$$I_1 = \frac{0.075}{15}$$

$$I_1 = 0.005 \text{ A}$$

$$I_2 = \frac{P_2}{E_2}$$

$$I_2 = \frac{0.45}{15}$$

$$I_2 = 0.03 \text{ A}$$

$$I_3 = \frac{P_3}{E_3}$$

$$I_3 = \frac{0.225}{15}$$

$$I_3 = 0.015 \text{ A}$$

All resistance values for the circuit can now be found using Ohm's law (Fig. 6-22).

$$R_1 = \frac{E_1}{I_1}$$

$$R_1 = \frac{15}{0.005}$$

$$R_1 = 3000\ \Omega$$

$$R_2 = \frac{E_2}{I_2}$$

Example 3 Continued

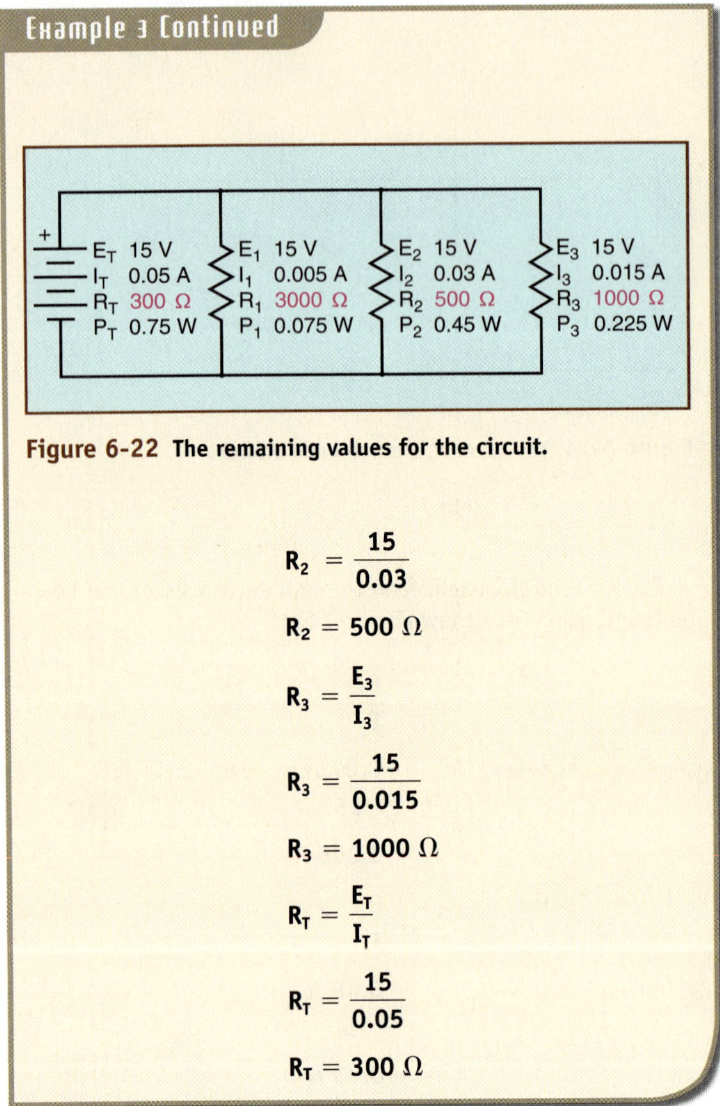

Figure 6-22 **The remaining values for the circuit.**

$$R_2 = \frac{15}{0.03}$$

$$R_2 = 500\ \Omega$$

$$R_3 = \frac{E_3}{I_3}$$

$$R_3 = \frac{15}{0.015}$$

$$R_3 = 1000\ \Omega$$

$$R_T = \frac{E_T}{I_T}$$

$$R_T = \frac{15}{0.05}$$

$$R_T = 300\ \Omega$$

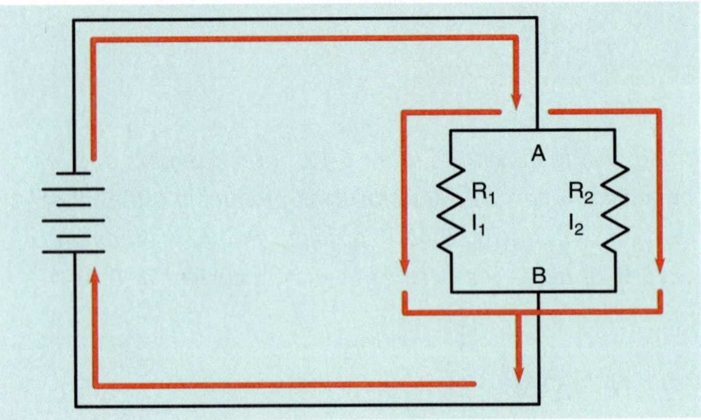

Figure 6-23 **Parallel circuits are current dividers.**

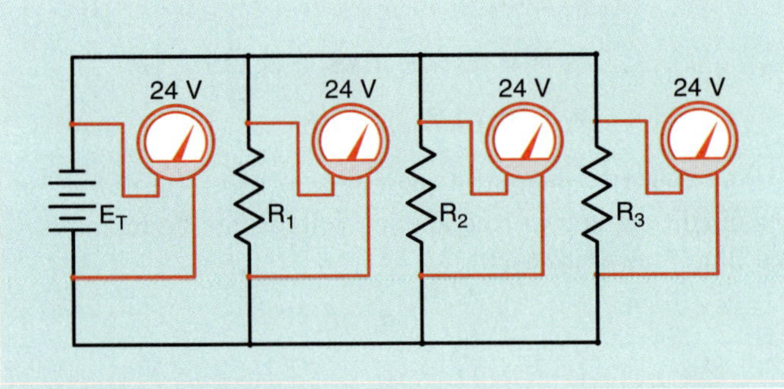

Figure 6-24 **The voltage is the same across all branches of a parallel circuit.**

Current Dividers

All parallel circuits are *current dividers (Fig. 6-23)*. As previously discussed in this chapter, the sum of the currents in a parallel circuit must equal the total current. Assume that a current of 1 ampere enters the circuit at point A. This 1 ampere of current will divide between resistors R_1 and R_2, and then recombine at point B. The amount of current that flows through each resistor is proportional to the resistance value. A greater amount of current will flow through a low value resistor and less current will flow through a high value resistor. In other words, the amount of current flowing through each resistor is inversely proportional to its resistance.

In a parallel circuit, the voltage across each branch must be equal (Fig. 6-24). Therefore, the current flow through any branch can be computed by dividing the source voltage (E_T) by the resistance of that branch. The current flow through branch #1 can be computed using the formula

$$I_1 = \frac{E_T}{R_1}$$

It is also true that the total circuit voltage is equal to the product of the total circuit current and the total circuit resistance.

$$E_T = I_T \times R_T$$

If the value of E_T is substituted for ($I_T \times R_T$) in the previous formula, it becomes

$$I_1 = \frac{I_T \times R_T}{R_1}$$

If the formula is rearranged, and the values of I_1 and R_1 are substituted for I_X and R_X, it becomes what is generally known as the **current divider formula.**

$$I_x = \left(\frac{R_T}{R_X}\right) I_T$$

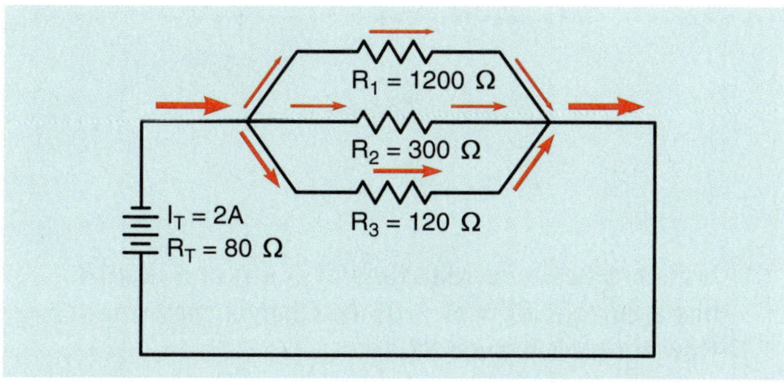

Figure 6-25 **The current divides through each branch of a parallel circuit.**

This formula can be used to compute the current flow through any branch by substituting the values of I_X and R_X for the branch values when the total circuit current and resistance are known. In the circuit shown in Figure 6-25, resistor R_1 has a value of 1200 Ω, resistor R_2 has a value of 300 Ω, and resistor R_3 has a value of 120 Ω, producing a total of resistance of 80 Ω for the circuit. It is assumed that a total current of 2 amps flows in the circuit. The amount of current flow through resistor R_1 can be found using the formula.

$$I_1 = \left(\frac{R_T}{R_1}\right) I_T$$

$$I_1 = \left(\frac{80}{1200}\right) 2$$

$$I_1 = 0.133 \text{ amp}$$

The current flow through each of the other resistors can be found by substituting in the same formula.

$$I_2 = \left(\frac{R_T}{R_2}\right) I_T$$

$$I_2 = \left(\frac{80}{300}\right) 2$$

$$I_2 = 0.533 \text{ amp}$$

$$I_3 = \left(\frac{R_T}{R_3}\right) I_T$$

$$I_3 = \left(\frac{80}{120}\right) 2$$

$$I_3 = 1.333 \text{ amp}$$

Summary

- A parallel circuit is characterized by the fact that it has more than one path for current flow.
- Three rules for solving parallel circuits are
  - A. The total current is the sum of the currents through all of the branches of the circuit.
  - B. The voltage across any part of the circuit is the same as the total voltage.
  - C. The total resistance is the reciprocal of the sum of the reciprocals of each individual branch.
- Circuits in homes are connected in parallel.
- The total power in a parallel circuit is equal to the sum of the power dissipation of all the components.
- Parallel circuits are current dividers.
- The current flowing through each branch of a parallel circuit can be computed when the total resistance and total current are known.
- The amount of current flow through each branch of a parallel circuit is inversely proportional to its resistance.

Review Questions

1. What characterizes a parallel circuit?
2. Why are circuits in homes connected in parallel?
3. State three rules concerning parallel circuits.
4. A parallel circuit contains four branches. One branch has a current flow of 0.8 A, another has a current flow of 1.2 A, the third has a current flow of 0.25 A, and the fourth has a current flow of 1.5 A. What is the total current flow in the circuit?
5. Four resistors having a value of 100 Ω each are connected in parallel. What is the total resistance of the circuit?
6. A parallel circuit has three branches. An ammeter is connected in series with the output of the power supply and indicates a total current flow of 2.8 A. If branch 1 has a current flow of 0.9 A and branch 2 has a current flow of 1.05 A, what is the current flow through branch 3?
7. Four resistors having values of 270 Ω, 330 Ω, 510 Ω, and 430 Ω are connected in parallel. What is the total resistance in the circuit?
8. A parallel circuit contains four resistors. The total resistance of the circuit is 120 Ω. Three of the resistors have values of 820 Ω, 750 Ω, and 470 Ω. What is the value of the fourth resistor?
9. A circuit contains a 1,200 Ω, a 2,200 Ω, and a 3,300 Ω resistor connected in parallel. The circuit has a total current flow of 0.25 A. How much current flows through each of the resistors?

Practice Problems

Parallel Circuits

Using the rules for parallel circuits and Ohm's law, solve for the missing values.

1

E_T	E_1	E_2	E_3	E_4
I_T 0.942	I_1	I_2	I_3	I_4
R_T	R_1 680 Ω	R_2 820 Ω	R_3 470 Ω	R_4 330 Ω
P_T	P_1	P_2	P_3	P_4

2

E_T	E_1	E_2	E_3	E_4
I_T 0.00639	I_1	I_2 0.00139	I_3 0.00154	I_4 0.00115
R_T	R_1	R_2	R_3	R_4
P_T	P_1 0.640	P_2	P_3	P_4

3

E_T	E_1	E_2	E_3	E_4
I_T	I_1	I_2	I_3 3.2	I_4
R_T 3.582 Ω	R_1 16 Ω	R_2 10 Ω	R_3	R_4 20 Ω
P_T	P_1	P_2	P_3	P_4

4

E_T	E_1	E_2	E_3	E_4
I_T	I_1	I_2	I_3	I_4
R_T	R_1 82 kΩ	R_2 75 kΩ	R_3 56 kΩ	R_4 62 kΩ
P_T 3.436	P_1	P_2	P_3	P_4

5 A parallel circuit contains the following resistor values:

$R_1 = 360\ \Omega$ $R_2 = 470\ \Omega$ $R_3 = 300\ \Omega$

$R_4 = 270\ \Omega$ $I_T = 0.05$ amp

Find the following missing values:

$R_T =$ _____ Ω $I_1 =$ _____ A $I_2 =$ _____ A

$I_3 =$ _____A $I_4 =$ _____ A

6 A parallel circuit contains the following resistor values:

$R_1 = 270\text{K}\ \Omega$ $R_2 = 360\text{K}\ \Omega$ $R_3 = 430\text{K}\ \Omega$

$R_4 = 100\text{K}\ \Omega$ $I_T = 0.006$ amp

Find the following missing values:

$R_T =$ _____ Ω $I_1 =$ _____ A $I_2 =$ _____ A

$I_3 =$ _____A $I_4 =$ _____ A

Chapter 7 Combination Circuits

Combination circuits contain both series and parallel elements. To determine which components are in parallel and which are in series, trace the flow of current through the circuit. Remember that a series circuit is one that has only one path for current flow, and a parallel circuit has more than one path for current flow.

OBJECTIVES

After studying this unit, you should be able to:

- **define a combination circuit.**
- **list the rules for parallel circuits.**
- **list the rules for series circuits.**
- **solve combination circuits using the rules for parallel circuits, the rules for series circuits, and Ohm's law.**

Glossary of Combination Circuit Terms

combination circuit a circuit that contains both series and parallel circuits

node a point of connection

parallel block a section of a combination circuit composed of parallel resistors

reduce a method employed to simplify parallel or series connected components into a single resistor

trace the current path a method used to determine if components are connected in series or parallel

Combination Circuits

A simple combination circuit is shown in Figure 7-1. It will be assumed that the current in Figure 7-1 will flow from point A to point B. To identify the series and parallel elements, **trace the current path.** All of the current in the circuit must flow through resistor R_1. Resistor R_1 is, therefore, in series with the rest of the circuit. When the current reaches the junction point of resistors R_2 and R_3, however, it splits. A junction point such as this is often referred to as a **node.** Part of the current flows through resistor R_2 and part flows through resistor R_3. These two resistors are in parallel. Because this circuit contains both series and parallel elements, it is a **combination circuit.**

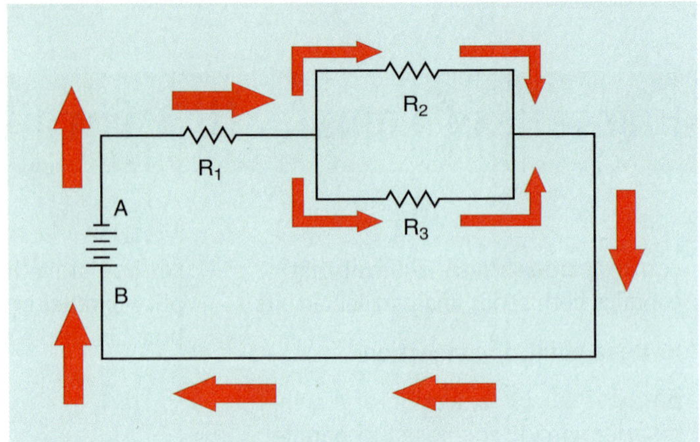

Figure 7-1 A simple combination circuit.

Solving Combination Circuits

The circuit shown in Figure 7-2 contains four resistors with values of 325 Ω, 275 Ω, 150 Ω, and 250 Ω. The circuit has a total current flow of 1 A. To determine which resistors are in series and which are in parallel, trace the path for current flow through the circuit. When the path of current flow is traced, it can be seen that current can flow by two separate paths from the negative terminal to the positive terminal. One path is through resistors R_1 and R_2, and the other path is through resistors R_3 and R_4. These two paths are, therefore, in parallel. However, the same current must flow through resistors R_1 and R_2. So these two resistors are in series. The same is true for resistors R_3 and R_4.

To solve the unknown values in a combination circuit, use series circuit rules for those sections of the circuit that are connected in series, and parallel circuit rules for sections connected in parallel. The circuit rules are as follows:

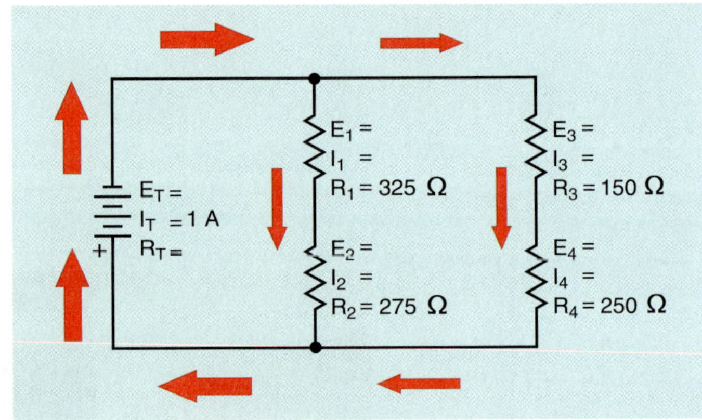

Figure 7-2 Tracing the current paths through a combination circuit.

Series Circuits

1. The current is the same at any point in the circuit.
2. The total resistance is the sum of the individual resistances.
3. The sum of the voltage drops of the individual resistors must equal the applied voltage.

Parallel Circuits

1. The voltage across any circuit branch is the same as the total voltage.
2. The total current is the sum of the current through all of the circuit paths.
3. The total resistance is equal to the reciprocal of the sum of the reciprocals of the branch resistances.

Simplifying the Circuit

The circuit shown in Figure 7-2 can be reduced or simplified to a *simple parallel circuit* (Fig. 7-3). Since resistors R_1 and R_2 are connected in series, their values can be added to form one equivalent resistor, $R_{c(1\&2)}$, which stands for a combination of resistors 1 and 2. The same is true for resistors R_3 and R_4. Their values are added to form resistor $R_{c(3\&4)}$. Now that the circuit has been reduced to a simple parallel circuit, the total resistance can be found.

$$R_T = \frac{1}{\frac{1}{R_{c(1\&2)}} + \frac{1}{R_{c(3\&4)}}}$$

$$R_T = \frac{1}{\frac{1}{600} + \frac{1}{400}}$$

$$R_T = \frac{1}{0.0016667 + 0.0025}$$

$$R_T = \frac{1}{0.0041667}$$

$$R_T = 240\ \Omega$$

Now that the total resistance has been found, the other circuit values can be computed. The applied voltage can be found using Ohm's law.

$$E_T = 1_T \times R_T$$

$$E_T = 1 \times 240$$

$$E_T = 240\ V$$

One of the rules for parallel circuits states that the voltage is the same across each branch of the circuit. For this reason, the voltage drops across resistors $R_{c(1\&2)}$ and $R_{c(3\&4)}$ are the same. Since the voltage drop and the resistance are known, Ohm's law can be used to find the current flow through each branch.

$$I_{c(1\&2)} = \frac{E_{c(1\&2)}}{R_{c(1\&2)}}$$

$$I_{c(1\&2)} = \frac{240}{600}$$

$$I_{c(1\&2)} = 0.4\ A$$

$$I_{c(3\&4)} = \frac{E_{c(3\&4)}}{R_{c(3\&4)}}$$

$$I_{c(3\&4)} = \frac{240}{400}$$

$$I_{c(3\&4)} = 0.6\ A$$

These values can now be used to solve the missing values in the original circuit. Resistor $R_{c(1\&2)}$ is actually a combination of resistors R_1 and R_2. The values of voltage and current that apply to $R_{c(1+2)}$ therefore apply to resistors R_1 and R_2. Resistors R_1 and R_2 are connected in series. One of the rules for a series circuit states that the current is the same at any point in the circuit. Since 0.4 A of current flows through resistor $R_{c(1+2)}$, the same amount of current flows through resistors R_1 and R_2. Now that the current flow through these two resistors is known, the voltage drop across each can be computed using Ohm's law.

$$E_1 = I_1 \times R_1$$

$$E_1 = 0.4 \times 325$$

$$E_1 = 130\ V$$

$$E_2 = I_2 \times R_2$$

$$E_2 = 0.4 \times 275$$

$$E_2 = 110\ V$$

These values of voltage and current can now be added to the circuit in Figure 7-3 to produce the circuit shown in Figure 7-4.

The values of voltage and current for resistor $R_{c(3\&4)}$ apply to resistors R_3 and R_4. The same amount of current that flows through resistor $R_{c(3\&4)}$ flows through resistors R_3 and R_4. The voltage across these two resistors can now be computed using Ohm's law.

$$E_3 = I_3 \times R_3$$

$$E_3 = 0.6 \times 150$$

$$E_3 = 90\ V$$

$$E_4 = I_4 \times R_4$$

$$E_4 = 0.6 \times 250$$

$$E_4 = 150\ V$$

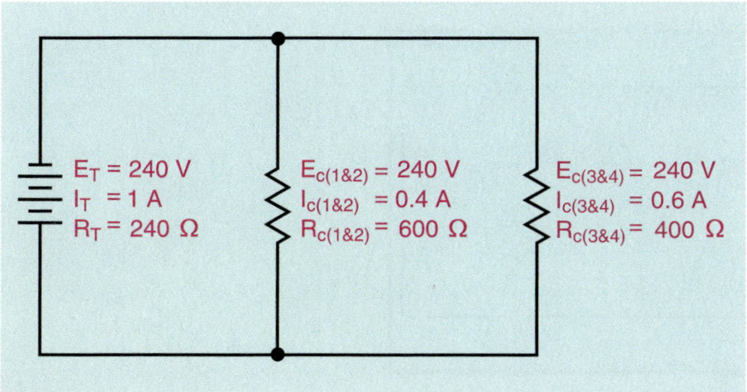

Figure 7-3 **Simplifying the combination circuit.**

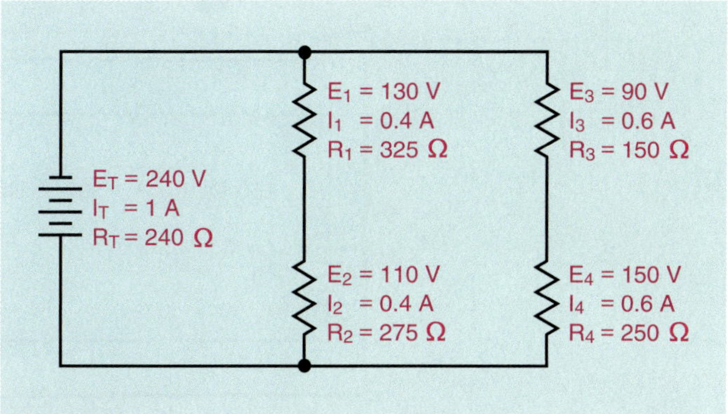

Figure 7-4 **All the missing values for the combination circuit.**

Example 1

Solve the combination circuit shown in Figure 7-5.

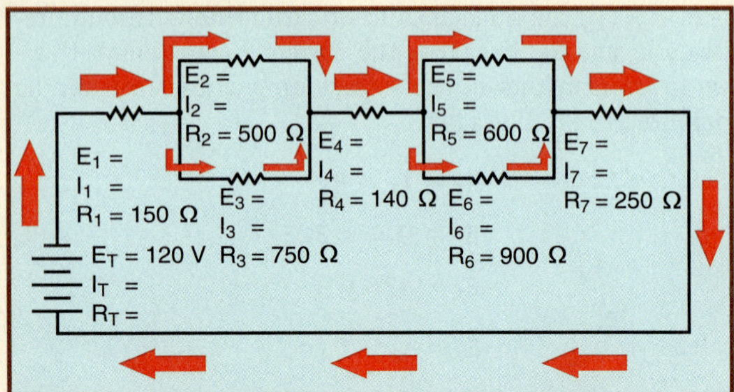

Figure 7-5 Tracing the flow of current through the combination circuit.

SOLUTION

The first step in finding the missing values is to trace the current path through the circuit to determine which resistors are in series and which are in parallel. All the current must flow through resistor R_1. Resistor R_1 is, therefore, in series with the rest of the circuit. When the current reaches the junction of resistors R_2 and R_3, it divides and part flows through each resistor. Resistors R_2 and R_3 are in parallel. All the current must then flow through resistor R_4, which is connected in series, to the junction of resistors R_5 and R_6. The current path is divided between these two resistors. Resistors R_5 and R_6 are connected in parallel. All the circuit current must then flow through resistor R_7.

The next step in solving this circuit is to **reduce** it to a simpler circuit. If the total resistance of the first **parallel block** formed by resistors R_2 and R_3 is found, this block can be replaced by a single resistor.

$$R_{(2,3)} = \frac{1}{\frac{1}{R_2} + \frac{1}{R_3}}$$

$$R_{(2,3)} = \frac{1}{\frac{1}{500} + \frac{1}{750}}$$

$$R_{(2,3)} = \frac{1}{0.002 + 0.0013333}$$

$$R_{(2,3)} = \frac{1}{0.0033333}$$

$$R_{(2,3)} = 300\ \Omega$$

The equivalent resistance of the second parallel block can be computed in the same way.

$$R_{(5,6)} = \frac{1}{\frac{1}{R_5} + \frac{1}{R_6}}$$

$$R_{(5,6)} = \frac{1}{\frac{1}{600} + \frac{1}{900}}$$

$$R_{(5,6)} = \frac{1}{0.0016667 + 0.0011111}$$

$$R_{(5,6)} = \frac{1}{0.0027778}$$

$$R_{(5,6)} = 360\ \Omega$$

Now that the total resistance of the second parallel block is known, you can *redraw* the circuit as a simple series circuit as shown in Figure 7-6. The first parallel block has been replaced with a single resistor of 300 Ω labeled $R_{c(2\&3)}$, and the second parallel block has been replaced with a single 360-Ω resistor labeled $R_{c(5\&6)}$. Ohm's law can be used to find the missing values in this series circuit.

One of the rules for series circuits states that the total resistance of a series circuit is equal to the sum of the individ-

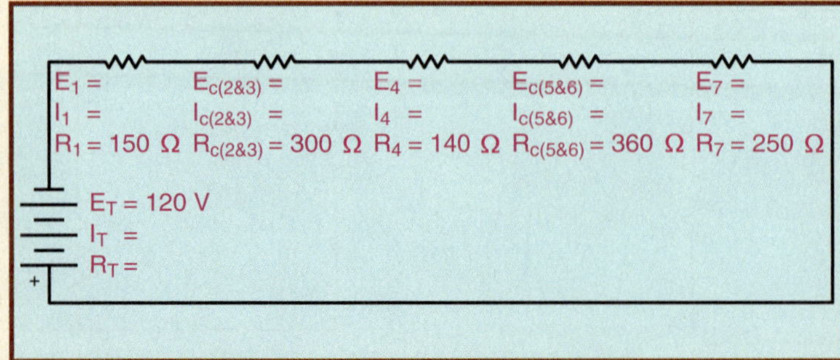

Figure 7-6 Simplifying the combination circuit.

Example 1 Continued

ual resistances. Therefore, R_T can be computed by adding the resistances of all the resistors.

$$R_T = R_1 + R_{c(2\&3)} + R_4 + R_{c(5\&6)} + R_7$$

$$R_T = 150 + 300 + 140 + 360 + 250$$

$$R_T = 1200\ \Omega$$

Since the total voltage and total resistance are known, the total current flow through the circuit can be computed.

$$I_T = \frac{E_T}{R_T}$$

$$I_T = \frac{120}{1200}$$

$$I_T = 0.1\ A$$

The first rule of series circuits states that the current is the same at any point in the circuit. The current flow through each resistor is, therefore, 0.1 A. The voltage drop across each resistor can now be computed using Ohm's law.

$$E_1 = I_1 \times R_1$$

$$E_1 = 0.1 \times 150$$

$$E_1 = 15\ V$$

$$E_{c(2\&3)} = I_{c(2\&3)} \times R_{c(2\&3)}$$

$$E_{c(2\&3)} = 0.1 \times 300$$

$$E_{c(2\&3)} = 30\ V$$

$$E_4 = I_4 \times R_4$$

$$E_4 = 0.1 \times 140$$

$$E_4 = 14\ V$$

$$E_{c(5\&6)} = I_{c(5\&6)} \times R_{c(5\&6)}$$

$$E_{c(5\&6)} = 0.1 \times 360$$

$$E_{c(5\&6)} = 36\ V$$

$$E_7 = I_7 \times R_7$$

$$E_7 = 0.1 \times 250$$

$$E_7 = 25\ V$$

The series circuit with all solved values is shown in Figure 7-7. These values can now be used to solve missing parts in the original circuit.

Resistor $R_{c(2\&3)}$ is actually the parallel block containing resistors R_2 and R_3. The values for $R_{c(2\&3)}$, therefore, apply to this parallel block. One of the rules for a parallel circuit states that the voltage drop of a parallel circuit is the same at any point in the circuit. Since 30 V is dropped across resistor $R_{c(2\&3)}$, the same 30 V is dropped across resistors R_2 and R_3 (Fig. 7-8). The current flow through these resistors can now be computed using Ohm's law.

$$I_2 = \frac{E_2}{R_2}$$

$$I_2 = \frac{30}{500}$$

$$I_2 = 0.06\ A$$

$$I_3 = \frac{E_3}{R_3}$$

$$I_3 = \frac{30}{750}$$

$$I_3 = 0.04\ A$$

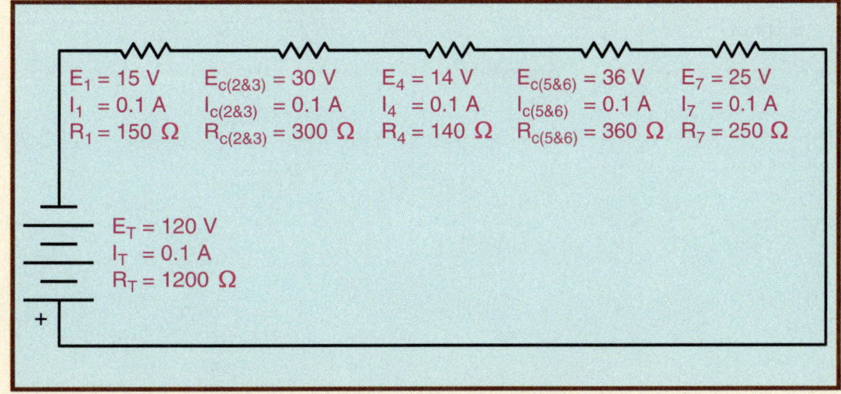

Figure 7-7 The simplified circuit with all values solved.

Example 1 Continued

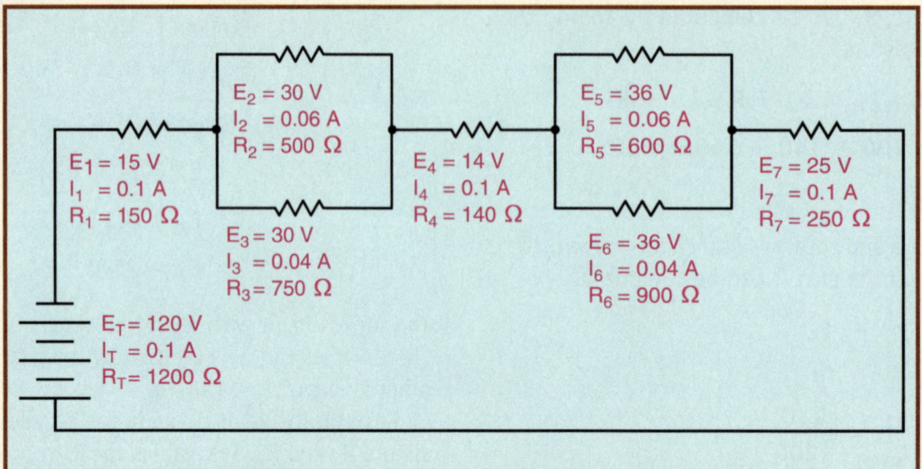

Figure 7-8 **All values solved for the combination circuit.**

The values of resistor $R_{c(5\&6)}$ can be applied to the parallel block composed of resistors R_5 and R_6. $E_{c(5\&6)}$ is 36 V. This is the voltage drop across resistors R_5 and R_6. The current flow through these two resistors can be computed using Ohm's law.

$$I_5 = \frac{E_5}{R_5}$$

$$I_5 = \frac{36}{600}$$

$$I_5 = 0.06 \text{ A}$$

$$I_6 = \frac{E_6}{R_6}$$

$$I_6 = \frac{36}{900}$$

$$I_6 = 0.04 \text{ A}$$

Example 2

Both of the preceding circuits were solved by first determining which parts of the circuit were in series and which were in parallel. The circuits were then reduced to a simple series or parallel circuit. This same procedure can be used for any combination circuit. The circuit shown in Figure 7-9 will be reduced to a simpler circuit first. Once the values of the simple circuit are found, they can be placed back in the original circuit to find other values.

SOLUTION

The first step will be to reduce the top part of the circuit to a single resistor. This part consists of resistors R_3 and R_4. Since these two resistors are connected in series with each other, their values can be added to form one single resistor. This combination will form R_{c1} (Fig. 7-10).

$$R_{c1} = R_3 + R_4$$

$$R_{c1} = 270 + 330$$

$$R_{c1} = 600\ \Omega$$

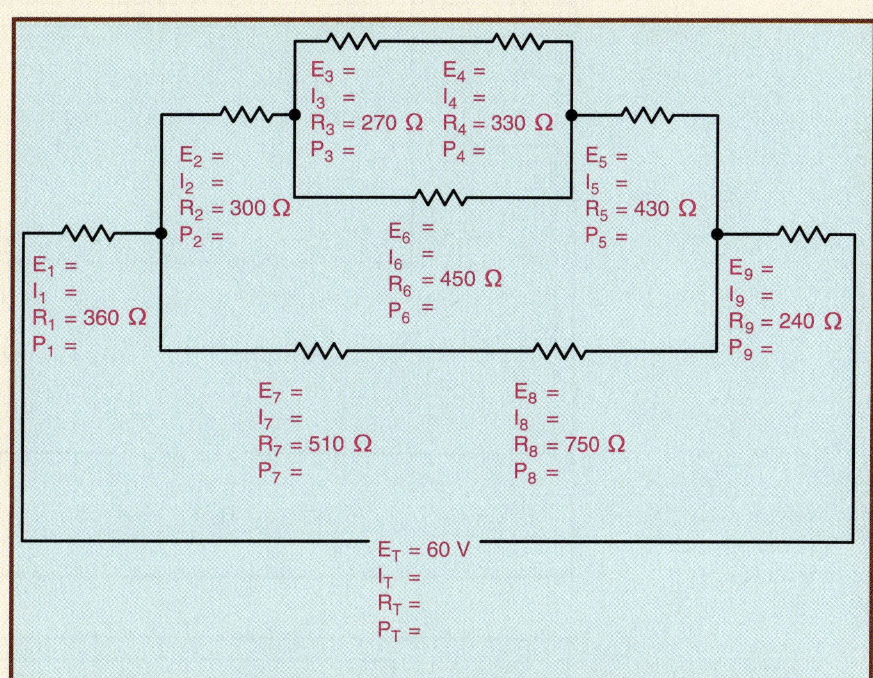

Figure 7-9 **A complex combination circuit.**

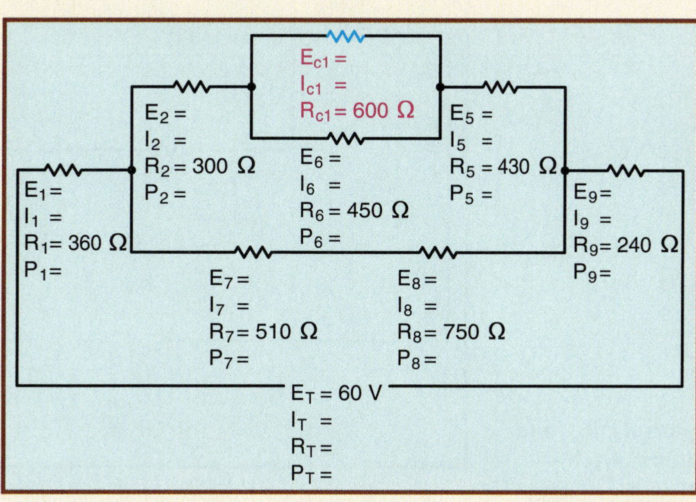

Figure 7-10 **Resistors R_3 and R_4 are combined to form R_{c1}.**

Example 2 Continued

The top part of the circuit is now formed by resistors R_{c1} and R_6. These two resistors are in parallel with each other. If their total resistance is computed, they can be changed into one single resistor with a value of 257.143 Ω. This combination will become resistor R_{c2} (Fig. 7-11).

$$R_T = \frac{1}{\frac{1}{600} + \frac{1}{450}}$$

$$R_T = 257.143\ \Omega$$

The top of the circuit now consists of resistors R_2, R_{c2}, and R_5. These three resistors are connected in series with each other. They can be combined to form resistor R_{c3} by adding their resistances together (Fig. 7-12).

$$R_{c3} = R_2 + R_{c2} + R_5$$

$$R_{c3} = 300 + 257.143 + 430$$

$$R_{c3} = 987.143\ \Omega$$

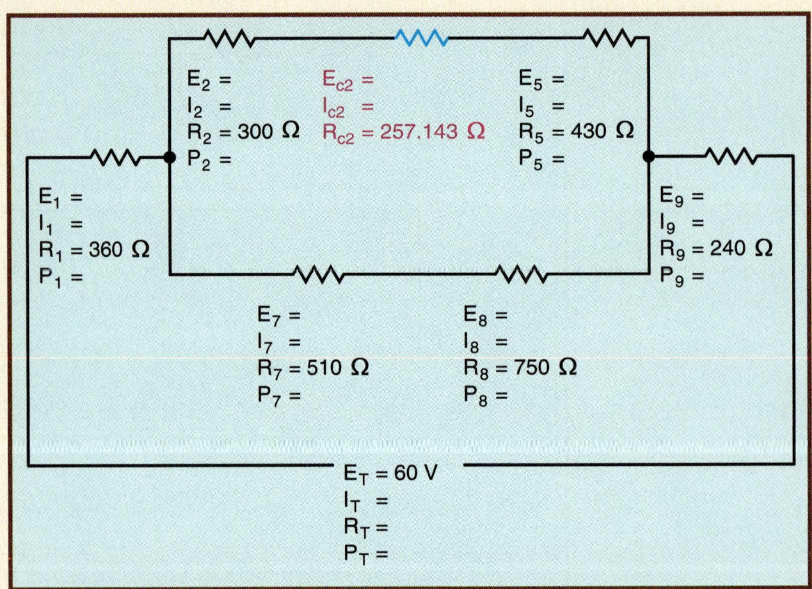

Figure 7-11 Resistors R_{c1} and R_6 are combined to form R_{c2}.

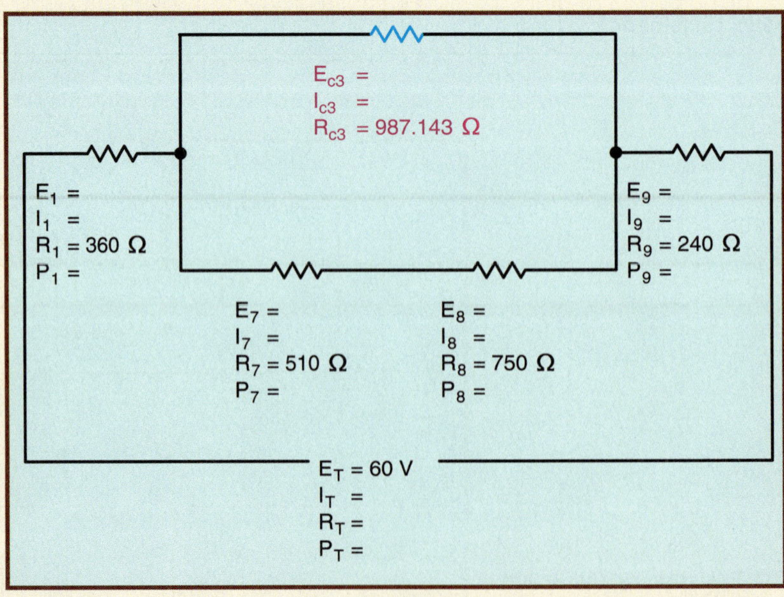

Figure 7-12 Resistors R_2, R_{c2}, and R_5 are combined to form R_{c3}.

Example 2 Continued

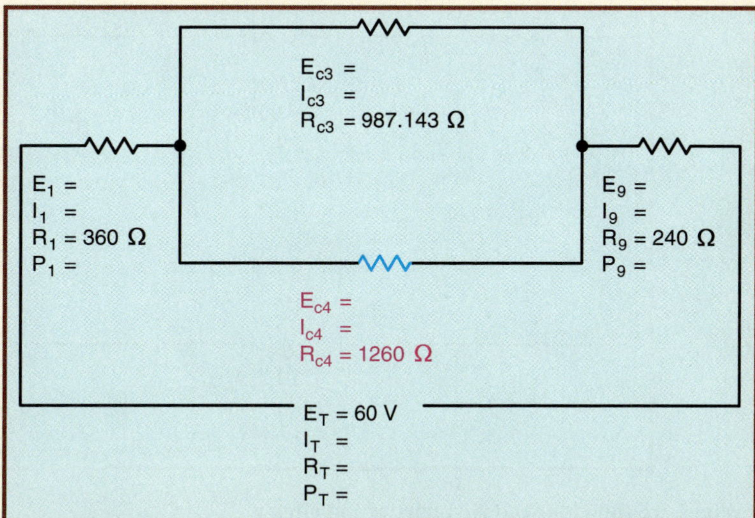

Figure 7-13 **Resistors R_7 and R_8 are combined to form R_{c4}.**

Resistors R_7 and R_8 are connected in series with each other also. These two resistors will be added to form resistor R_{c4} (Fig. 7-13).

$$R_{c4} = R_7 + R_8$$

$$R_{c4} = 510 + 750$$

$$R_{c4} = 1260\ \Omega$$

Resistors R_{c3} and R_{c4} are connected in parallel with each other. Their total resistance can be computed to form resistor R_{c5} (Fig. 7-14).

$$R_{c5} = \frac{1}{\dfrac{1}{987.143} + \dfrac{1}{1260}}$$

$$R_{c5} = 553.503\ \Omega$$

The circuit has now been reduced to a simple series circuit containing three resistors. The total resistance of the circuit can be computed by adding resistors R_1, R_{c5}, and R_9.

$$R_T = R_1 + R_{c5} + R_9$$

$$R_T = 360 + 553.503 + 240$$

$$R_T = 1153.503\ \Omega$$

Now that the total resistance and total voltage are known, the total circuit current and total circuit power can be computed using Ohm's law.

$$I_T = \frac{E_T}{R_T}$$

E_1 =
I_1 =
R_1 = 360 Ω
P_1 =
E_{c5} =
I_{c5} =
R_{c5} = 553.503 Ω
E_9 =
I_9 =
R_9 = 240 Ω
P_9 =
E_T = 60 V
I_T =
R_T =
P_T =

Figure 7-14 **Resistors R_{c3} and R_{c4} are combined to form R_{c5}.**

$$I_T = \frac{60}{1153.503}$$

$$I_T = 0.052\ \text{A}$$

$$P_T = E_T \times I_T$$

$$P_T = 60 \times 0.052$$

$$P_T = 3.12\ \text{W}$$

Ohm's law can now be used to find the missing values for resistors R_1, R_{c5}, and R_9 (Fig. 7-15).

$$E_1 = I_1 \times R_1$$

$$E_1 = 0.052 \times 360$$

$$E_1 = 18.72\ \text{V}$$

Example 2 Continued

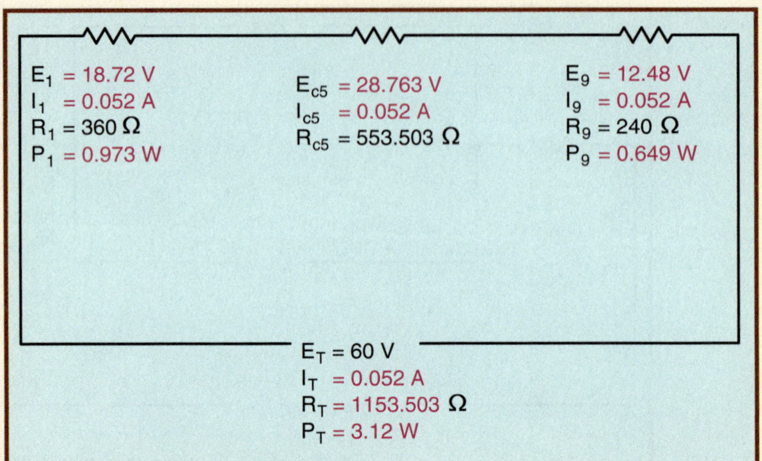

Figure 7-15 Missing values are found for the first part of the circuit.

$$P_1 = E_1 \times I_1$$
$$P_1 = 18.72 \times 0.052$$
$$P_1 = 0.973 \text{ W}$$
$$E_{c5} = I_{c5} \times R_{c5}$$
$$E_{c5} = 0.052 \times 553.503$$
$$E_{c5} = 28.763 \text{ V}$$
$$E_9 = I_9 \times R_9$$
$$E_9 = 0.052 \times 240$$
$$E_9 = 12.48 \text{ V}$$
$$P_9 = E_9 \times I_9$$
$$P_9 = 1.48 \times 0.052$$
$$P_9 = 0.649 \text{ W}$$

Resistor R_{c5} is actually the combination of resistors R_{c3} and R_{c4}. The values of R_{c5}, therefore, apply to resistors R_{c3} and R_{c4}. Since these two resistors are connected in parallel with each other, the voltage drop across them will be the same. Each will have the same voltage drop as resistor R_{c5} (Fig. 7-16). Ohm's law can now be used to find the remaining values of R_{c3} and R_{c4}.

$$I_{c4} = \frac{E_{c4}}{R_{c4}}$$
$$I_{c4} = \frac{28.763}{1260}$$

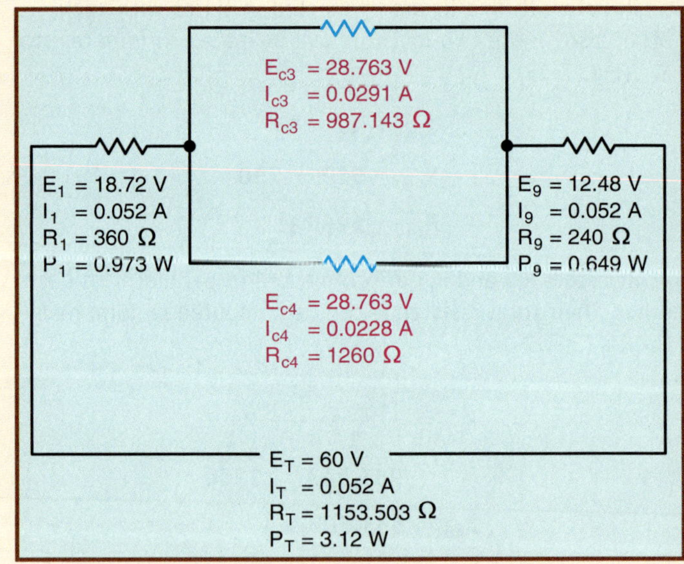

Figure 7-16 The values for resistors R_{c3} and R_{c4}.

$$I_{c4} = 0.0228 \text{ A}$$
$$I_{c3} = \frac{E_{c3}}{R_{c3}}$$
$$I_{c3} = \frac{28.763}{987.143}$$
$$I_{c3} = 0.0291 \text{ A}$$

Resistor R_{c4} is the combination of resistors R_7 and R_8. The values of resistor R_{c4} apply to resistors R_7 and R_8. Since

Example 2 Continued

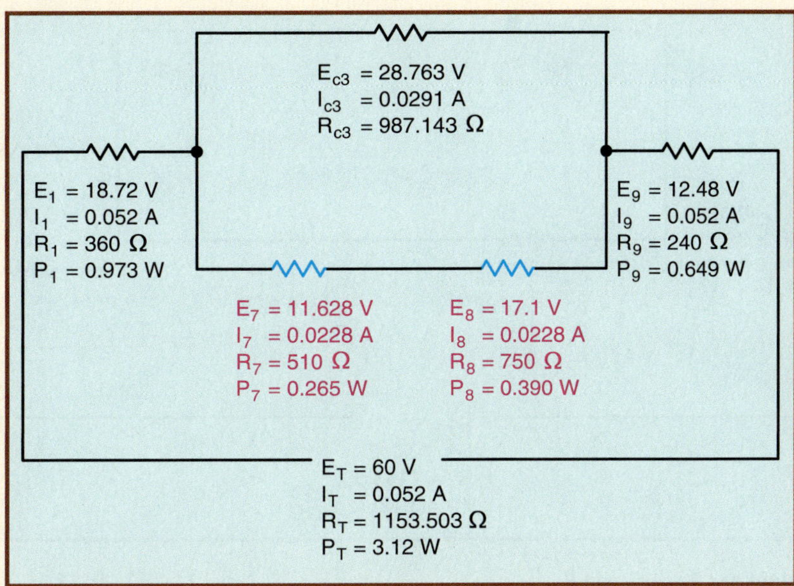

Figure 7-17 **The values for resistors R_7 and R_8.**

resistors R_7 and R_8 are connected in series with each other, the current flow will be the same through both (Fig. 7-17). Ohm's law can now be used to compute the remaining values for these two resistors.

$$E_7 = I_7 \times R_7$$
$$E_7 = 0.0228 \times 510$$
$$E_7 = 11.268 \text{ V}$$

$$P_7 = E_7 \times I_7$$
$$P_7 = 11.268 \times 0.0228$$
$$P_7 = 0.265 \text{ W}$$

$$E_8 = I_8 \times R_8$$
$$E_8 = 0.0228 \times 750$$
$$E_8 = 17.1 \text{ V}$$

$$P_8 = E_8 \times I_8$$
$$P_8 = 17.1 \times 0.0228$$
$$P_8 = 0.390 \text{ W}$$

Resistor R_{c3} is the combination of resistors R_2, R_{c2}, and R_5. Since these resistors are connected in series with each other, the current flow through each will be the same as the current flow through R_{c3}. The remaining values can now be computed using Ohm's law (Fig. 7-18).

$$E_{c2} = I_{c2} \times R_{c2}$$
$$E_{c2} = 0.0291 \times 257.143$$
$$E_{c2} = 7.483 \text{ V}$$

$$E_2 = I_2 \times R_2$$
$$E_2 = 0.0291 \times 300$$
$$E_2 = 8.73 \text{ V}$$

$$P_2 = E_2 \times I_2$$
$$P_2 = 8.73 \times 0.291$$
$$P_2 = 0.254 \text{ W}$$

$$E_5 = I_5 \times R_5$$
$$E_5 = 0.0291 \times 430$$
$$E_5 = 12.513 \text{ V}$$

$$P_5 = E_5 \times I_5$$
$$P_5 = 12.513 \times 0.0291$$
$$P_5 = 0.364 \text{ W}$$

Example 2 Continued

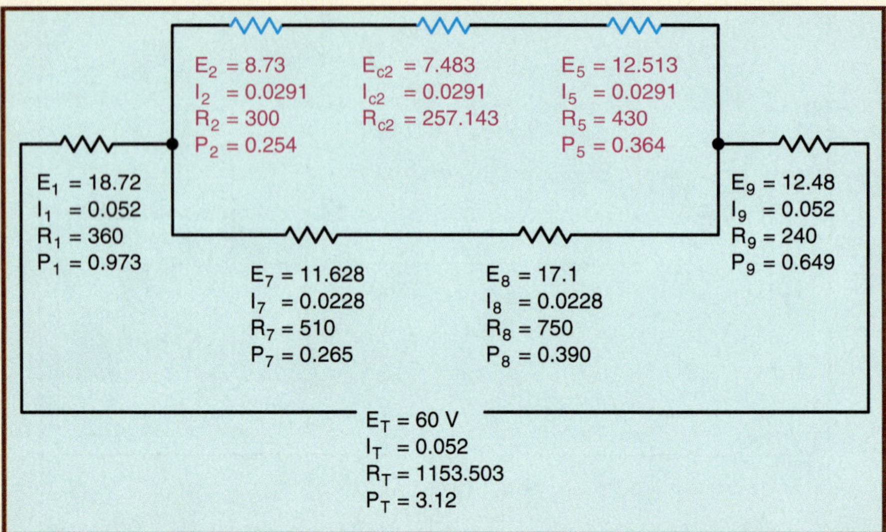

Figure 7-18 Determining values for R_2, R_{c2}, and R_5.

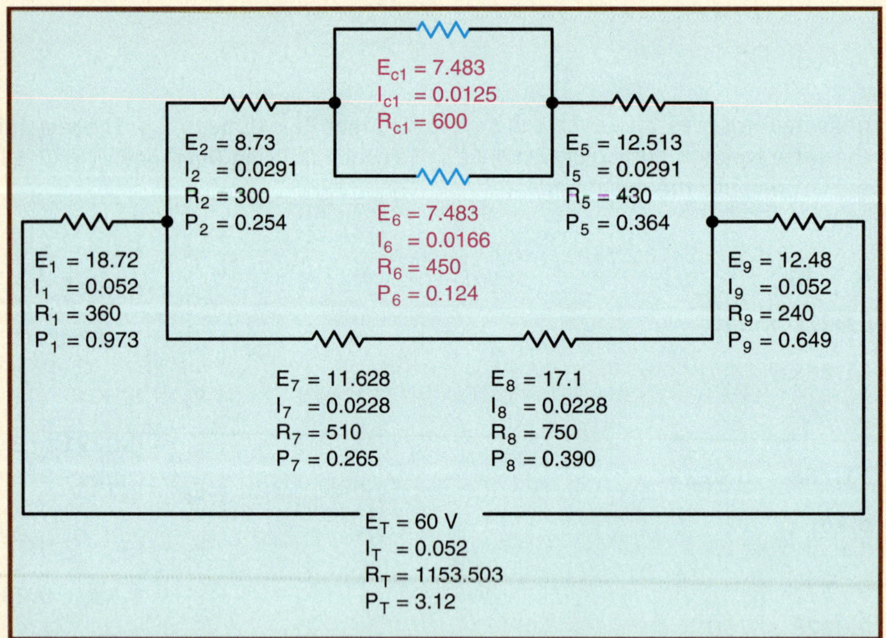

Figure 7-19 The values of R_{c1} and R_6.

Resistor R_{c2} is the combination of resistors R_{c1} and R_6. Resistors R_{c1} and R_6 are connected in parallel and will, therefore, have the same voltage drop as resistor R_{c2}. Ohm's law can be used to compute the remaining values for R_{c2} and R_6 (Fig. 7-19).

$$I_{c1} = \frac{E_{c1}}{R_{c1}}$$

$$I_{c1} = \frac{7.483}{600}$$

Example 2 Continued

$$I_{c1} = 0.0125 \text{ A}$$

$$I_6 = \frac{E_6}{R_6}$$

$$I_6 = \frac{7.483}{450}$$

$$I_6 = 0.0166 \text{ A}$$

$$P^6 = E_6 \times I_6$$

$$P_6 = 7.483 \times 0.0166$$

$$P_6 = 0.0124 \text{ A}$$

Resistor R_{c1} is the combination of resistors R_3 and R_4. Since these two resistors are connected in series, the amount of current flow through resistor R_{c1} will be the same as the flow through R_3 and R_4. The remaining values of the circuit can now be found using Ohm's law (Fig. 7-20).

$$E_3 = I_3 \times R_3$$

$$E_3 = 0.0125 \times 270$$

$$E_3 = 3.375 \text{ V}$$

$$P_3 = E_3 \times I_3$$

$$P_3 = 3.375 \times 0.0125$$

$$P_3 = 0.0423 \text{ W}$$

$$E_4 = I_4 \times R_4$$

$$E_4 = 0.0125 \times 330$$

$$E_4 = 4.125 \text{ V}$$

$$P_4 = E_4 \times I_4$$

$$P_4 = 4.125 \times 0.0125$$

$$P_4 = 0.0516 \text{ W}$$

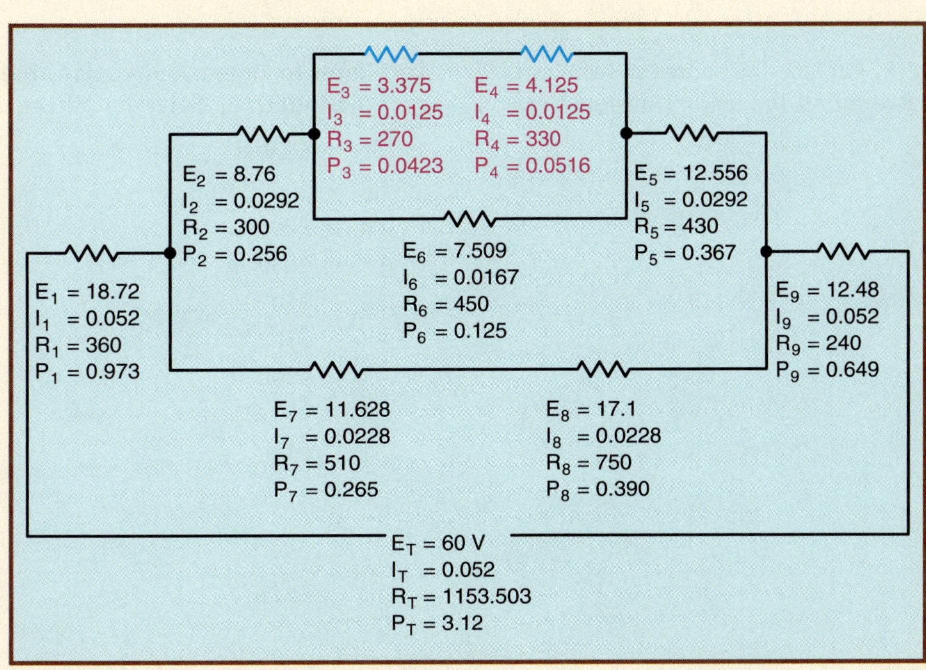

Figure 7-20 **The values for R_3 and R_4.**

Summary

- Combination circuits contain both series and parallel branches.
- The three rules for series circuits are
  - A. The current is the same at any point in the circuit.
  - B. The total resistance is the sum of the individual resistances.
  - C. The applied voltage is equal to the sum of the voltage drops across the individual components.
- The three rules for parallel circuits are
  - A. The total voltage is the same as the voltage across any branch of a parallel circuit.
  - B. The total current is the sum of the individual currents through each path in the circuit.
  - C. The total resistance is the reciprocal of the sum of the reciprocals of the branch resistances.
- When solving combination circuits, it is generally easier if the circuit is reduced to simpler circuits.

Review Questions

1. **Refer to Figure 7-2. Replace the values shown with the following. Solve for all the unknown values.**

   $I_T = 0.6$ A
   $R_1 = 470\ \Omega$
   $R_2 = 360\ \Omega$
   $R_3 = 510\ \Omega$
   $R_4 = 430\ \Omega$

2. **Refer to Figure 7-5. Replace the values shown with the following. Solve for all the unknown values.**

   $E_T = 63$ V
   $R_1 = 1000\ \Omega$
   $R_2 = 2200\ \Omega$
   $R_3 = 1800\ \Omega$
   $R_4 = 910\ \Omega$
   $R_5 = 3300\ \Omega$
   $R_6 = 4300\ \Omega$
   $R_7 = 860\ \Omega$

Practice Problems

Series-Parallel Circuits

Refer to the circuit shown in Figure 7-21 to solve the following problems.

1. Find the unknown values in the circuit if the applied voltage is 75 V and the resistors have the following values:

   $R_1 = 1.5\ K\Omega$; $R_2 = 910\ \Omega$; $R_3 = 2\ K\Omega$; and $R_4 = 3.6\ K\Omega$.

   I_T _____ E_1 _____ E_2 _____ E_3 _____ E_4 _____

   R_T _____ I_1 _____ I_2 _____ I_3 _____ I_4 _____

2. Find the unknown values in the circuit if the applied voltage is 350 V and the resistors have the following values:

   $R_1 = 22\ K\Omega$; $R_2 = 18\ K\Omega$; $R_3 = 12\ K\Omega$; and $R_4 = 30\ K\Omega$.

   I_T _____ E_1 _____ E_2 _____ E_3 _____ E_4 _____

   R_T _____ I_1 _____ I_2 _____ I_3 _____ I_4 _____

3. Find the unknown values in the circuit if the applied voltage is 18 V and the resistors have the following values:

   $R_1 = 82\ \Omega$; $R_2 = 160\ \Omega$; $R_3 = 220\ \Omega$; and $R_4 = 470\ \Omega$.

   I_T _____ E_1 _____ E_2 _____ E_3 _____ E_4 _____

   R_T _____ I_1 _____ I_2 _____ I_3 _____ I_4 _____

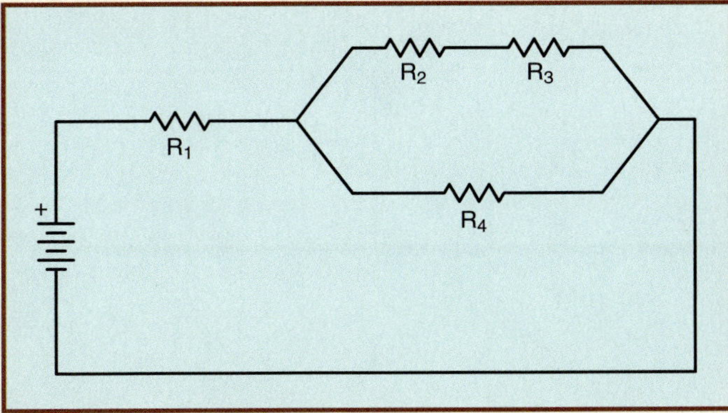

Figure 7-21 **Series-parallel circuit.**

Parallel-Series Circuits

Refer to the circuit shown in Figure 7-22 to solve the following problems.

4. Find the unknown values in the circuit if the total current is 0.8 A and the resistors have the following values:

   $R_1 = 1.5\ K\Omega$; $R_2 = 910\ \Omega$; $R_3 = 2\ K\Omega$; and $R_4 = 3.6\ K\Omega$.

   E_T _____ E_1 _____ E_2 _____ E_3 _____ E_4 _____

   R_T _____ I_1 _____ I_2 _____ I_3 _____ I_4 _____

5. Find the unknown values in the circuit if the total current is 0.65 A and the resistors have the following values:

   $R_1 = 22\ K\Omega$; $R_2 = 15\ K\Omega$; $R_3 = 22\ K\Omega$; and $R_4 = 33\ K\Omega$.

   E_T _____ E_1 _____ E_2 _____ E_3 _____ E_4 _____

   R_T _____ I_1 _____ I_2 _____ I_3 _____ I_4 _____

6. Find the unknown values in the circuit if the total current is 1.2 A and the resistors have the following values:

   $R_1 = 75\ \Omega$; $R_2 = 47\ \Omega$; $R_3 = 220\ \Omega$; and $R_4 = 160\ \Omega$.

   E_T _____ E_1 _____ E_2 _____ E_3 _____ E_4 _____

   R_T _____ I_1 _____ I_2 _____ I_3 _____ I_4 _____

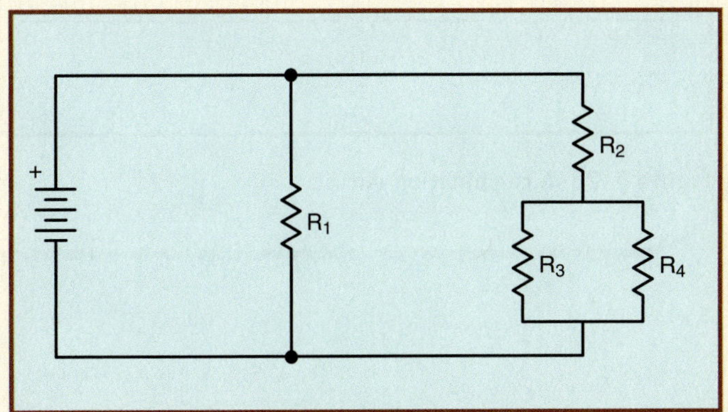

Figure 7-22 **A parallel-series circuit.**

Combination Circuits

Refer to the circuit shown in Figure 7-23 to solve the following problems.

E_T 250	E_1	E_2	E_3	E_4
I_T	I_1	I_2	I_3	I_4
R_T	R_1 220 Ω	R_2 500 Ω	R_3 470 Ω	R_4 280 Ω
P_T	P_1	P_2	P_3	P_4

E_5	E_6	E_7	E_8	E_9
I_5	I_6	I_7	I_8	I_9
R_5 400 Ω	R_6 500 Ω	R_7 350 Ω	R_8 450 Ω	R_9 300 Ω
P_5	P_6	P_7	P_8	P_9

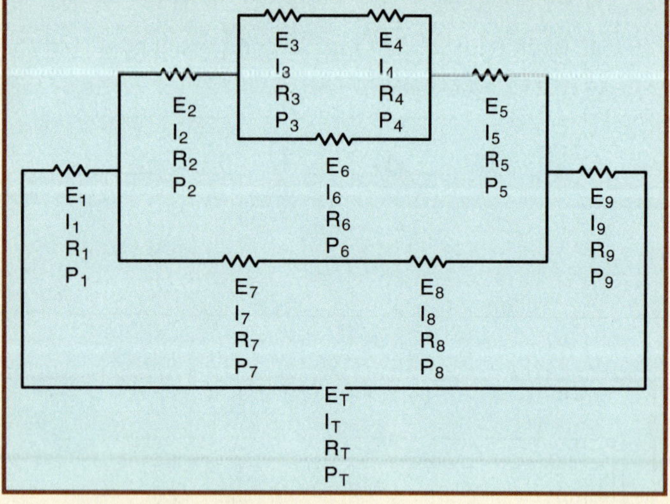

Figure 7-23 A combination circuit.

E_T	E_1	E_2	E_3	E_4 1.248
I_T	I_1	I_2	I_3	I_4
R_T	R_1	R_2	R_3	R_4
P_T 0.576	P_1 0.0806	P_2 0.0461	P_3 0.00184	P_4

E_5	E_6	E_7	E_8	E_9
I_5	I_6	I_7	I_8	I_9
R_5	R_6	R_7	R_8	R_9
P_5 0.0203	P_6 0.00995	P_7 0.0518	P_8 0.0726	P_9 0.288

Chapter 8 Measuring Instruments

Anyone desiring to work in the electrical and electronics field must become proficient with the common instruments used to measure electrical quantities. These instruments are the voltmeter, ammeter, and ohmmeter. Without meters it would be impossible to make meaningful interpretations of what is happening in a circuit. Meters can be divided into two general types: analog and digital.

OBJECTIVES

After studying this unit, you should be able to:

- **discuss the operation of a d'Arsonval meter movement.**
- **connect a voltmeter to a circuit.**
- **connect and read an analog multimeter.**
- **connect an ammeter.**
- **measure resistance using an ohmmeter.**

Glossary of Measuring Instrument Terms

ammeter a device for measuring current flow

ammeter shunt a device that allows an ammeter to measure large amounts of current; the shunt is connected in series with the load and the ammeter is connected in parallel with the shunt

analog meters meters that employ a moving pointer to indicate a value

clamp-on ammeter an ammeter with a movable jaw that can be clamped around a conductor to measure current flow

current transformer a transformer used to measure large values of AC current

d'Arsonval movement a galvanometer that employs a moving coil suspended inside a permanent magnet to move a pointer

galvanometers very sensitive meters that require only a few microamperes of current to operate

multirange voltmeters voltmeters that employ more than one full range value

ohmmeter a meter used to measure values of resistance

voltmeter a meter used to measure voltage

Analog Meters

Analog meters are characterized by the fact that they use a pointer and scale to indicate their value (Fig. 8-1). There are different types of analog meter movements. One of the most common is the **d'Arsonval movement** shown in Figure 8-2. This type of movement is often referred to as a *moving coil meter*. A coil of wire is suspended between the poles of a permanent magnet, either by jeweled movements similar to those used in watches or by taut bands. The taut band type offers less turning friction than the jeweled movement. These meters can be made to operate on very small amounts of current and often are referred to as **galvanometers.**

Principle of Operation

Analog meters operate on the principle that like magnetic poles repel each other. As current passes through the coil, a magnetic field is created around the coil. The direction of current flow through the meter is such that the same polarity of magnetic pole is created around the coil as that of the permanent magnet. This like polarity causes the coil to be deflected away from the pole of the magnet. A spring is used to retard the turning of the coil. The distance the coil turns against the spring is proportional to the strength of the magnetic field developed in the coil. If a pointer is added to the coil and a scale is placed behind the pointer, a meter movement is created.

Since the turning force of this meter depends on the repulsion of magnetic fields, it will operate on DC current only. If an AC current is connected to the moving coil, the magnetic polarity will change 60 times per second and the net turning force will be zero. For this reason, a DC voltmeter will indicate zero if connected to an AC line. When this type of movement is to be used to measure AC values, the current must be rectified, or changed into DC, before it is applied to the meter (Fig. 8-3).

Figure 8-1 **An analog meter.** *Courtesy of Simpson Electric.*

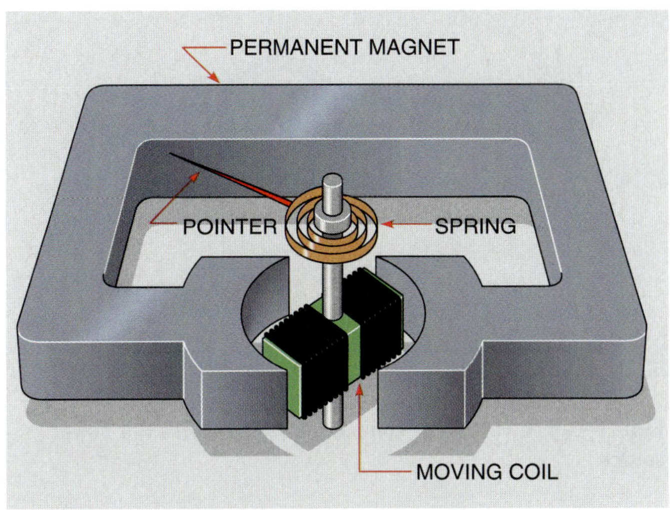

Figure 8-2 **Basic d'Arsonval meter movement.**

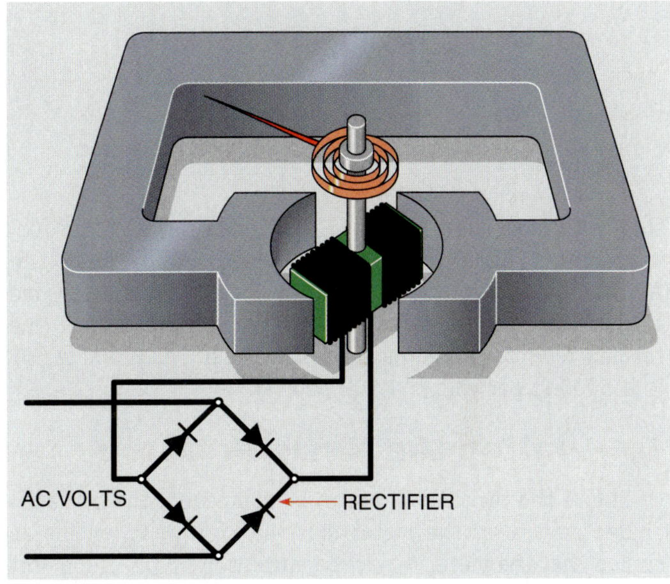

Figure 8-3 **Rectifier changes AC voltage into DC voltage.**

The Voltmeter

The **voltmeter** is designed to be connected directly across the source of power. Figure 8-4 shows a voltmeter being used to test the voltage of a battery. Notice that the leads of the meter are connected directly across the source of voltage. A voltmeter can be connected directly across the power source because it has a very high resistance connected in series with

Figure 8-4 **A voltmeter connects directly across the power source.**

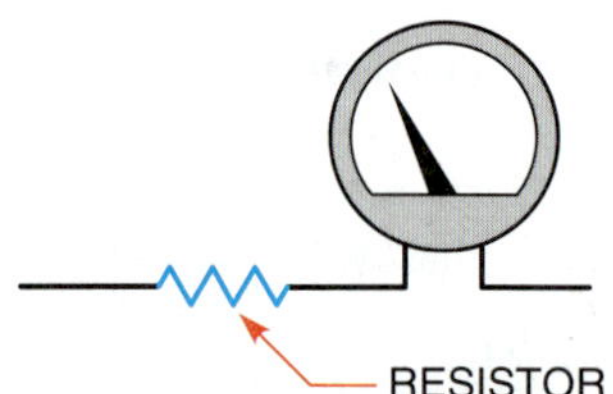

Figure 8-5 **A resistor connects in series with the meter.**

the meter movement (Fig. 8-5). The industrial standard for a voltmeter is 20,000 Ω per volt for DC and 5000 Ω per volt for AC. Assume the voltmeter shown in Figure 8-5 is an AC meter and has a full scale range of 300 V. The meter circuit (meter plus resistor) would have a resistance of 1,500,000 Ω (300 V × 5000 Ω per volt = 1,500,000 Ω).

Calculating the Resistor Value

Before the resistor value can be computed, the operating characteristics of the meter must be known. It will be assumed that the meter requires a current of 50 μA and a voltage of 1 V to deflect the pointer full scale. These are known as the *full scale values* of the meter.

When the meter and resistor are connected to a source of voltage, their combined voltage drop must be 300 V. Since the meter has a voltage drop of 1 V, the resistor must have a drop of 299 V. The resistor and meter are connected in series with each other. In a series circuit, the current flow must be the same in all parts of the circuit. If 50 μA of current flow is required to deflect the meter full scale, then the resistor must have a current of 50 μA flowing through it when it has a voltage drop of 299 V. The value of resistance can now be computed using Ohm's law.

$$R = \frac{E}{I}$$

$$R = \frac{299}{0.000050}$$

$$R = 5.98\ M\Omega\ (5{,}980{,}000\ \Omega)$$

Multirange Voltmeters

Most voltmeters are **multirange voltmeters,** which means that they are designed to use one meter movement to measure several ranges of voltage. For example, one meter may have a selector switch that permits full scale ranges to be selected. These ranges may be 3 V full scale, 12 V full scale, 30 V full scale, 60 V full scale, 120 V full scale, 300 V full scale, and 600 V full scale. Meters are made with that many scales so that they will be as versatile as possible. If it is necessary to check for a voltage of 480 V, the meter can be set on the 600-V range. It would be very difficult, however, to check a 24-V system on the 600-V range. If the meter is set on the 30-V range, it is simple to test for a voltage of 24 V. The meter shown in Figure 8-6 has multirange selection for voltage.

When the selector switch of this meter is turned, steps of resistance are inserted in the circuit to increase the range or removed from the circuit to decrease the range. The meter shown in Figure 8-7 has four range settings for full scale voltage: 30 V, 60 V, 300 V, and 600 V. Notice that when the higher voltage settings are selected, more resistance is inserted in the circuit.

Calculating the Resistor Values

The values of the four resistors shown in Figure 8-7 can be determined using Ohm's law. Assume that the full scale values of the meter are 50 μA and 1 V. The first step is to determine the value for resistor R_1, which provides a full scale

Figure 8-6 **Volt-ohm-milliam meter with multi-range selection.** *Courtesy of Triplett Corp.*

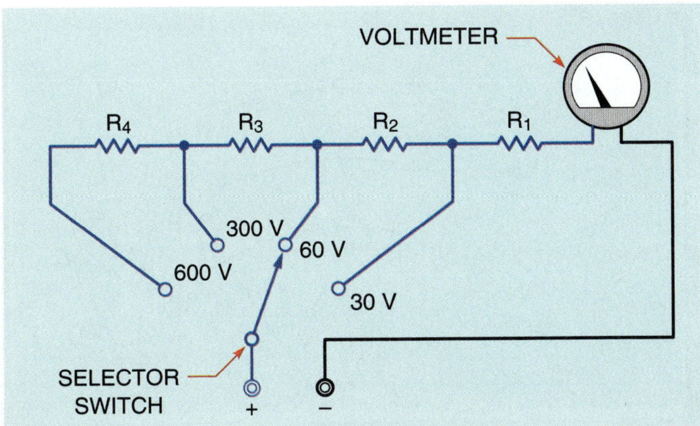

Figure 8-7 A rotary selector switch is used to change the full range setting.

value of 30 V. Resistor R_1, therefore, must have a voltage drop of 29 V when a current of 50 μA is flowing through it.

$$R = \frac{E}{I}$$

$$R = \frac{29}{0.000050}$$

$$R = 580\ k\Omega\ (580{,}000\ \Omega)$$

When the selector switch is moved to the second position, the meter circuit should have a total voltage drop of 60 V. The meter movement and resistor R_1 have a total voltage drop of 30 V, so resistor R_2 must have a voltage drop of 30 V when 50 μA of current flow through it. This will provide a total voltage drop of 60 V for the entire circuit.

$$R = \frac{E}{I}$$

$$R = \frac{30}{50\ \mu A\ (0.000050\ A)}$$

$$R = 600\ k\Omega\ (600{,}000\ \Omega)$$

When the selector switch is moved to the third position, the circuit must have a total voltage drop of 300 V. Resistors R_1 and R_2 plus the meter movement have a combined voltage drop of 60 V at rated current. Resistor R_3, therefore, must have a voltage drop of 240 V at 50 μA.

$$R = \frac{E}{I}$$

$$R = \frac{240}{50\ \mu A}$$

$$R = 4.8\ M\Omega\ (4{,}800{,}000\ \Omega)$$

When the selector switch is moved to the fourth position, the circuit must have a total voltage drop of 600 V at rated current. Since resistors R_1, R_2, and R_3 plus the meter movement produce a voltage drop of 300 V at rated current, resistor R_4 must have a voltage drop of 300 V when 50 μA of current flow through it.

$$R = \frac{E}{I}$$

$$R = \frac{300}{50\ \mu A}$$

$$R = 6\ M\Omega\ (6{,}000{,}000\ \Omega)$$

Reading a Meter

Learning to read the scale of a multimeter takes time and practice. Most people use meters every day without thinking about it. A common type of meter used daily by most people is shown in Figure 8-8. It is a speedometer similar to those seen in automobiles. This meter is designed to measure speed. It is calibrated in miles per hour. The speedometer shown has a full scale value of 80 mph. If the pointer is positioned as shown in Figure 8-8, most people would know instantly that the speed of the automobile is 55 mph.

Figure 8-9 illustrates another common meter used by most people. This meter measures the amount of fuel in the tank of the automobile. Most people can glance at the pointer and know that the meter is indicating that there is one-quarter of a tank of fuel remaining. Now assume that the tank has a capacity of 20 gallons. The meter is indicating that 5 gallons of fuel remain in the tank.

Figure 8-8 A speedometer.

Figure 8-9 A fuel gauge.

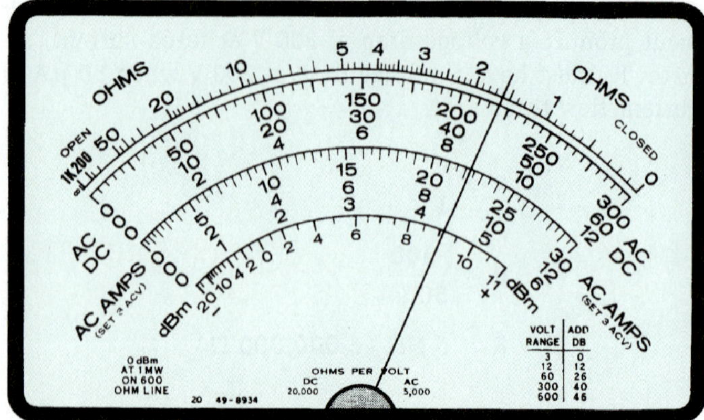

Figure 8-10 A typical multimeter. *From Herman and Sparkman,* Electricity and Controls for HVAC, *2nd ed., copyright 1991 by Delmar Publishers Inc.*

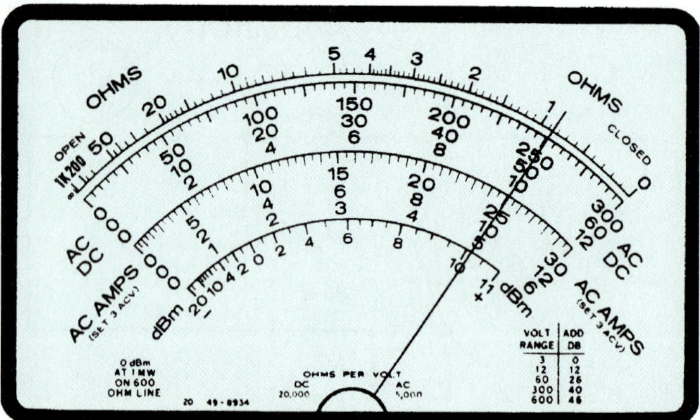

Figure 8-11 Reading the meter. *From Herman and Sparkman,* Electricity and Controls for HVAC, *2nd ed., copyright 1991 by Delmar Publishers Inc.*

Learning to read the scale of a multimeter is similar to learning to read a speedometer or fuel gauge. The meter scale shown in Figure 8-10 has several scales used to measure different quantities and values. The top of the scale measures resistance, or ohms. Notice that the scale begins on the left at infinity and ends at zero on the right. Ohmmeters will be covered later in this chapter. The second scale is labeled AC-DC and is used to measure voltage. Notice that this scale has three different full scale values. The top scale is 0–300, the second scale is 0–60, and the third scale is 0–12. The scale used is determined by the setting of the range control switch. The third set of scales is labeled AC amps. This scale is used with a clamp-on ammeter attachment that can be used with some meters. The last scale is labeled dBm, which is used to measure decibels.

Reading a Voltmeter

Notice that the three voltmeter scales use the primary numbers 3, 6, and 12, and are in multiples of 10 of these numbers. Since the numbers are in multiples of 10, it is easy to multiply or divide the readings in your head by moving a decimal point. Remember that any number can be multiplied by 10 by moving the decimal point one place to the right, and any number can be divided by 10 by moving the decimal point one place to the left. For example, if the selector switch were set to permit the meter to indicate a voltage of 3 V full scale, the 300-V scale would be used, and the reading would be divided by 100. The reading can be divided by 100 by moving the decimal point two places to the left. In Figure 8-11, the pointer is indicating a value of 250. If the selector switch is set for 3 V full scale, moving the decimal point two places to the left will give a reading of 2.5 V. If the selector switch were set for a full scale value of 30 V, the meter shown in Figure 8-11 would be indicating a value of 25 V. That reading is obtained by dividing the scale by 10 and moving the decimal point one place to the left.

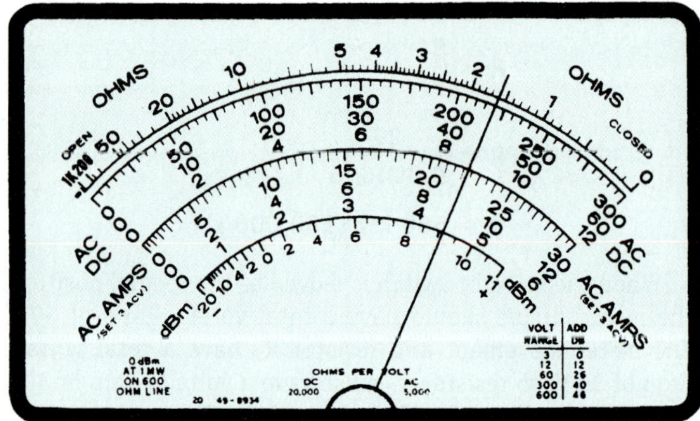

Figure 8-12 Reading the meter. *From Herman and Sparkman.* Electricity and Controls for HVAC, *2nd ed., copyright 1991 by Delmar Publishers Inc.*

Now assume that the meter has been set to have a full scale value of 600 V. The pointer in Figure 8-12 is indicating a value of 44. Since the full scale value of the meter is set for 600 V, use the 60-V range and multiply the reading on the meter by 10. By moving the decimal point one place to the right, the correct reading becomes 440 V.

Three distinct steps should be followed when reading a meter. These steps are especially helpful for someone who has not had a great deal of experience reading a multimeter. The steps are

1. *Determine what the meter indicates.* Is the meter set to read a value of DC voltage, DC current, AC voltage, AC current, or ohms? It is impossible to read a meter if you don't know what the meter is used to measure.
2. *Determine the full scale value of the meter.* The advantage of a multimeter is that it can measure a wide range of values and quantities. After determining what

quantity the meter is set to measure, it must be determined what the range of the meter is. There is a great deal of difference in reading when the meter is set to indicate a value of 600 V full scale and when it is set for 30 V full scale.

3. *Read the meter.* The last step is to determine what the meter is indicating. It may be necessary to determine the value of the hash marks on the meter face for the range for which the selector switch is set. If the meter in Figure 8-10 is set for 300 V full scale, each hash mark has a value of 5 V. If the full scale value of the meter is 60 volts, however, each hash mark has a value of 1 V.

The Ammeter

The **ammeter,** unlike the voltmeter, is a very low-impedance device. The ammeter is used to measure current and must be connected in series with the load to permit the load to limit the current flow (Fig. 8-13). An ammeter has a typical impedance of less than 0.1 Ω. If this meter is connected in parallel with the power supply, the impedance of the ammeter is the only thing to limit the amount of current flow in the circuit. Assume that an ammeter with a resistance of 0.1 Ω is connected across a 240-V AC line. The current flow in this circuit would be 2400 A (240/0.1 = 2400). A blinding flash of light would be followed by the destruction of the ammeter. Ammeters connected directly into the circuit as shown in Figure 8-13 are referred to as in-line ammeters. Figure 8-14 shows an ammeter of this type.

Ammeter Shunts

DC ammeters are constructed by connecting a common moving coil type of meter across a shunt. An **ammeter shunt** is a low-resistance device used to conduct most of the circuit current away from the meter movement. Since the meter movement is connected in parallel with the shunt, the voltage drop across the shunt is the voltage applied to the meter. Most ammeter shunts are manufactured to have a voltage drop of 50 mV (millivolts). If a 50-mV meter movement is connected across the shunt as shown in Figure 8-15 the pointer will move to the full scale value when the rated current of the shunt is flowing. In the example shown, the ammeter shunt is rated to have a 50-mV drop when a 10-A current is flowing in the circuit. Since the meter movement has a full scale voltage of 50 mV, it will indicate the full scale value when 10 A of current are flowing through the shunt. An ammeter shunt is shown in Figure 8-16.

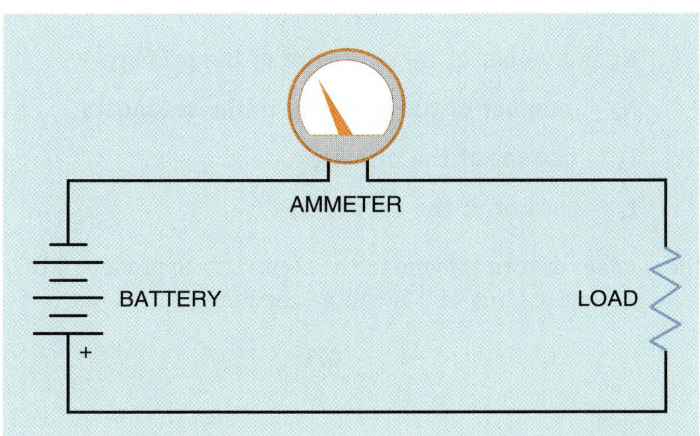

Figure 8-13 An ammeter connects in series with the load.

Figure 8-14 In-line ammeter.

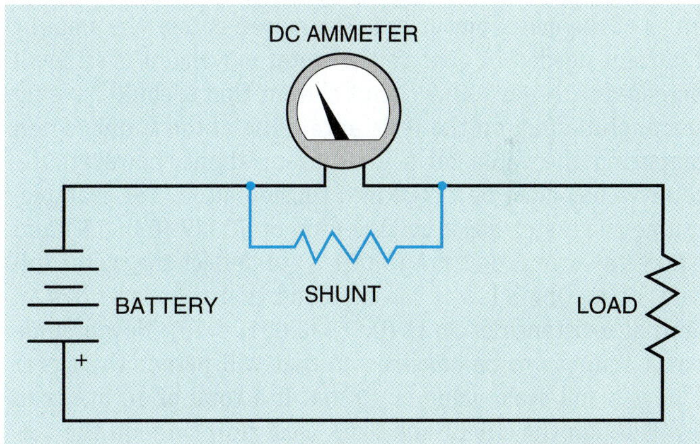

Figure 8-15 A shunt is used to set the value of the ammeter.

Figure 8-16 Ammeter shunt.

Ammeter shunts can be purchased to indicate different values. If the same 50-mV movement is connected across a shunt designed to drop 50 mV when 100 A of current flow through it, the meter will now have a full scale value of 100 A.

The resistance of an ammeter shunt can be computed using Ohm's law. The resistance of a shunt designed to have a voltage drop of 50 mV when 100 A of current flow through it is

$$R = \frac{E}{I}$$

$$R = \frac{0.050}{100}$$

$$R = 0.0005\ \Omega\text{, or } 0.5\ \text{m}\Omega$$

In this problem, no consideration is given to the electrical values of the meter movement. The reason is that the amount of current needed to operate the meter movement is so small compared with the 100-A circuit current that it could have no meaningful effect on the resistance value of the shunt. When computing the value for a low-current shunt, however, the meter values must be taken into consideration. For example, assume the meter has a voltage drop of 50 mV (0.050 V) and requires a current of 1 mA (0.001 A) to deflect the meter full scale. Using Ohm's law it can be found that the meter has an internal resistance of 50 Ω (0.050/0.001 = 50). Now assume that a shunt is to be constructed that will permit the meter to have a full scale value of 10 mA. If a total of 10 mA is to flow through the circuit and 1 mA must flow through the meter, then 9 mA must flow through the shunt (Fig. 8-17). Since the shunt must have a voltage drop of 50 mV when 9 mA of current are flowing through it, its resistance must be 5.555 Ω (0.050/0.009 = 5.555).

AC Ammeters

Shunts can be used with AC ammeters to increase their range, but cannot be used to decrease their range. Most AC ammeters use a **current transformer** instead of shunts to change scale values. This type of ammeter is shown in Figure 8-18. The primary of the transformer is connected in series with the load, and the ammeter is connected to the secondary of the transformer. Notice that the range of the meter is changed by selecting different taps on the secondary of the current transformer. The different taps on the transformer provide different turns ratios between the primary and secondary of the transformer. The turns ratio is the ratio of the number of turns of wire in the primary as compared with the number of turns of wire in the secondary.

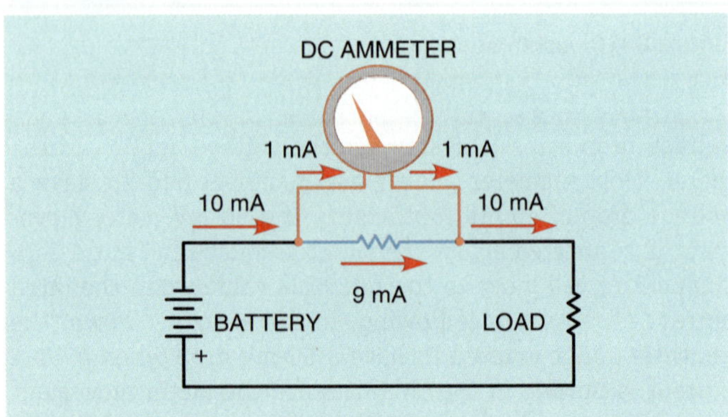

Figure 8-17 **The total current is divided between the meter and the shunt.**

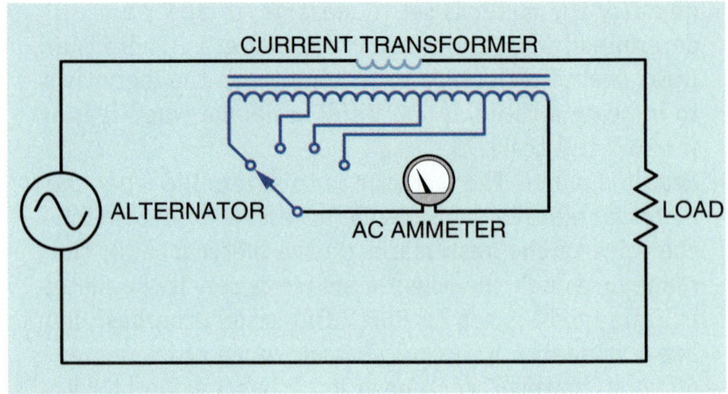

Figure 8-18 **A current transformer is used to change the range of an AC ammeter.**

Computing the Turns Ratio

In this example, it will be assumed that an AC meter movement requires a current flow of 100 mA to deflect the meter full scale. It is also assumed that the primary of the current transformer contains 5 turns of wire. A transformer will be designed to provide full scale current readings of 1 A, 5 A, and 10 A. To find the number of turns required in the secondary winding, the following formula can be used.

$$\frac{N_p}{N_s} = \frac{I_s}{I_p}$$

where

N_p = **number of turns of wire in the primary**

N_s = **number of turns of wire in the secondary**

I_p = **current of the primary**

I_s = **current of the secondary**

The number of turns of wire in the secondary to produce a full scale current reading of 1 A can be computed as follows:

$$\frac{5}{N_s} = \frac{0.1}{1}$$

Cross-multiplication is used to solve the problem. Cross-multiplication is accomplished by multiplying the bottom half of the equation on one side of the equals sign by the top half of the equation on the other side of the equals sign.

$$0.1\ N_s = 5$$

$$N_s = 50$$

The transformer secondary must contain 50 turns of wire if the ammeter is to indicate a full scale reading when 1 A of current flows through the primary winding.

The number of secondary turns can be found for the other values of primary current in the same way.

$$\frac{5}{N_s} = \frac{0.1}{5}$$

$$0.1\ N_s = 25$$

$$N_s = 250 \text{ turns}$$

$$\frac{5}{N_s} = \frac{0.1}{10}$$

$$0.1\ N_s = 50$$

$$N_s = 500 \text{ turns}$$

Current Transformers (CTs)

When a large amount of AC current must be measured, a different type of current transformer is connected in the power line. These transformers have ratios that start at 200:5 and can have ratios of several thousand to five. These current transformers, generally referred to in industry as CTs, have a standard secondary current rating of 5 A AC. They are designed to be operated with a 5-A AC ammeter connected directly to their secondary winding, which produces a short circuit. CTs are designed to operate with the secondary winding shorted. **The secondary winding of a CT should never be opened when there is power applied to the primary. This will cause the transformer to produce a step-up in voltage that could be high enough to kill anyone who comes in contact with it.**

A current transformer is basically a toroid transformer. A toroid transformer is constructed with a hollow core similar to a doughnut (Fig. 8-19). When current transformers are used, the main power line is inserted through the opening in the transformer (Fig. 8-20). The power line acts as the primary of the transformer and is considered to be 1 turn.

The turns ratio of the transformer can be changed by looping the power wire through the opening in the transformer to produce a primary winding of more than 1 turn. For example, assume a current transformer has a ratio of 600:5. If the primary power wire is inserted through the opening, it will require a current of 600 A to deflect the meter full scale. If the primary power conductor is looped around and inserted through the window a second time, the primary now contains 2 turns of wire instead of 1 (Fig. 8-21). It now requires 300 A of current flow in the primary to de-

Figure 8-19 **A toroid current transformer.** *Courtesy of SQUARE D COMPANY.*

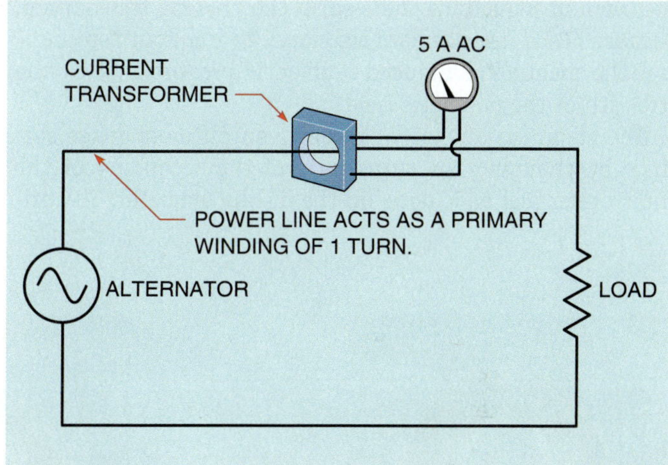

Figure 8-20 **Toroid transformer used to change the scale factor of an AC ammeter.**

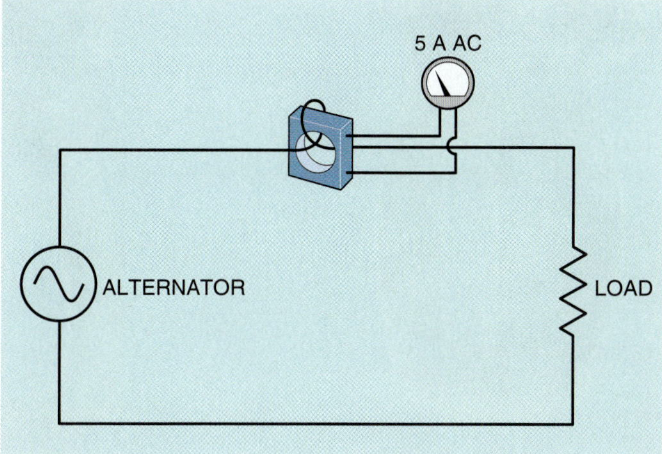

Figure 8-21 **The primary conductor loops through the CT to produce a second turn, which changes the ratio.**

flect the meter full scale. If the primary conductor is looped through the opening a third time, it would require only 200 A of current flow to deflect the meter full scale.

Clamp-on Ammeters

Many electricians use the **clamp-on** type of AC **ammeter** (Fig. 8-22). The jaw of the meter is clamped around one of the conductors supplying power to the load (Fig. 8-23). The meter is clamped around only one of the lines. If it is clamped around more than one line, the magnetic fields of the wires cancel each other and the meter indicates zero.

The clamp-on meter also uses a current transformer to operate. The jaw of the meter is part of the core material of the transformer. When the meter is connected around the current-carrying wire, the changing magnetic field produced by the AC current induces a voltage into the current transformer. The strength and frequency of the magnetic field determine the amount of voltage induced in the current transformer. Because 60 Hz is a standard frequency throughout the country, the amount of induced voltage is proportional to the strength of the magnetic field.

The clamp-on ammeter can be given different range settings by changing the turns ratio of the secondary of the transformer just as is done on the in-line ammeter. The primary of the transformer is the conductor around which the movable jaw is connected. If the ammeter is connected around one wire, the primary has one turn of wire compared with the turns of the secondary. The turns ratio can be changed in the same manner that the ratio of the CT is changed. If 2 turns of wire are wrapped around the jaw of the ammeter (Fig. 8-24), the primary winding now contains 2 turns instead of 1, and the turns ratio of the transformer is changed. The ammeter will now indicate double the amount of current in the circuit. The reading on the scale of the meter would have to be divided by 2 to get the correct reading. The ability to change the turns ratio of a clamp-on ammeter can be useful for measuring low currents. Changing the turns ratio is not limited to wrapping 2 turns of wire around the jaw of the ammeter. Any number of turns can be wrapped around the jaw of the ammeter, and the reading will be divided by that number.

DC-AC Clamp-on Ammeters

Most clamp-on ammeters with the ability to measure both direct and alternating current do not operate on the principle of the current transformer. Current transformers depend

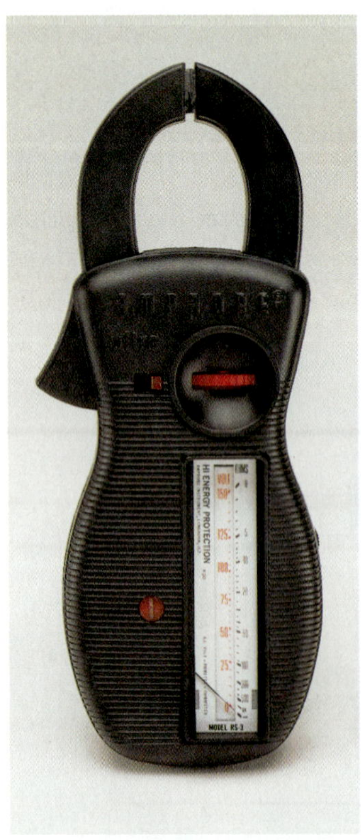

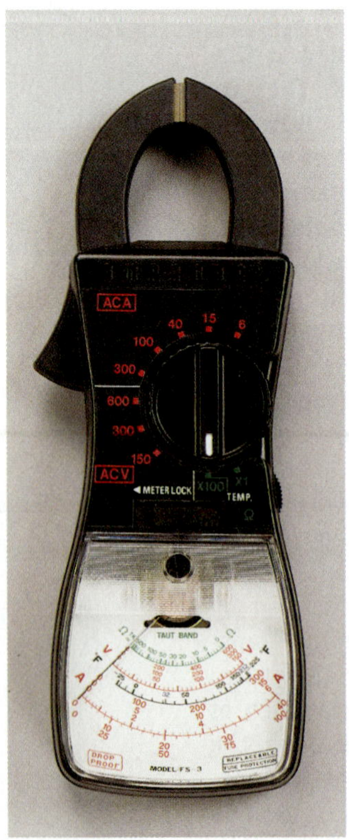

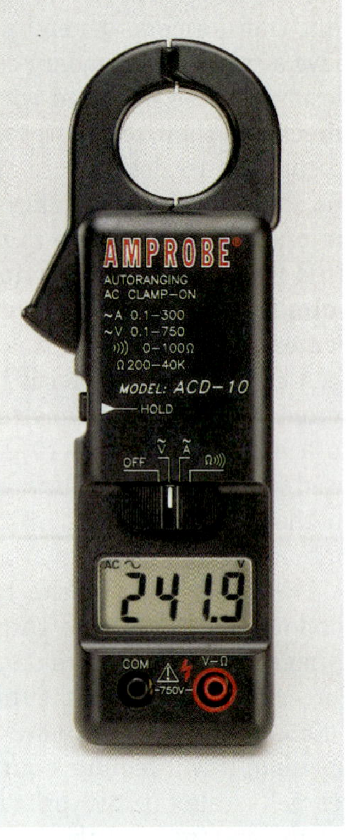

Figure 8-22 (A) Analog type clamp-on ammeter with vertical scale. (B) Analog type clamp-on ammeter with flat scale. (C) Clamp-on ammeter with digital scale. *Courtesy of Amprobe Instrument.*

Figure 8-23 The clamp-on ammeter connects around only one conductor.

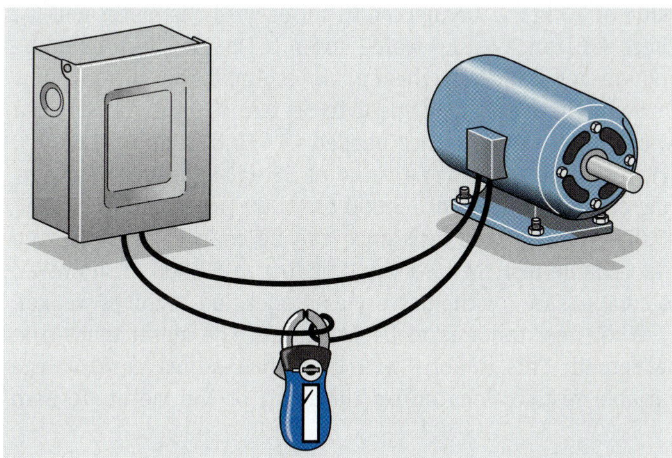

Figure 8-24 Looping the conductor around the jaw of the ammeter changes the ratio.

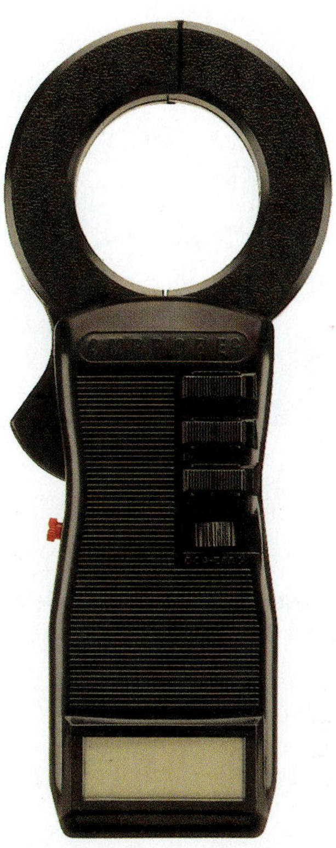

Figure 8-25 DC-AC clamp-on ammeter. *Courtesy of Amprobe Instrument.*

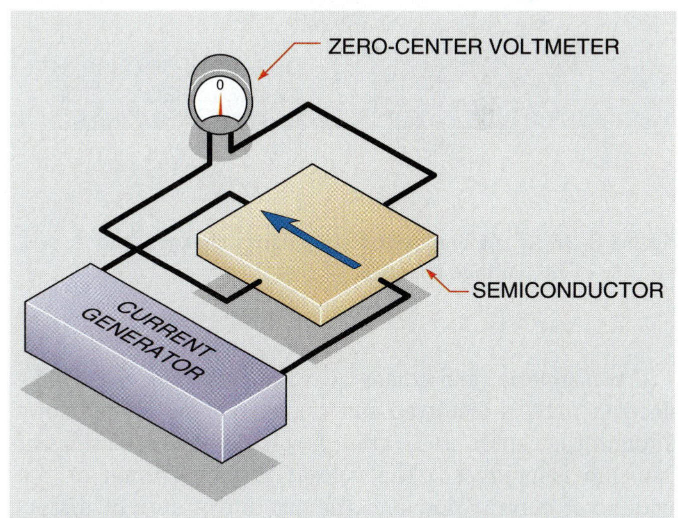

Figure 8-26 Basic Hall generator.

on induction, which means that the current in the line must change direction periodically to provide a change of magnetic field polarity. It is the continuous change of field strength and direction that permits the current transformer to operate. The current in a DC circuit is unidirectional and does not change polarity, which would not permit the current transformer to operate.

DC-AC clamp-on ammeters (Fig. 8-25) use the Hall effect as the basic principle of operation. The Hall effect was discovered by Edward H. Hall at Johns Hopkins University in 1879. Hall originally used a piece of pure gold to produce the Hall effect, but today a semiconductor material is used because it has better operating characteristics and is less expensive. The device is often referred to as a Hall generator. Figure 8-26 illustrates the operating principle of the Hall generator. A constant-current generator supplies a continuous current to the semiconductor chip. The leads of a zero-center voltmeter are connected across the opposite sides of the chip. As long as the current flows through the center of the semiconductor chip, no potential difference or voltage develops across the chip.

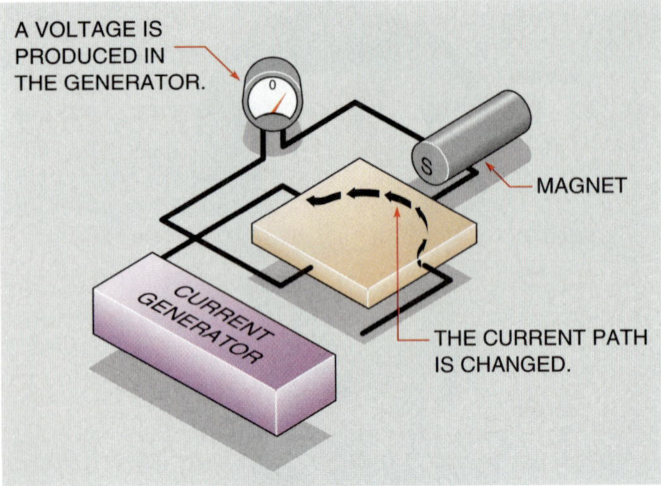

Figure 8-27 The presence of a magnetic field causes the Hall generator to produce a voltage.

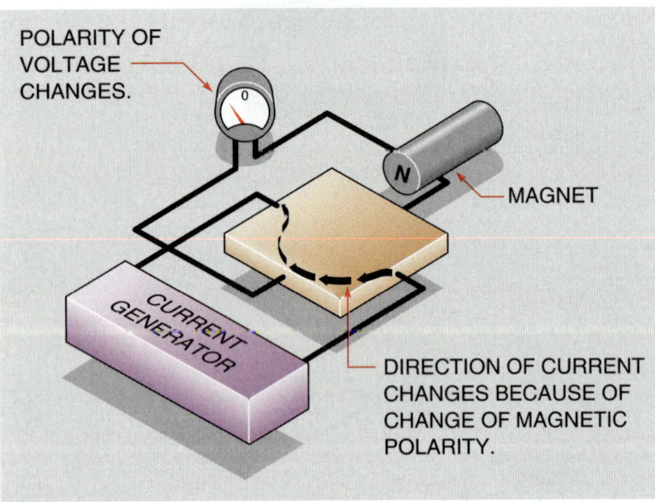

Figure 8-28 If the magnetic field polarity changes, the polarity of the voltage changes.

If a magnetic field comes near the chip (Fig. 8-27), the electron path is distorted and the current no longer flows through the center of the chip. A voltage across the sides of the chip is produced. The voltage is proportional to the amount of current flow and the amount of current distortion. Since the current remains constant and the amount of distortion is proportional to the strength of the magnetic field, the voltage produced across the chip is proportional to the strength of the magnetic field.

If the polarity of the magnetic field were reversed (Fig. 8-28), the current path would be distorted in the opposite direction, producing a voltage of the opposite polarity. Notice that the Hall generator produces a voltage in the presence of a magnetic field. It makes no difference whether the field is moving or stationary. The Hall effect can, therefore, be used to measure direct or alternating current.

The Ohmmeter

The **ohmmeter** is used to measure resistance. The common VOM (volt-ohm-milliammeter) contains an ohmmeter. The ohmmeter has the only scale on a VOM that is nonlinear. The scale numbers increase in value as they progress from right to left. There are two basic types of analog ohmmeters, the series and the shunt. The series ohmmeter is used to measure high values of resistance, and the shunt type is used to measure low values of resistance. Regardless of the type used, the meter must provide its own power source to measure resistance. The power is provided by batteries located inside the instrument.

The Series Ohmmeter

A schematic for a basic series ohmmeter is shown in Figure 8-29. It is assumed that the meter movement has a resistance of 1000 Ω and requires a current of 50 μA to deflect the meter full scale. The power source will be a 3-V battery. R_1, a fixed resistor with a value of 54 kΩ, is connected in series with the meter movement, and R_2, a variable resistor with a value of 10 kΩ, is connected in series with the meter and R_1. These resistance values were chosen to ensure there would be enough resistance in the circuit to limit the current flow through the meter movement to 50 μA. If Ohm's law is used to compute the resistance needed (3 V/0.000050 A = 60,000 Ω), it will be seen that a value of 60 kΩ is needed. This circuit contains a total of 65,000 Ω (1000 [meter] + 54,000 + 10,000). The circuit resistance can be changed by adjusting the variable resistor to a value as low as 55,000 Ω, however, to compensate for the battery as it ages and becomes weaker.

When resistance is to be measured, the meter must first be zeroed. This is done with the ohms-adjust control, the variable resistor located on the front of the meter. To zero

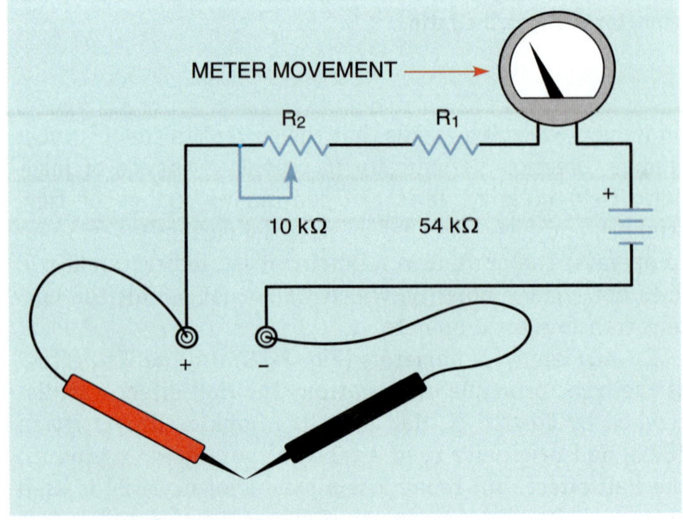

Figure 8-29 Basic series ohmmeter.

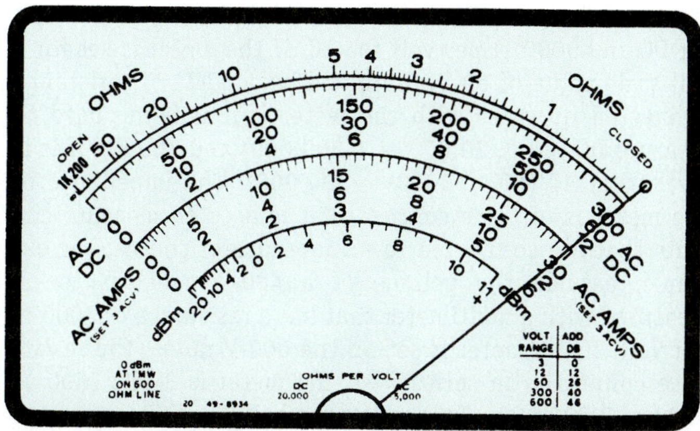

Figure 8-30 Adjusting the ohmmeter to zero. *From Herman and Sparkman,* Electricity and Controls for HVAC, *2nd ed., copyright 1991 by Delmar Publishers Inc.*

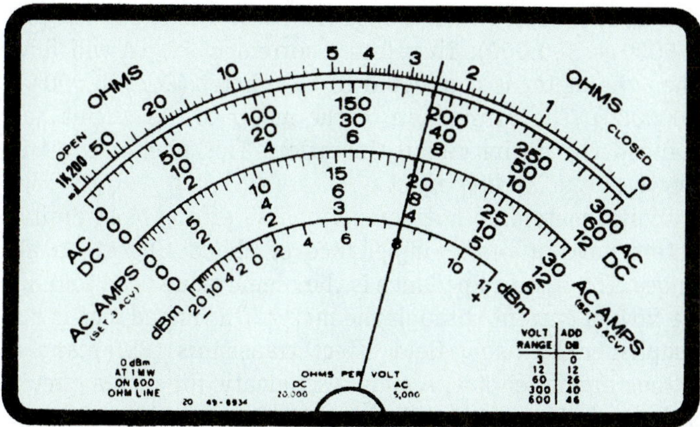

Figure 8-31 Reading the ohmmeter. *From Herman and Sparkman,* Electricity and Controls for HVAC, *2nd ed. copyright 1991 by Delmar Publishers Inc.*

the meter, connect the leads (Fig. 8-29) and turn the ohms-adjust knob until the meter indicates zero at the far right end of the scale (Fig. 8-30). When the leads are separated, the meter will again indicate infinity resistance at the left side of the scale. When the leads are connected across a resistance, the meter will again go up the scale. Since resistance has been added to the circuit, less than 50 μA of current will flow, and the meter will indicate some value other than zero. Figure 8-31 shows a meter indicating a resistance of 2.5 Ω, assuming the range setting is Rx1.

Ohmmeters can have different range settings such as R×1, R×100, R×1000, or R×10,000. These different scales can be obtained by adding different values of resistance in the meter circuit and resetting the meter to zero. **An ohmmeter should always be readjusted to zero when the scale is changed.** On the R×1 setting, the resistance is measured straight off the resistance scale located at the top of the meter. If the range is set for R×1000, however, the reading must be multiplied by 1000. The ohmmeter reading shown in Figure 8-31 would be indicating a resistance of 2,500 Ω if the range had been set for R×1000. Notice that the ohmmeter scale is read backward from the other scales. Zero ohms is located on the far right side of the scale, and maximum ohms is located at the far left side. It generally takes a little time and practice to read the ohmmeter properly.

Digital Meters

Digital Ohmmeters

Digital ohmmeters display the resistance in figures instead of using a meter movement. When using a digital ohmmeter, care must be taken to notice the scale indication on the meter. For example, most digital meters will display a *K* on the scale to indicate kilohms or an *M* to indicate megohms (kilo means 1000 and mega means 1,000,000). If the meter is showing a resistance of 0.200 K, it means 0.200 × 1000, or 200 Ω. If the meter indicates 1.65 M, it means 1.65 × 1,000,000, or 1,650,000 Ω.

Appearance is not the only difference between analog and digital ohmmeters. Their operating principle is different also. Analog meters operate by measuring the amount of current change in the circuit when an unknown value of resistance is added. Digital ohmmeters measure resistance by measuring the amount of voltage drop across an unknown resistance. In the circuit shown in Figure 8-32 a constant-current generator supplies a known amount of current to a resistor, R_x. It will be assumed that the amount of current supplied is 1 mA. The voltage dropped across the resistor is proportional to the resistance of the resistor and the amount of current flow. For example, assume the value of the unknown resistor is 4700 Ω. The voltmeter would indicate a drop of 4.7 V when 1 mA of current flowed through the resistor. The scale factor of the ohmmeter can be changed by changing the amount of current flow through the resistor. Digital ohmmeters generally exhibit an accuracy of about 1%.

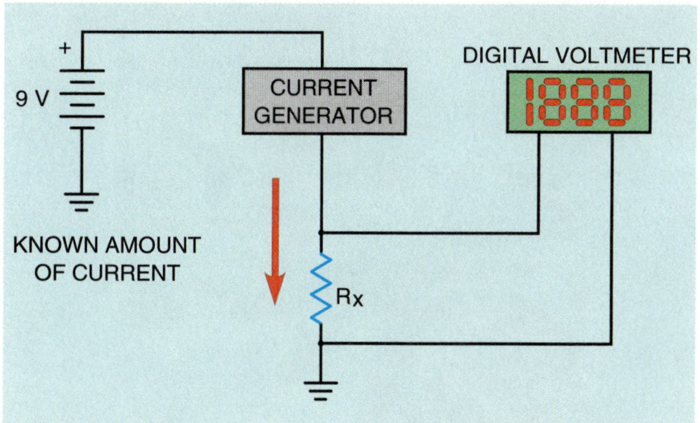

Figure 8-32 Digital ohmmeters operate by measuring the voltage drop across a resistor when a known amount of current flows through it.

The ohmmeter, whether digital or analog, must never be connected to a circuit when the power is turned on. Since the ohmmeter uses its own internal power supply, it has a very low operating voltage. Connecting a meter to power when it is set in the ohms position will probably damage or destroy the meter.

Digital Multimeters

Digital multimeters have become increasingly popular in the past few years. The most apparent difference between digital meters and analog meters is that digital meters display their reading in discrete digits instead of with a pointer and scale. A digital multimeter is shown in Fig. 8-33. Some digital meters have a range switch similar to the range switch used with analog meters. This switch sets the full range value of the meter. Many digital meters have voltage range settings from 200 mV to 2000 V. The lower ranges are used for accuracy. For example, assume it is necessary to measure a voltage of 16 V. The meter will be able to make a more accurate measurement when set on the 20-V range than when set on the 2000-V range.

Some digital meters do not contain a range setting control. These meters are known as autoranging meters. They have a function control switch that permits selection of the electrical quantity to be measured, such as AC volts, DC volts, ohms, and so on. When the meter probes are connected to the object to be tested, the meter automatically selects the proper range and displays the value.

Analog meters change scale value by inserting or removing resistance from the meter circuit (Fig. 8-7). The typical resistance of an analog meter is 20,000 Ω per volt for DC and 5000 Ω per volt for AC. If the meter is set for a full scale value of 60 V, there will be 1.2 MΩ of resistance connected in series with the meter if it is being used to measure DC (60 × 20,000 = 1,200,000) and 300 kΩ if it is measuring AC (60 × 5000 = 300,000). The impedance of the meter is of little concern if it is used to measure circuits that are connected to a high-current source. For example, assume the voltage of a 480-V panel is to be measured with a multimeter that has a resistance of 5000 Ω per volt. If the meter is set on the 600-V range, the resistance connected in series with the meter is 3 MΩ (600 × 5000 = 3,000,000). This resistance will permit a current of 160 μA to flow in the meter circuit (480/3,000,000 = 0.000160). This 160 μA of current is not enough to affect the circuit being tested.

Now assume that this meter is to be used to test a 24-V circuit that has a current flow of 100 μA. If the 60-V range is used, the meter circuit contains a resistance of 300 kΩ (60 × 5000 = 300,000). Therefore a current of 80 μA will flow when the meter is connected to the circuit (24/300,000 = 0.000080). The connection of the meter to the circuit has changed the entire circuit operation. This phenomenon is known as the *loading effect.*

Digital meters do not have a loading effect. Most digital meters have an input impedance of about 10 MΩ on all ranges. The input impedance is the ohmic value used to limit the flow of current through the meter. This impedance is accomplished by using field effect transistors (FETs) and a voltage divider circuit. A simple schematic for such a circuit is shown in Figure 8-34. Notice that the meter input is connected across 10 MΩ of resistance regardless of the range setting of the meter. If this meter is used to measure the voltage of the 24-V circuit, a current of 2.4 μA will flow through the meter. This is not enough current to upset the rest of the circuit, and voltage measurements can be made accurately.

Figure 8-33 Digital multimeter. *Courtesy of Amprobe Instrument.*

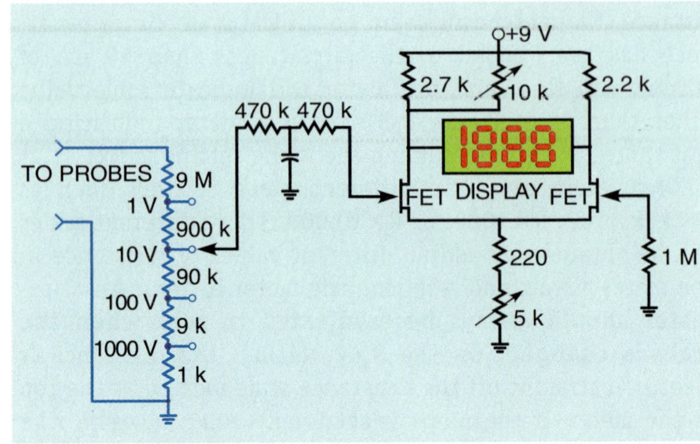

Figure 8-34 Digital voltmeter.

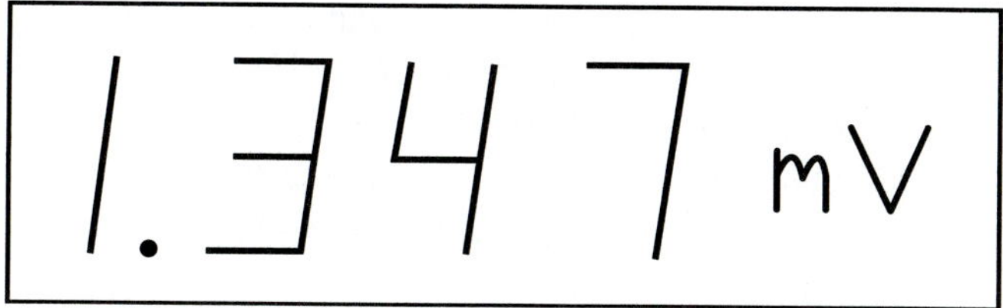

Figure 8-35 Display indicates a voltage of 1.347 millivolts.

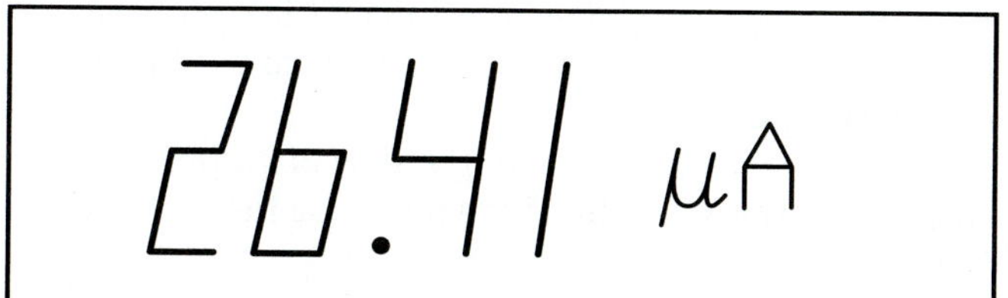

Figure 8-36 Display indicates a current of 26.41 microamperes.

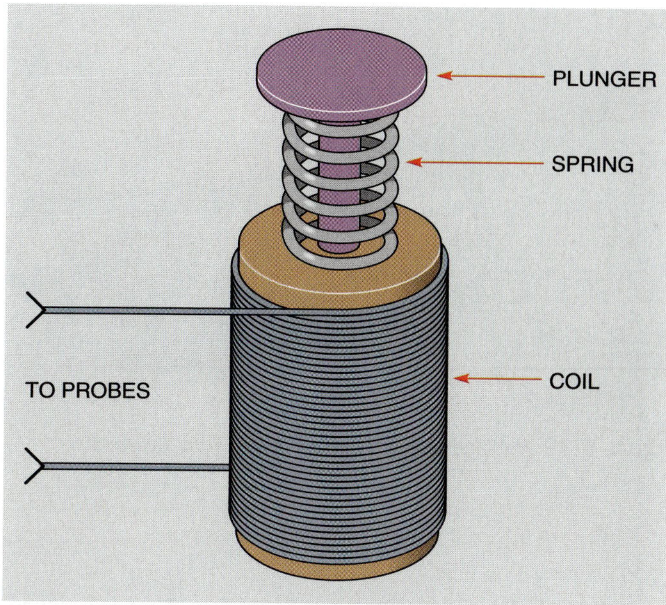

Figure 8-37 Low-impedance voltage tester.

Reading a Digital Meter

It would seem that reading a digital meter would be a simple matter of looking at the numbers on the display. This may not be the case, however. The reading on the display of autoranging digital meters, for example, may represent anything from ohms to megohms, or volts to millivolts. These meters usually display a notation beside the numerical digits to indicate the meter scale. Figure 8-35 illustrates the display of a typical digital meter. It shows the numbers 1.347 with a notation beside the numbers of mV, or millivolts. The actual meter reading is 1.347 millivolts or 0.001347 volts.

The display in Figure 8-36 indicates a current of 26.41 μA. The μ symbol means micro or one millionth and the A means amps. The meter is indicating a current of 26.41 microamperes or 0.00002641 amps. It is very important to pay attention to the notation symbols following the digits on the display of a digital meter.

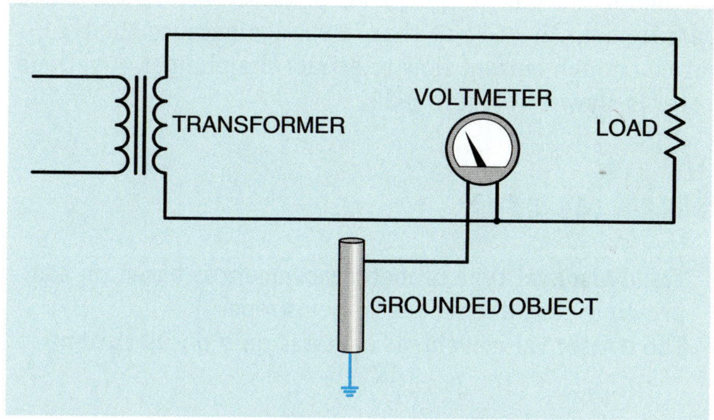

Figure 8-38 High-impedance ground paths can produce misleading voltage reading.

The Low-Impedance Voltage Tester

Another device used to test voltage is often referred to as a voltage tester. This device does measure voltage, but it does not contain a meter movement or digital display. It contains a coil and a plunger. The coil produces a magnetic field that is proportional to the amount of voltage to which the coil of the tester is connected. The higher the voltage, the stronger the magnetic field becomes. The plunger must overcome the force of a spring as it is drawn into the coil (Fig. 8-37). The plunger acts as a pointer to indicate the amount of voltage to which the tester is connected. The tester has an impedance of approximately 5000 Ω and can generally be used to measure voltages as high as 600 V. **The low-impedance voltage tester has a very large current draw compared with other types of voltmeters and should never be used to test low-power circuits.**

The relatively high current draw of the voltage tester can be an advantage when testing certain types of circuits, however, because it is not susceptible to giving the misleading voltage readings caused by high-impedance ground paths or feedback voltages that affect other types of voltmeters. An example of this advantage is shown in Figure 8-38. A transformer is used to supply power to a load. Notice that neither the output side of the transformer nor the load is connected to ground. If a high-impedance voltmeter measures between one side of the transformer and a grounded point, it will

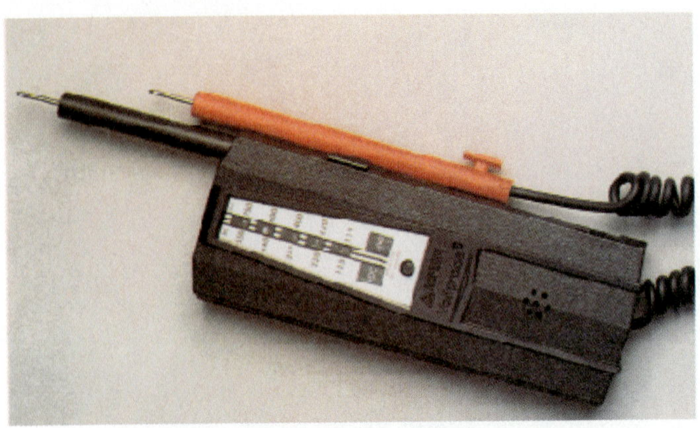

Figure 8-39 Voltage tester. *Courtesy of Amprobe Instrument.*

most likely indicate some amount of voltage. That is because ground can act as a large capacitor and can permit a small amount of current to flow through the circuit created by the meter. This high-impedance ground path can support only a few microamps of current flow, but it is enough to operate the meter movement. If a voltage tester makes the same measurement, it will not show a voltage because there cannot be enough current flow to attract the plunger. A voltage tester is shown in Figure 8-39.

Summary

- The d'Arsonval type of meter movement is based on the principle that like magnetic fields repel.
- The d'Arsonval movement operates only on DC current.
- Voltmeters have a high resistance and are designed to be connected directly across the power line.
- The steps to reading a meter are
  - A. Determine what quantity the meter is set to measure.
  - B. Determine the full range value of the meter.
  - C. Read the meter.
- Ammeters have a low resistance and must be connected in series with a load to limit the flow of current.
- Shunts are used to change the value of DC ammeters.
- AC ammeters use a current transformer to change the range setting.
- Clamp-on ammeters measure the flow of current by measuring the strength of the magnetic field around a conductor.
- Ohmmeters are used to measure the resistance in a circuit.
- Ohmmeters contain an internal power source, generally batteries.
- Ohmmeters must never be connected to a circuit that has power applied to it.
- Digital multimeters display their value in digits instead of a meter movement.
- Digital multimeters generally have an input impedance of 10 MΩ on all ranges.
- Digital ohmmeters measure resistance by measuring the voltage drop across an unknown resistor when a known amount of current flows through it.
- Low-impedance voltage testers are not susceptible to indicating a voltage caused by a high-impedance ground or a feedback.

Review Questions

1. To what is the turning force of a d'Arsonval meter movement proportional?
2. What type of voltage must be connected to a d'Arsonval meter movement?
3. A DC voltmeter has a resistance of 20,000 Ω per volt. What is the resistance of the meter if the range selection switch is set on the 250-V range?
4. What is the purpose of an ammeter shunt?
5. How is an ammeter connected into a circuit?
6. How is a voltmeter connected into a circuit?
7. An ammeter shunt has a voltage drop of 50 mV when 50 A of current flow through it. What is the resistance of the shunt?
8. What type of meter contains its own separate power source?

Practice Problems

Measuring Instruments

1. A d'Arsonval meter movement has a full scale current value of 100 μA (0.000100 A) and a resistance of 5 kΩ (5000 Ω). What size resistor must be placed in series with this meter to permit it to indicate 10 V full scale?
2. The meter movement described in question 1 is to be used to construct a multirange voltmeter. The meter is to have voltage ranges of 15 V, 60 V, 150 V, and 300 V (Fig. 8-40). Find the values of resistors R_1, R_2, R_3, and R_4.
3. A meter movement has a full scale value of 500 μA (0.000500 A) and 50 mV (0.050 V). A shunt is to be connected to the meter that permits it to have a full scale current value of 2 A. What is the resistance of the shunt?
4. A digital voltmeter indicates a voltage of 2.5 V when 10 μA of current flows through a resistor. What is the resistance of the resistor?

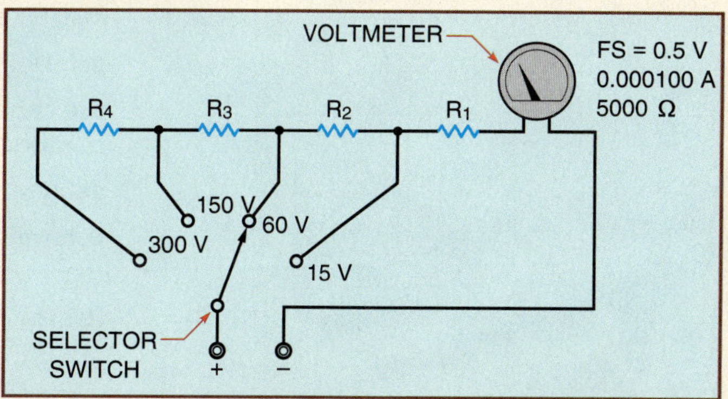

Figure 8-40 **The multirange voltmeter operates by connecting different values of resistance in series with the meter movement.**

Chapter 9 Using Wire Tables and Determining Conductor Sizes

The size of the conductor needed for a particular application can be determined by several methods. The **National Electrical Code** *is used throughout the industry to determine the conductor size for most applications. It is imperative that an electrician be familiar with code tables and correction factors. In some circumstances, however, wire tables cannot be used, as in the case of extremely long wire runs or for windings of a transformer or motor. In these instances the electrician should know how to determine the conductor size needed by computing maximum voltage drop and resistance of the conductor.*

OBJECTIVES

After studying this unit, you should be able to:

- select a conductor from the proper wire table.
- discuss the different types of wire insulation.
- determine insulation characteristics.
- use correction factors to determine the proper ampacity rating of conductors.
- determine the resistance of long lengths of conductors.
- determine the proper wire size for loads located long distances from the power source.
- list the requirements for using parallel conductors.
- discuss the use of a MEGGER® for testing insulation.

Glossary of Using Wire Tables and Determining Conductor Size Terms

ambient air temperature the temperature of the surrounding air

American Wire Gauge (AWG) an English standard for measuring the diameter of wire

ampacity the amount of current that a wire can carry

circular mil an English measure for determining the cross-sectional area of round wire; circular mils is the diameter in mils (1 mil = 0.001 in.) squared

correction factor a decimal fraction employed to determine the amount of current a conductor is permitted to carry over a range of ambient temperature

damp location a location that is covered and not likely to be saturated by water

dry location a location that is not normally subject to dampness or wetness

insulation the plastic or rubber coating around the conductor

maximum operating temperature the maximum temperature in an area where the wire is to be used

MEGGER® a device used to measure the insulation resistance; meggers measure resistance in millions of ohms

mil foot an English standard of measure for wire; a mil foot is a piece of wire one thousandth of an inch in diameter and one foot long

National Electrical Code a national standard for installation of electrical apparatus and equipment published by the National Fire Protection Association

parallel conductors conductors connected in parallel with each other for the purpose of carrying more electric current

voltage drop the amount of voltage employed to push current through the resistance of the wire

wet location areas that are subject to saturation by water; some of these areas include direct burial in the earth, underground conduit installations, or installation in masonry or concrete slabs

Using the National Electrical Code Charts

Section 310 of the *National Electrical Code* deals with conductors for general wiring. Tables 310.16 through 310.19 are generally used to select a wire size according to the requirements of the circuit. Each of these tables lists different conditions. The table used is determined by the wiring conditions. Table 310.16 (Fig. 9-1) lists **ampacities** (current-carrying ability) of not more than three single insulated conductors in raceway or cable, or buried in the earth based on an **ambient** (surrounding) **air temperature** of 30°C (86°F). Table 310.17 (Fig. 9-2) lists ampacities of single insulated conductors in free air based on an ambient temperature of 30°C. Table 310.18 (Fig. 9-3) lists the ampacities of three single insulated conductors in raceway or cable based on an ambient temperature of 40°C (104°F). The conductors listed in Tables 310.18 and 310.19 are generally used for high-temperature locations. The heading at the top of each table lists a different set of conditions.

Factors That Determine Ampacity

Conductor Material

One factor that determines the resistivity of wire is the material from which the wire is made. The wire tables list the current-carrying capacity of both copper and aluminum or copper-clad aluminum conductors. The currents listed in the left half of Table 310.16, for example, are for copper wire. The currents listed in the right half of the table are for aluminum or copper-clad aluminum. The table indicates that a copper conductor is permitted to carry more current than an aluminum conductor of the same size and insulation type. A #8 **AWG (American Wire Gauge)** copper conductor with type TW insulation is rated to carry a maximum of 40 A. A #8 AWG aluminum conductor with type TW insulation is rated to carry only 30 A. One of the columns of Tables 310.18 and 310.19 gives the ampacity rating of nickel or nickel-coated copper conductors.

Insulation Type

Another factor that determines the amount of current a conductor is permitted to carry is the type of insulation used. This is because some types of insulation can withstand more heat than others. The **insulation** is the nonconductive covering around the wire, as shown in Figure 9-4. The voltage rating of the conductor is also determined by the type of insulation. The amount of voltage a particular type of insulation can withstand without breaking down is determined by the type of material it is made of and its thickness. Table 310.13 lists different types of insulation and certain specifications about each one. The table is divided into seven columns. The first column lists the trade name of the insulation; the second, its identification code letter; the third, its **maximum operating temperature**; the fourth, its applications and where it is permitted to be used; the fifth, the material from which the insulation is made; the sixth, the thickness of the insulation; and the last, the type of outer covering over the insulation.

Example 1

Find the maximum operating temperature of type RHW insulation. (Note: Refer to the National Electrical Code.)

SOLUTION

Find type RHW in the second column of Table 310.13. The third column lists a maximum operating temperature of 75°C or 167°F.

Example 2

Can type THHN insulation be used in wet locations?

SOLUTION

Locate type THHN insulation in the second column. The fourth column indicates that this insulation can be used in **dry** and **damp locations.** It cannot be used in **wet locations.** For an explanation of the difference between damp and wet locations, consult "locations" in article 100 of the *National Electrical Code.*

Table 310.16 Allowable Ampacities of Insulated Conductors Rated 0 Through 2000 Volts, 60°C Through 90°C (140°F Through 194°F), Not More Than Three Current-Carrying Conductors in Raceway, Cable, or Earth (Directly Buried), Based on Ambient Temperature of 30°C (86°F)

Size AWG or kcmil	Temperature Rating of Conductor (See Table 310.13.)						Size AWG or kcmil
	60°C (140°F)	75°C (167°F)	90°C (194°F)	60°C (140°F)	75°C (167°F)	90°C (194°F)	
	Types TW, UF	Types RHW, THHW, THW, THWN, XHHW, USE, ZW	Types TBS, SA, SIS, FEP, FEPB, MI, RHH, RHW-2, THHN, THHW, THW-2, THWN-2, USE-2, XHH, XHHW, XHHW-2, ZW-2	Types TW, UF	Types RHW, THHW, THW, THWN, XHHW, USE	Types TBS, SA, SIS, THHN, THHW, THW-2, THWN-2, RHH, RHW-2, USE-2, XHH, XHHW, XHHW-2, ZW-2	
	COPPER			ALUMINUM OR COPPER-CLAD ALUMINUM			
18	—	—	14	—	—	—	—
16	—	—	18	—	—	—	—
14*	20	20	25	—	—	—	—
12*	25	25	30	20	20	25	12*
10*	30	35	40	25	30	35	10*
8	40	50	55	30	40	45	8
6	55	65	75	40	50	60	6
4	70	85	95	55	65	75	4
3	85	100	110	65	75	85	3
2	95	115	130	75	90	100	2
1	110	130	150	85	100	115	1
1/0	125	150	170	100	120	135	1/0
2/0	145	175	195	115	135	150	2/0
3/0	165	200	225	130	155	175	3/0
4/0	195	230	260	150	180	205	4/0
250	215	255	290	170	205	230	250
300	240	285	320	190	230	255	300
350	260	310	350	210	250	280	350
400	280	335	380	225	270	305	400
500	320	380	430	260	310	350	500
600	355	420	475	285	340	385	600
700	385	460	520	310	375	420	700
750	400	475	535	320	385	435	750
800	410	490	555	330	395	450	800
900	435	520	585	355	425	480	900
1000	455	545	615	375	445	500	1000
1250	495	590	665	405	485	545	1250
1500	520	625	705	435	520	585	1500
1750	545	650	735	455	545	615	1750
2000	560	665	750	470	560	630	2000

CORRECTION FACTORS

Ambient Temp. (°C)	For ambient temperatures other than 30°C (86°F), multiply the allowable ampacities shown above by the appropriate factor shown below.						Ambient Temp. (°F)
21–25	1.08	1.05	1.04	1.08	1.05	1.04	70–77
26–30	1.00	1.00	1.00	1.00	1.00	1.00	78–86
31–35	0.91	0.94	0.96	0.91	0.94	0.96	87–95
36–40	0.82	0.88	0.91	0.82	0.88	0.91	96–104
41–45	0.71	0.82	0.87	0.71	0.82	0.87	105–113
46–50	0.58	0.75	0.82	0.58	0.75	0.82	114–122
51–55	0.41	0.67	0.76	0.41	0.67	0.76	123–131
56–60	—	0.58	0.71	—	0.58	0.71	132–140
61–70	—	0.33	0.58	—	0.33	0.58	141–158
71–80	—	—	0.41	—	—	0.41	159–176

* See 240.4(D).

Figure 9-1 ***NEC*** **Table 310.16.** *Reprinted with permission from NFPA 70–2002, the* National Electrical Code® *Copyright© 2002, National Fire Protection Association, Quincy, MA 02260. This reprinted material is not the referenced subject which is represented only by the standard in its entirety.*

Table 310.17 Allowable Ampacities of Single-Insulated Conductors Rated 0 Through 2000 Volts in Free Air, Based on Ambient Air Temperature of 30°C (86°F)

	Temperature Rating of Conductor (See Table 310.13.)						
	60°C (140°F)	75°C (167°F)	90°C (194°F)	60°C (140°F)	75°C (167°F)	90°C (194°F)	
	Types TW, UF	Types RHW, THHW, THW, THWN, XHHW, ZW	Types TBS, SA, SIS, FEP, FEPB, MI, RHH, RHW-2, THHN, THHW, THW-2, THWN-2, USE-2, XHH, XHHW, XHHW-2, ZW-2	Types TW, UF	Types RHW, THHW, THW, THWN, XHHW	Types TBS, SA, SIS, THHN, THHW, THW-2, THWN-2, RHH, RHW-2, USE-2, XHH, XHHW, XHHW-2, ZW-2	
Size AWG or kcmil	COPPER			ALUMINUM OR COPPER-CLAD ALUMINUM			Size AWG or kcmil
18	—	—	18	—	—	—	—
16	—	—	24	—	—	—	—
14*	25	30	35	—	—	—	—
12*	30	35	40	25	30	35	12*
10*	40	50	55	35	40	40	10*
8	60	70	80	45	55	60	8
6	80	95	105	60	75	80	6
4	105	125	140	80	100	110	4
3	120	145	165	95	115	130	3
2	140	170	190	110	135	150	2
1	165	195	220	130	155	175	1
1/0	195	230	260	150	180	205	1/0
2/0	225	265	300	175	210	235	2/0
3/0	260	310	350	200	240	275	3/0
4/0	300	360	405	235	280	315	4/0
250	340	405	455	265	315	355	250
300	375	445	505	290	350	395	300
350	420	505	570	330	395	445	350
400	455	545	615	355	425	480	400
500	515	620	700	405	485	545	500
600	575	690	780	455	540	615	600
700	630	755	855	500	595	675	700
750	655	785	885	515	620	700	750
800	680	815	920	535	645	725	800
900	730	870	985	580	700	785	900
1000	780	935	1055	625	750	845	1000
1250	890	1065	1200	710	855	960	1250
1500	980	1175	1325	795	950	1075	1500
1750	1070	1280	1445	875	1050	1185	1750
2000	1155	1385	1560	960	1150	1335	2000
	CORRECTION FACTORS						
Ambient Temp. (°C)	For ambient temperatures other than 30°C (86°F), multiply the allowable ampacities shown above by the appropriate factor shown below.						Ambient Temp. (°F)
21–25	1.08	1.05	1.04	1.08	1.05	1.04	70–77
26–30	1.00	1.00	1.00	1.00	1.00	1.00	78–86
31–35	0.91	0.94	0.96	0.91	0.94	0.96	87–95
36–40	0.82	0.88	0.91	0.82	0.88	0.91	96–104
41–45	0.71	0.82	0.87	0.71	0.82	0.87	105–113
46–50	0.58	0.75	0.82	0.58	0.75	0.82	114–122
51–55	0.41	0.67	0.76	0.41	0.67	0.76	123–131
56–60	—	0.58	0.71	—	0.58	0.71	132–140
61–70	—	0.33	0.58	—	0.33	0.58	141–158
71–80	—	—	0.41	—	—	0.41	159–176

* See 240.4(D).

Figure 9-2 ***NEC*** **Table 310.17.** *Reprinted with permission from NFPA 70–2002, the* National Electrical Code® *Copyright© 2002, National Fire Protection Association, Quincy, MA 02260. This reprinted material is not the referenced subject which is represented only by the standard in its entirety.*

Table 310.18 Allowable Ampacities of Insulated Conductors Rated 0 Through 2000 Volts, 150°C Through 250°C (302°F Through 482°F). Not More Than Three Current-Carrying Conductors in Raceway or Cable, Based on Ambient Air Temperature of 40°C (104°F)

	Temperature Rating of Conductor (See Table 310.13.)				
	150°C (302°F)	200°C (392°F)	250°C (482°F)	150°C (302°F)	
	Type Z	Types FEP, FEPB, PFA	Types PFAH, TFE	Type Z	
Size AWG or kcmil	COPPER		NICKEL OR NICKEL-COATED COPPER	ALUMINUM OR COPPER-CLAD ALUMINUM	**Size AWG or kcmil**
14	34	36	39	—	14
12	43	45	54	30	12
10	55	60	73	44	10
8	76	83	93	57	8
6	96	110	117	75	6
4	120	125	148	94	4
3	143	152	166	109	3
2	160	171	191	124	2
1	186	197	215	145	1
1/0	215	229	244	169	1/0
2/0	251	260	273	198	2/0
3/0	288	297	308	227	3/0
4/0	332	346	361	260	4/0
	CORRECTION FACTORS				
Ambient Temp. (°C)	For ambient temperatures other than 40°C (104°F), multiply the allowable ampacities shown above by the appropriate factor shown below.				**Ambient Temp. (°F)**
41–50	0.95	0.97	0.98	0.95	105–122
51–60	0.90	0.94	0.95	0.90	123–140
61–70	0.85	0.90	0.93	0.85	141–158
71–80	0.80	0.87	0.90	0.80	159–176
81–90	0.74	0.83	0.87	0.74	177–194
91–100	0.67	0.79	0.85	0.67	195–212
101–120	0.52	0.71	0.79	0.52	213–248
121–140	0.30	0.61	0.72	0.30	249–284
141–160	—	0.50	0.65	—	285–320
161–180	—	0.35	0.58	—	321–356
181–200	—	—	0.49	—	357–392
201–225	—	—	0.35	—	393–437

Figure 9-3 ***NEC*** **Table 310.18.** *Reprinted with permission from NFPA 70–2002.*

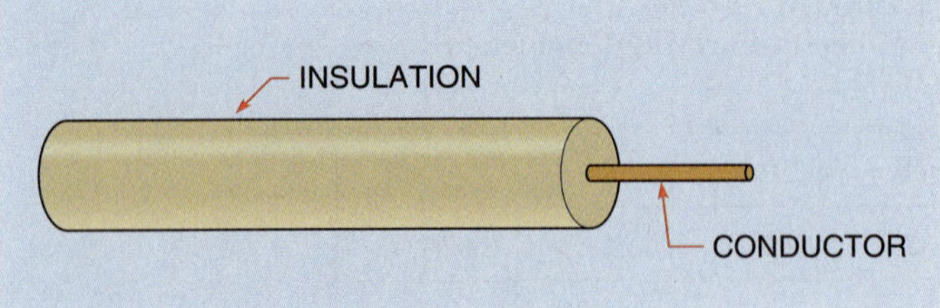

Figure 9-4 **Insulation around conductor.**

Correction Factors

One of the main factors that determines the amount of current a conductor is permitted to carry is the ambient, or surrounding, air temperature. Table 310.16, for example, lists the ampacity of not more than three conductors in raceway in free air. These ampacities are based on an ambient air temperature of 30°C or 86°F. If these conductors are to be used in a location that has a higher ambient temperature, the ampacity of the conductor must be reduced, because the resistance of copper or aluminum increases with an increase of temperature. The **correction factor** chart located at the bottom of the table is used to make this amendment. The chart is divided into the same number of columns as the wire table directly above it. The correction factors in each column are used for the conductors listed in the same column of the wire table.

Example 3

What is the maximum ampacity of a #4 AWG copper conductor with type THWN insulation used in an area with an ambient temperature of 43°C?

SOLUTION

Determine the ampacity of a #4 AWG copper conductor with type THWN insulation from the wire table. Type THWN insulation is located in the second column of Table 310.16. The table lists an ampacity of 85 A for this conductor. Follow the second column down to the correction factor chart. Locate 43°C in the far left column of the correction factor chart; 43°C falls between 41° and 45°. The chart lists a correction factor of 0.82. The ampacity of the conductor in the wire table is multiplied by the correction factor.

$$85 \times 0.82 = 69.7 \text{ A}$$

Example 4

What is the maximum ampacity of a #1/0 AWG copper-clad aluminum conductor with type RHH insulation if the conductor is to be used in an area with an ambient air temperature of 100°F?

SOLUTION

Locate the column that contains type RHH insulation in the copper-clad aluminum section of Table 310.16. The table indicates a maximum ampacity of 135 A for this conductor. Follow this column down to the correction factor chart and locate 100°F. Fahrenheit degrees are in the far right column of the chart. One hundred degrees falls between 97° and 104°F. The correction factor for this temperature is 0.91. Multiply the ampacity of the conductor by this factor.

$$135 \times 0.91 = 122.85 \text{ A}$$

More Than Three Conductors in a Raceway

Tables 310.16 and 310.18 list three conductors in a raceway. If a raceway is to contain more than three conductors, the ampacity of the conductors must be derated because the heat from each conductor combines with the heat dissipated by the other conductors to produce a higher temperature inside the raceway. NEC Table 310.15 (B)(2)(a) (Fig. 9-5) lists these correction factors. If the raceway is used in an area with a greater ambient temperature than that listed in the appropriate wire table, the temperature correction factor must be applied also.

Table 310.15(B)(2)(a) Adjustment Factors for More Than Three Current-Carrying Conductors in a Raceway or Cable

Number of Current-Carrying Conductors	Percent of Values in Tables 310.16 through 310.19 as Adjusted for Ambient Temperature if Necessary
4–6	80
7–9	70
10–20	50
21–30	45
31–40	40
41 and above	35

Figure 9-5 **NEC Table 310.15 (B)(2)(a).** *Reprinted with permission from NFPA 70–2002.*

Example 5

Twelve #14 AWG copper conductors with type RHW insulation are to be run in a conduit. The conduit is used in an area that has an ambient temperature of 110°F. What is the maximum ampacity of these conductors?

SOLUTION

Find the ampacity of a #14 AWG copper conductor with type RHW insulation. Type RHW insulation is located in the second column of Table 310.16. A #14 AWG copper conductor has an ampacity of 20 A. Next, use the correction factor for ambient temperature. A correction factor of 0.82 will be used.

$$\mathbf{20 \times 0.82 = 16.4\ A}$$

The correction factor located in Table 310.15(B)(2)(a) must now be used. The table indicates a correction factor of 50% when 10 through 20 conductors are run in a raceway.

$$\mathbf{16.4 \times 0.50 = 8.2\ A}$$

Each #14 AWG conductor has a maximum current rating of 8.2 A.

Duct Banks

Duct banks are often used when it becomes necessary to bury cables in the ground. An electrical duct can be a single metallic or nonmetallic conduit. An electrical duct bank is a group of electrical ducts buried together, as shown in Figure 9-6. When a duct bank is used, the center points of individual ducts should be separated by a distance of not less than 7.5 inches.

Computing Conductor Sizes and Resistance

Although the wire tables in the *National Electrical Code* are used to determine the proper size wire for most installations, there are instances in which these tables are not used. The formula in 310.15(c) of the *NEC* is used for ampacities not listed in the wire tables. The formula is

$$I = \sqrt{\frac{T_C - (T_A + \Delta_{TD})}{R_{DC}(1 + Y_C)R_{CA}}}$$

where

T_C = **conductor temperature in °C**

T_A = **ambient temperature in °C**

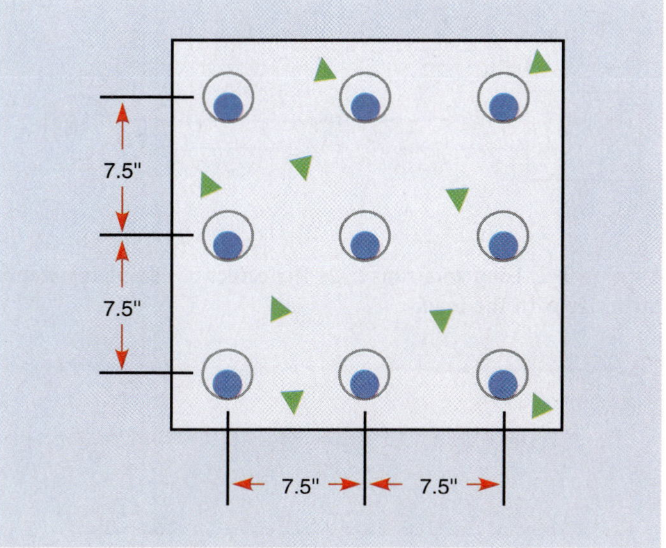

Figure 9-6 **Electrical duct banks.**

Δ_{TD} = **dielectric loss temperature rise**

R_{DC} = **DC resistance of conductor at temperature TC**

Y_C = **component AC resistance resulting from skin effect and proximity effect**

R_{CA} = **effective thermal resistance between conductor and surrounding ambient**

Although this formula is seldom used by electricians, the *National Electrical Code* does permit its use under the supervision of an electrical engineer.

Long Wire Lengths

Another situation in which it becomes necessary to compute wire sizes instead of using the tables in the code is when the conductor becomes excessively long. The listed ampacities in the code tables assume that the length of the conductor will not increase the resistance of the circuit by a significant amount. When the wire becomes extremely long, however, it is necessary to compute the size of wire needed.

All wire contains resistance. As wire is added to a circuit, it adds resistance in series with the load (Fig. 9-7). Four factors determine the resistance of a length of wire.

1. *The type of material from which the wire is made.* Different types of material have different wire resistances. A copper conductor will have less resistance than an aluminum conductor of the same size and length. An aluminum conductor will have less resistance than a piece of iron wire the same size and length.
2. *The diameter of the conductor.* The larger the diameter, the less resistance it will have. A large-diameter pipe, for example, will have less resistance to the flow of water than will a small-diameter pipe (Fig. 9-8). The diameter of round wire is measured in circular mils

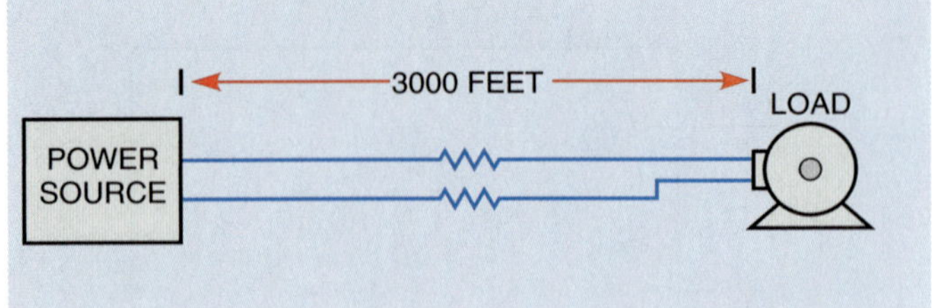

Figure 9-7 Long wire runs have the effect of adding resistance in series with the load.

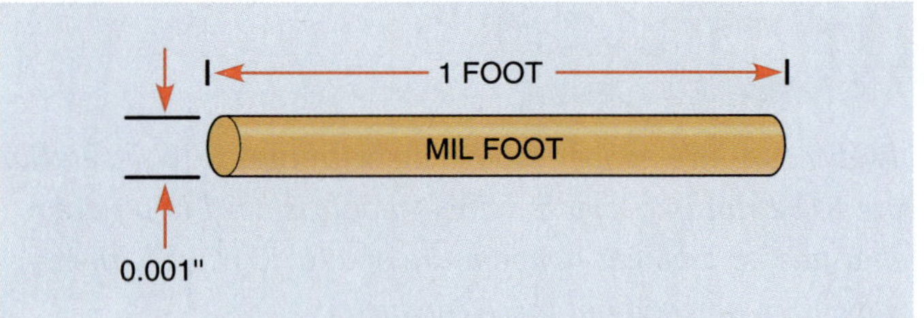

Figure 9-9 A mil foot is equal to a piece of wire one foot long and one thousandth of an inch in diameter.

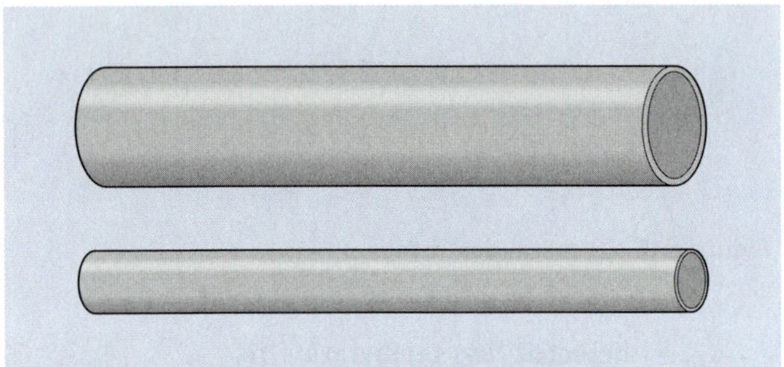

Figure 9-8 A large pipe has less resistance to the flow of water than a small pipe.

(CM). One mil equals 0.001 inch. A **circular mil** is the diameter of the wire in mils squared. For example, assume a wire has a diameter of 0.064 inch. Sixty-four thousandths should be written as a whole number, not as a decimal or a fraction (64^2 [64×64] = 4096 CM).

3. *The length of the conductor.* The longer the conductor, the more resistance it will have. Adding length to a conductor has the same effect as connecting resistors in series.
4. *The temperature of the conductor.* As a general rule, most conductive materials will increase their resistance with an increase of temperature. Some exceptions to this rule are carbon, silicon, and germanium. If the coefficient of temperature for a particular material is known, its resistance at different temperatures can be computed. Materials that increase their resistance with an increase of temperature have a positive coefficient of temperature. Materials that decrease their resistance with an increase of temperature have a negative coefficient of temperature.

In the English system of measure, a standard value of resistance, called the mil foot, is used to determine the resistance of different lengths and sizes of wire. A **mil foot** is a piece of wire 1 foot long and 1 mil in diameter (Fig. 9-9). A chart showing the resistance of a mil foot of wire at 20°C is shown in Figure 9-10. Notice the wide range of resistances for different materials. The temperature coefficient of the different types of conductors is listed also.

RESISTIVITY (K) OF MATERIALS

MATERIAL	Ω PER MIL FOOT AT 20°C	TEMP. COEFF. (Ω PER °C)
ALUMINUM	17	0.004
CARBON	22,000 APRX.	−0.0004
CONSTANTAN	295	0.000002
COPPER	10.4	0.0039
GOLD	14	0.004
IRON	60	0.0055
LEAD	126	0.0043
MANGANIN	265	0.000000
MERCURY	590	0.00088
NICHROME	675	0.0002
NICKEL	52	0.005
PLATINUM	66	0.0036
SILVER	9.6	0.0038
TUNGSTEN	33.8	0.005

Figure 9-10 Resistivity of materials.

Computing Resistance

Now that a standard measure of resistance for different types of materials is known, the resistance of different lengths and sizes of these materials can be computed. The formula for computing resistance of a certain length, size, and type of wire is

$$R = \frac{K \times L}{CM}$$

where

R = resistance of the wire
K = ohms per mil foot
L = length of wire in feet
CM = circular mil area of the wire

This formula can be converted to compute other values in the formula, such as:

to find the SIZE of wire to use:

$$CM = \frac{K \times L}{R}$$

to find the LENGTH of wire to use:

$$L = \frac{R \times CM}{K}$$

to find the TYPE of wire to use:

$$K = \frac{R \times CM}{L}$$

Example 6

Find the resistance of a piece of #6 AWG copper wire 550 ft long. Assume a temperature of 20°C. The formula to be used is

$$R = \frac{K \times L}{CM}$$

SOLUTION

The value for K can be found in the table in Figure 9-10. The table indicates a value of 10.4 Ω per mil foot for a copper conductor. The length, L, was given as 550 ft, and the circular mil area of #6 AWG wire is listed as 26,250 in the table shown in Figure 9-11.

$$R = \frac{10.4 \times 550}{26{,}250}$$

$$R = \frac{5720}{26{,}250}$$

$$R = 0.218\ \Omega$$

Example 7

An aluminum wire 2250 ft long cannot have a resistance greater than 0.2 Ω. What size aluminum wire must be used?

SOLUTION

To find the size of wire, use

$$CM = \frac{K \times L}{R}$$

$$CM = \frac{17 \times 2250}{0.2}$$

$$CM = \frac{38{,}250}{0.2}$$

$$CM = 191{,}250$$

The nearest standard size conductor for this installation can be found in the American Wire Gauge table. Since the resistance cannot be greater than 0.2 Ω, the conductor cannot be smaller than 191,250 CM. The nearest standard conductor size is 0000 AWG.

Good examples of when it becomes necessary to compute the wire size for a particular installation can be seen in the following problems.

AMERICAN WIRE GAUGE TABLE							
B & S Gauge No.	Diam. in Mils	Area in Circular Mils	Ohms per 1000 Ft. (ohms per 100 m)			Pounds per 1,000 Ft. (kg per 100 m)	
			Copper* 68°F (20°C)	Copper* 167°F (75°C)	Aluminum 68°F (20°C)	Copper	Aluminum
0000	460	211 600	0.049 (0.016)	0.0596 (0.0195)	0.0804 (0.0263)	640 (95.2)	195 (29.0)
000	410	167 800	0.0618 (0.020)	0.0752 (0.0246)	0.101 (0.033)	508 (75.5)	154 (22.9)
00	365	133 100	0.078 (0.026)	0.0948 (0.031)	0.128 (0.042)	403 (59.9)	122 (18.1)
0	325	105 500	0.0983 (0.032)	0.1195 (0.0392)	0.161 (0.053)	320 (47.6)	97 (14.4)
1	289	83 690	0.1239 (0.0406)	0.151 (0.049)	0.203 (0.066)	253 (37.6)	76.9 (11.4)
2	258	66 370	0.1563 (0.0512)	0.190 (0.062)	0.256 (0.084)	201 (29.9)	61.0 (9.07)
3	229	52 640	0.1970 (0.0646)	0.240 (0.079)	0.323 (0.106)	159 (23.6)	48.4 (7.20)
4	204	41 740	0.2485 (0.0815)	0.302 (0.099)	0.408 (0.134)	126 (18.7)	38.4 (5.71)
5	182	33 100	0.3133 (0.1027)	0.381 (0.125)	0.514 (0.168)	100 (14.9)	30.4 (4.52)
6	162	26 250	0.395 (1.29)	0.481 (0.158)	0.648 (0.212)	79.5 (11.8)	24.1 (3.58)
7	144	20 820	0.498 (0.163)	0.606 (0.199)	0.817 (0.268)	63.0 (9.37)	19.1 (2.84)
8	128	16 510	0.628 (0.206)	0.764 (0.250)	1.03 (0.338)	50.0 (7.43)	15.2 (2.26)
9	114	13 090	0.792 (0.260)	0.963 (0.316)	1.30 (0.426)	39.6 (5.89)	12.0 (1.78)
10	102	10 380	0.999 (0.327)	1.215 (0.398)	1.64 (0.538)	31.4 (4.67)	9.55 (1.42)
11	91	8 234	1.260 (0.413)	1.532 (0.502)	2.07 (0.678)	24.9 (3.70)	7.57 (1.13)
12	81	6 530	1.588 (0.520)	1.931 (0.633)	2.61 (0.856)	19.8 (2.94)	6.00 (0.89)
13	72	5 178	2.003 (0.657)	2.44 (0.80)	3.29 (1.08)	15.7 (2.33)	4.8 (0.71)
14	64	4 107	2.525 (0.828)	3.07 (1.01)	4.14 (1.36)	12.4 (1.84)	3.8 (0.56)
15	57	3 257	3.184 (0.043)	3.98 (1.27)	5.22 (1.71)	9.86 (1.47)	3.0 (0.45)
16	51	2 583	4.016 (0.316)	4.88 (1.60)	6.59 (2.16)	7.82 (1.16)	2.4 (0.36)
17	45.3	2 048	5.06 (1.66)	6.16 (2.02)	8.31 (2.72)	6.20 (0.922)	1.9 (0.28)
18	40.3	1 624	6.39 (2.09)	7.77 (2.55)	10.5 (3.44)	4.92 (0.731)	1.5 (0.22)
19	35.9	1 288	8.05 (2.64)	9.79 (3.21)	13.2 (4.33)	3.90 (0.580)	1.2 (0.18)
20	32.0	1 022	10.15 (3.33)	12.35 (4.05)	16.7 (5.47)	3.09 (0.459)	0.94 (0.14)
21	28.5	810	12.8 (4.2)	15.6 (5.11)	21.0 (6.88)	2.45 (0.364)	0.745 (0.110)
22	25.4	642	16.1 (5.3)	19.6 (6.42)	26.5 (8.69)	1.95 (0.290)	0.591 (0.09)
23	22.0	510	20.4 (0.7)	24.0 (8.13)	33.4 (10.9)	1.54 (0.229)	0.408 (0.07)
24	20.1	404	25.7 (8.4)	31.2 (10.2)	42.1 (13.8)	1.22 (0.181)	0.371 (0.05)
25	17.9	320	32.4 (10.6)	39.4 (12.9)	53.1 (17.4)	0.97 (0.14)	0.295 (0.04)
26	15.9	254	40.8 (13.4)	49.6 (16.3)	67.0 (22.0)	0.77 (0.11)	0.234 (0.03)
27	14.2	202	51.5 (16.9)	62.6 (20.5)	84.4 (27.7)	0.61 (0.09)	0.185 (0.03)
28	12.6	160	64.9 (21.3)	78.9 (25.9)	106 (34.7)	0.48 (0.07)	0.147 (0.02)
29	11.3	126.7	81.8 (26.8)	99.5 (32.6)	134 (43.9)	0.384 (0.06)	0.117 (0.02)
30	10.0	100.5	103.2 (33.8)	125.5 (41.1)	169 (55.4)	0.304 (0.04)	0.092 (0.01)
31	8.93	79.7	130.1 (42.6)	158.2 (51.9)	213 (69.8)	0.241 (0.04)	0.073 (0.01)
32	7.95	63.2	164.1 (53.8)	199.5 (65.4)	269 (88.2)	0.191 (0.03)	0.058 (0.01)
33	7.08	50.1	207 (68)	252 (82.6)	339 (111)	0.152 (0.02)	0.046 (0.01)
34	6.31	39.8	261 (86)	317 (104)	428 (140)	0.120 (0.02)	0.037 (0.01)
35	5.62	31.5	329 (108)	400 (131)	540 (177)	0.095 (0.01)	0.029
36	5.00	25.0	415 (136)	505 (165)	681 (223)	0.076 (0.01)	0.023
37	4.45	19.8	523 (171)	636 (208)	858 (281)	0.0600 (0.01)	0.0182
38	3.96	15.7	660 (216)	802 (263)	1080 (354)	0.0476 (0.01)	0.0145
39	3.53	12.5	832 (273)	1012 (332)	1360 (446)	0.0377 (0.01)	0.0115
40	3.15	9.9	1049 (344)	1276 (418)	1720 (564)	0.0299 (0.01)	0.0091
41							
42	2.50	6.3					
43							
44	1.97	3.9					

*Resistance figures are given for standard annealed copper. For hard-drawn copper add 2%

Figure 9-11 American Wire Gauge table.

Example 8

A manufacturing plant has a cooling pond located 4000 feet from the plant. Six pumps are used to circulate water between the pond and the plant. In cold weather, however, the pumps can freeze and fail to supply water. The plant owner decides to connect electric-resistance heaters to each pump to prevent this problem. The six heaters are connected to a two-conductor cable from the plant at a junction box, Figure 9-12. The heaters operate on 480 volts and have a total current draw of 50 amperes. What size copper conductors should be used to supply power to the heaters if the voltage drop at the junction box is to be kept to 3% of the applied voltage? Assume an average ambient temperature of 20°C.

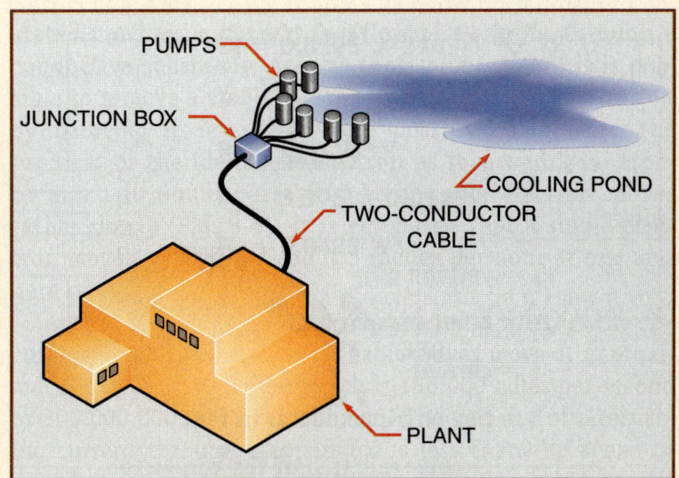

Figure 9-12 **Computing long wire lengths.**

SOLUTION

The first step in the solution of this problem will be to determine the maximum amount of resistance the conductors can have without producing a voltage drop greater than 3% of the applied voltage. The maximum amount of voltage drop can be computed by multiplying the applied voltage by 3%.

$$480 \times 0.03 = 14.4 \text{ V}$$

The maximum amount of resistance can now be computed using Ohm's law.

$$R = \frac{E}{I}$$

$$R = \frac{14.4}{50}$$

$$R = 0.288\ \Omega$$

The distance to the pond is 4000 feet. Since two conductors are used, the resistance of both conductors must be considered. Two conductors 4000 ft long will have the same resistance as one conductor 8000 ft long. For this reason, a length of 8000 ft will be used in the formula.

$$CM = \frac{K \times L}{R}$$

$$CM = \frac{10.4 \times 8000}{0.288}$$

$$CM = \frac{83{,}200}{0.288}$$

$$CM = 288{,}888.9$$

300-kcmil (thousand circular mils) cable will be used for this installation.

Computing Voltage Drop

Sometimes it is necessary to compute the **voltage drop** of an installation when the length, size of wire, and current are known. The following formula can be used to find the voltage drop of conductors used on a single-phase system.

$$E_D = \frac{2\ KIL}{CM}$$

where

E_D = voltage drop

K = ohms per mil foot

I = current

L = length of conductor in feet

CM = circular mil area of the conductor

Example 9

A single-phase motor is located 250 ft from its power source. The conductors supplying power to the motor are #10 AWG copper. The motor has a full load current draw of 24 A. What is the voltage drop across the conductors when the motor is in operation?

SOLUTION

$$E_D = \frac{2 \times 10.4 \times 24 \times 250}{10,380}$$

$$E_D = 12.02\ V$$

Parallel Conductors

Under certain conditions it may become necessary or advantageous to connect conductors in parallel. One such condition for **parallel conductors** is when the conductor is very large. For example, conductors supplying a motor 2500 feet from its source would have to be 500 kcmil. A 500-kcmil conductor is very large and difficult to handle. Therefore, it may be preferable to use parallel conductors for this installation. The *National Electrical Code* lists five conditions that must be met when conductors are connected in parallel (Section 310.4). These conditions are

1. The conductors must be the same length.
2. The conductors must be made of the same material. For example, all parallel conductors must be either copper or aluminum. It is not permissible to use copper for one of the conductors and aluminum for the other.
3. The conductors must have the same circular mil area.
4. The conductors must use the same type of insulation.
5. The conductors must be terminated or connected in the same manner.

In the example, the actual conductor size needed was computed to be 443,433.7 CM. This circular mil area could be obtained by connecting two 250-kcmil conductors in parallel for each phase, or three 000 (3/0) conductors in parallel for each phase. (Note: Each 000 [3/0] conductor has an area of 167,800 CM. This is a total of 503,400 CM.)

It might also be necessary to connect wires in parallel when conductors of a large size must be run in a conduit. Conductors of a single phase are not permitted to be run in metallic conduits, as shown in Figure 9-13 (*NEC* Sections 300.5(I) and 300.20(A) to (B)), because when current flows through a conductor, a magnetic field is produced around the conductor. In an alternating current circuit, the current continuously changes direction and magnitude, which causes the magnetic field to cut through the wall of the metal conduit (Fig. 9-14). This cutting action of the mag-

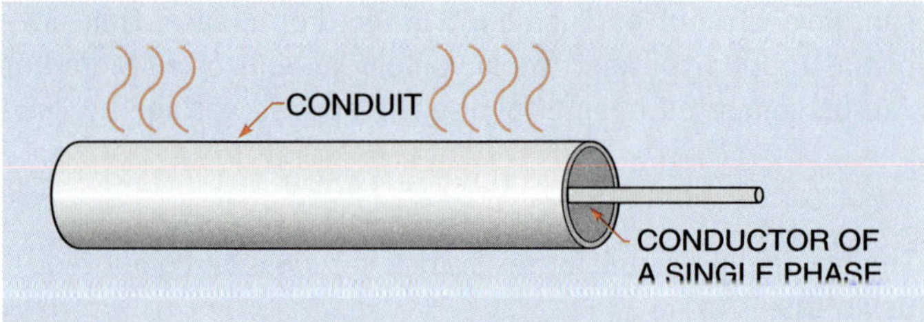

Figure 9-13 **A single-phase conductor causes heat to be produced in the conduit.**

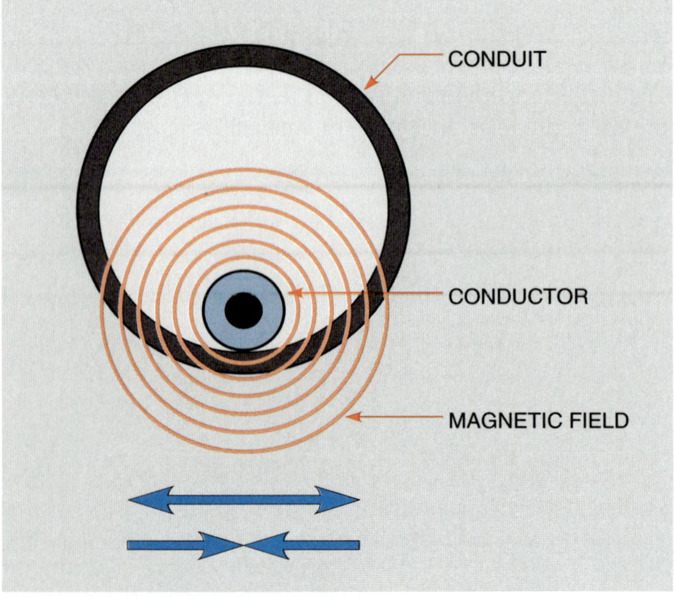

Figure 9-14 **The magnetic field expands and contracts.**

netic field induces a current, called an eddy current, into the metal of the conduit. Eddy currents are currents that are induced into metals. They tend to move in a circular fashion similar to the eddies of a river, hence the name eddy currents (Fig. 9-15). Eddy currents can produce enough heat in high-current circuits to melt the insulation surrounding the conductors. All metal conduits can have eddy current induction, but conduits made of magnetic materials such as steel have an added problem with hysteresis loss. Hysteresis loss is caused by molecular friction (Fig. 9-16). As the direction of the magnetic field reverses, the molecules of the metal are magnetized with the opposite polarity and swing to realign themselves. This continuous aligning and realigning of the molecules produces heat caused by friction. Hysteresis losses become greater with an increase in frequency.

To correct this problem, a conductor of each phase must be run in each conduit (Fig. 9-17). When all three phases are contained in a single conduit, the magnetic fields of the separate conductors cancel each other resulting in no current being induced in the walls of the conduit.

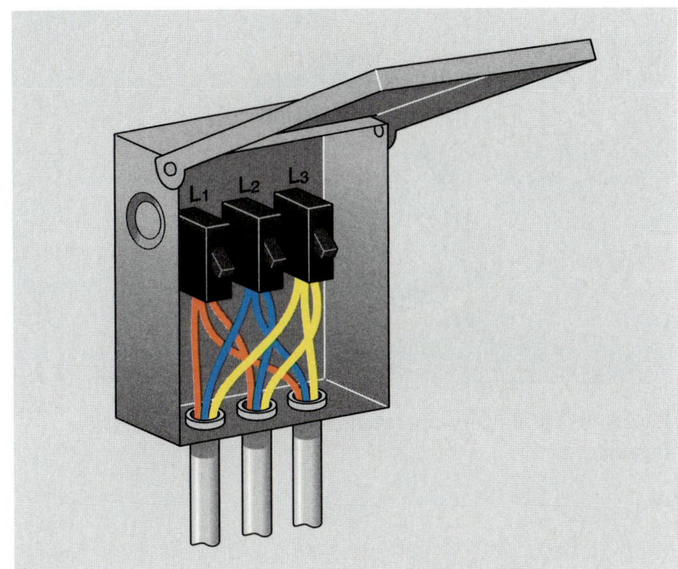

Figure 9-17 **Each conduit contains a conductor from each phase. This permits the magnetic fields to cancel each other.**

Testing Wire Installations

After the conductors have been installed in conduits or raceways, it is accepted practice to test the installation for grounds and shorts. This test requires an ohmmeter, which cannot only measure resistance in millions of ohms, but can also provide a high enough voltage to ensure that the insulation will not break down when rated line voltage is applied to the conductors. Most ohmmeters operate with a maximum voltage ranging from 1.5 V to about 9 V depending on the type of ohmmeter and the setting of the range scale. To test wire insulation, a special type of ohmmeter, called a MEGGER®, is used. The term **MEGGER®** is a registered trademark of Biddle Instruments. The MEGGER® is a megohmmeter that can produce voltages ranging from about 250 to 5000 V depending on the model of the meter and the range setting. One model of a MEGGER® is shown in Figure 9-18. This instrument contains a hand crank that is connected to the rotor of a brushless AC generator. The advantage of this instrument is that it does not require batteries. A range selector switch permits the meter to be used as a standard ohmmeter or as a megohmmeter. When it is used as a megohmmeter, the selector switch permits the test voltage to be selected. Test voltages of 100 V, 250 V, 500 V, and 1000 V can be obtained.

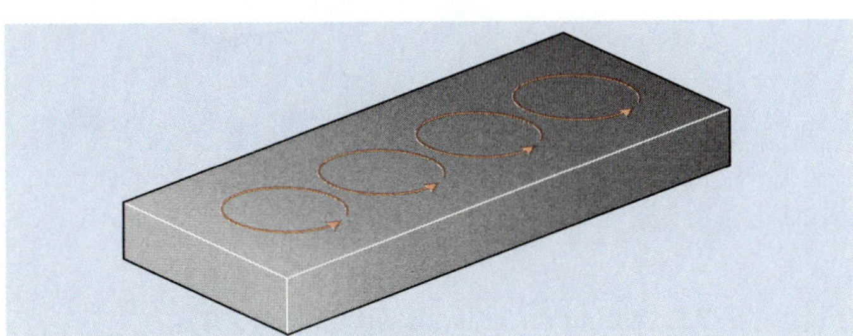

Figure 9-15 **Eddy currents are currents induced in metals.**

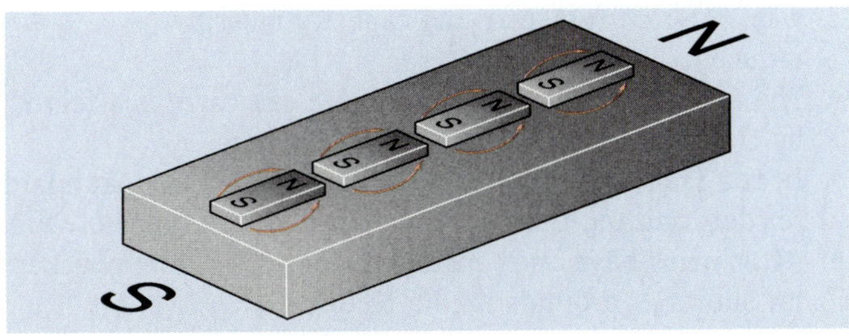

Figure 9-16 **The molecules reverse direction each time the magnetic field changes direction.**

Figure 9-18 **A hand-crank MEGGER®.** *Courtesy of Biddle Instruments.*

Figure 9-19 Battery-operated MEGGER®. *Courtesy of Biddle Instruments.*

MEGGER®s can also be obtained in battery-operated models as shown in Figure 9-19. These models are small, lightweight, and particularly useful when it becomes necessary to test the dielectric of a capacitor.

Wire installations are generally tested for two conditions, shorts and grounds. Shorts are current paths that exist between conductors. To test an installation for shorts, the MEGGER® is connected across two conductors at a time as shown in Figure 9-20. The circuit is tested at rated voltage or slightly higher. The MEGGER® indicates the resistance between the two conductors. Since both conductors are insulated, the resistance between them should be extremely high. Each conductor should be tested against every other conductor in the installation.

To test the installation for grounds, one lead of the MEGGER® is connected to the conduit or raceway as shown in Figure 9-21. The other meter lead is connected to one of the conductors. The conductor should be tested at rated voltage or slightly higher. Each conductor should be tested.

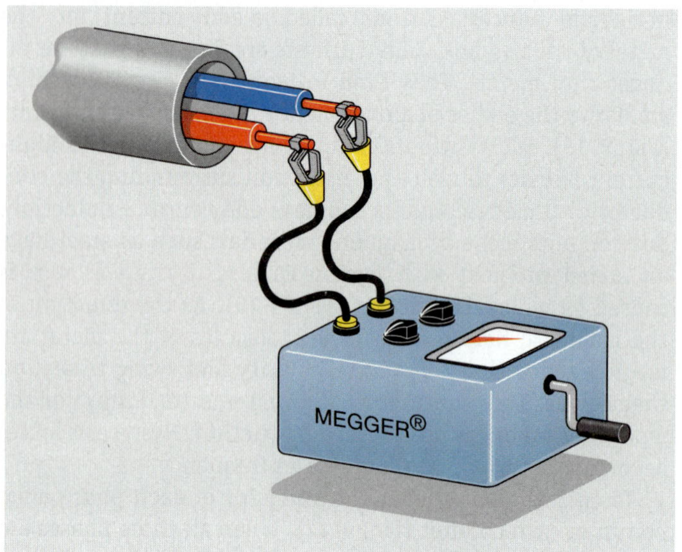

Figure 9-20 Testing for shorts with a MEGGER®.

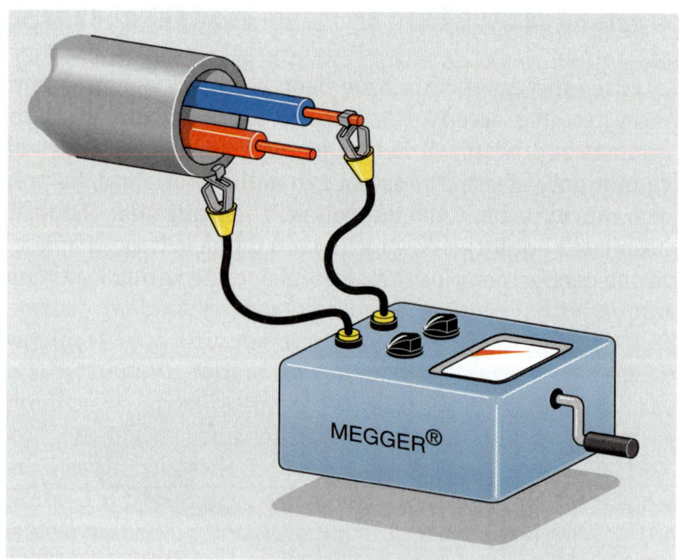

Figure 9-21 Testing for grounds with a MEGGER®.

Summary

- The National Electrical Code is used to determine the wire size for most installations.
- The resistance of a wire is determined by four factors:
  - A. the type of material from which the conductor is made
  - B. the length of the conductor
  - C. the area of the conductor
  - D. the temperature of the conductor
- When conductors are used in high-temperature locations, their current capacity must be reduced.
- When there are more than three conductors in a raceway, their current-carrying capacity must be reduced.
- The amount of current a conductor can carry is affected by the type of insulation around the wire.
- In the English system, the mil foot is used as a standard for determining the resistance of different types of wire.
- After wires have been installed, they should be checked for shorts or grounds with a MEGGER®.

Review Questions

Use of the National Electrical Code will be required to answer some of the following questions.

1. **What is the maximum temperature rating of type XHHW insulation when used in a wet location?**
2. **Name two types of conductor insulation designed to be used underground.**
3. **A #10 AWG copper conductor with type THW insulation is to be run in free air. What is the maximum ampacity of this conductor if the ambient air temperature is 40°C?**
4. **Six #1/0 aluminum conductors are to be run in a conduit. Each conductor has type THWN insulation, and the ambient air temperature is 30°C. What is the ampacity of each conductor?**
5. **Name five conditions that must be met for running conductors in parallel.**
6. **What is the largest solid (nonstranded) conductor listed in the wire tables?**
7. **Can type TW cable be used in an area that has an ambient temperature of 65°C?**
8. **How is the grounded conductor in a flat multiconductor cable #4 or larger identified?**
9. **What three colors are ungrounded conductors not permitted to be?**
10. **Twenty-five #12 AWG copper conductors are run in conduit. Each conductor has type THHN insulation. The conduit is located in an area that has an ambient temperature of 95°F. What is the ampacity of each conductor?**
11. **A single-phase load is located 2800 ft from its source. The load draws a current of 86 A and operates on 480 V. The maximum voltage drop at the load cannot be greater than 3%. What size aluminum conductors should be installed to operate this load?**
12. **It is decided to use parallel 0000 conductors to supply the load in question 11. How many 0000 conductors will be needed?**

Practice Problems

Using Wire Tables and Determining Conductor Sizes

1. A #2 AWG copper conductor is 450 ft long. What is the resistance of this wire? Assume the ambient temperature to be 20°C.
2. A #8 AWG conductor is 500 ft long and has a resistance of 1.817 Ω. The ambient temperature is 20°C. Of what material is the wire made?
3. Three 500-kCM copper conductors with type RHH insulation are to be used in an area with an ambient temperature of 58°C. What is the maximum current-carrying capacity of these conductors?
4. Eight #10 AWG aluminum conductors with type THWN insulation are installed in a single conduit. What is the maximum current-carrying capacity of these conductors? Assume an ambient temperature of 30°C.

Chapter 10 Magnetic Induction

Magnetic induction is one of the most important concepts in the electrical field. It is the basic operating principle underlying alternators, transformers, and most alternating current motors. It is imperative that anyone desiring to work in the electrical field have an understanding of the principles involved.

OBJECTIVES

After studying this unit, you should be able to:

- discuss magnetic induction.
- list factors that determine the amount and polarity of an induced voltage.
- discuss Lenz's law.
- discuss an exponential curve.
- list devices used to help prevent inductive voltage spikes.

Glossary of Magnetic Induction Terms

eddy current electric currents induced in metal; generally core material of motors and transformers

exponential curve a rate at which current increases or decreases in an inductor

henry the standard measure of inductance

hysteresis loss the amount of power loss in the form of heat caused by molecular friction

Lenz's law an electrical law stated by a scientist named Lenz that says, basically, that an induced voltage will always oppose the force that induces it

magnetic induction using magnetic field to induce a voltage into a conductive material, generally wire

metal oxide varistor (MOV) an electronic device that exhibits a rapid change of resistance with a change of temperature

R-L time constant the amount of time required for current to increase or decrease in value

strength of magnetic field the number of magnetic flux lines per square inch or flux density of a magnetic field

turns of wire the number of turns of wire on a coil

voltage spike a large amount of voltage that exists for a very short period of time

weber a measure of magnetism equal to 100,000,000 lines of magnetic flux

Magnetic Induction

One of the basic laws of electricity is that whenever current flows through a conductor, a magnetic field is created around the conductor (Fig. 10-1). The direction of the current flow determines the polarity of the magnetic field, and the amount of current determines the strength of the magnetic field.

That basic law in reverse is the principle of **magnetic induction,** which states that **whenever a conductor cuts through magnetic lines of flux, a voltage is induced into the conductor.** The conductor in Figure 10-2 is connected to a zero-center microammeter, creating a complete circuit. When the conductor is moved downward through the magnetic lines of flux, the induced voltage will cause electrons to flow in the direction indicated by the arrows. This flow of electrons causes the pointer of the meter to be deflected from the center-zero position.

If the conductor is moved upward, the polarity of induced voltage will be reversed and the current will flow in the opposite direction (Fig. 10-3). The pointer will be deflected in the opposite direction.

The polarity of the induced voltage can also be changed by reversing the polarity of the magnetic field (Fig. 10-4). In this example, the conductor is again moved downward through the lines of flux, but the polarity of the magnetic field has been reversed. Therefore the polarity of the induced

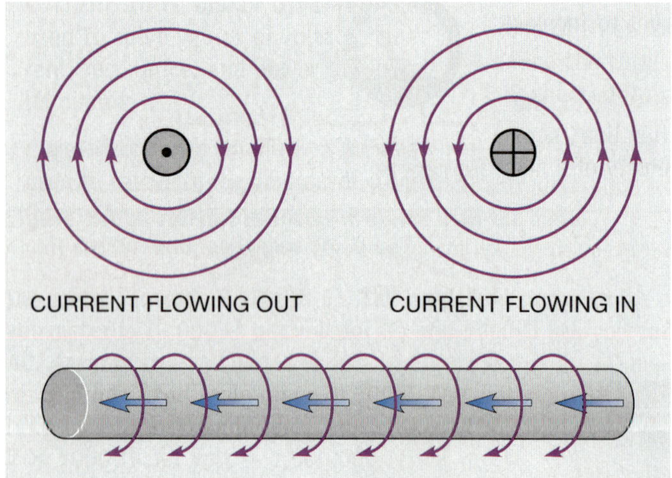

Figure 10-1 Current flowing through a conductor produces a magnetic field around the conductor.

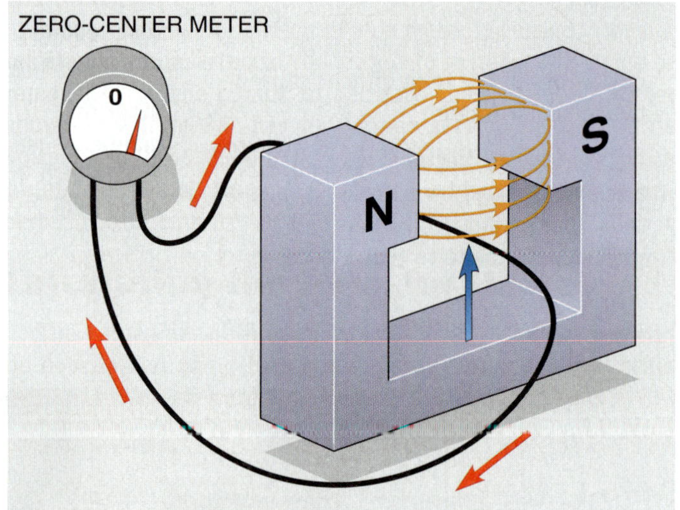

Figure 10-3 Reversing the direction of movement reverses the polarlty of the voltage.

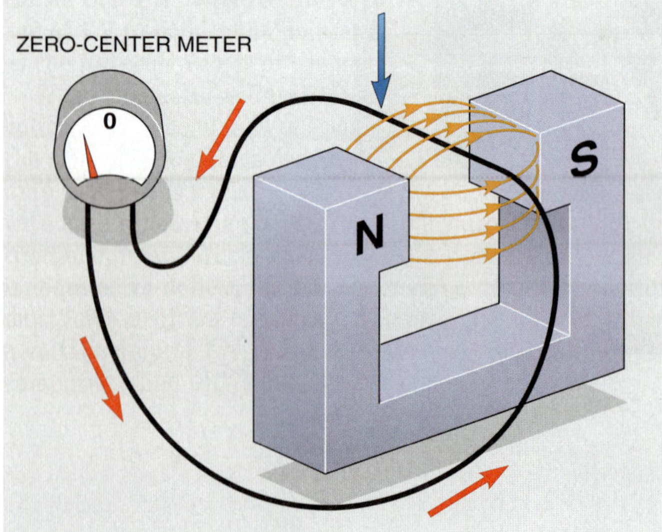

Figure 10-2 A voltage is induced when a conductor cuts magnetic lines of flux.

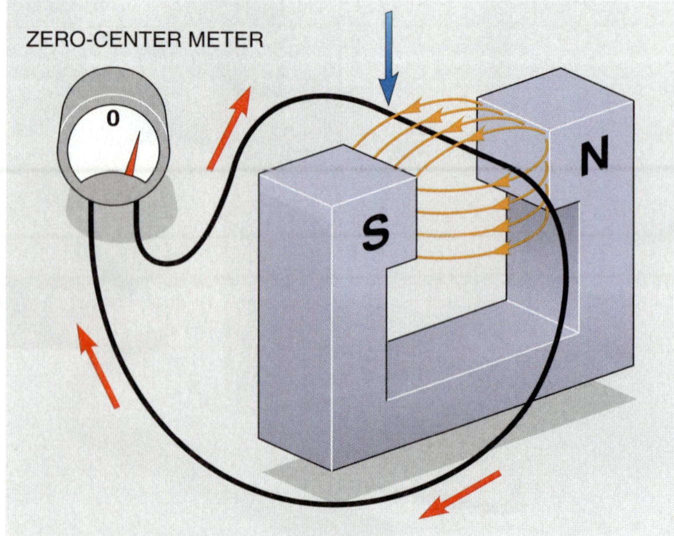

Figure 10-4 Reversing the polarity of the magnetic field reverses the polarity of the voltage.

voltage will be the opposite of that in Figure 10-2, and the pointer of the meter will be deflected in the opposite direction. It can be concluded that **the polarity of the induced voltage is determined by the polarity of the magnetic field in relation to the direction of movement.**

Fleming's Left-Hand Generator Rule

Fleming's *left-hand generator rule* can be used to determine the relationship of the motion of the conductor in a magnetic field to the direction of the induced current. To use the left-hand rule, place the thumb, forefinger, and center finger at right angles to each other as shown in Figure 10-5. **The forefinger points in the direction of the field flux,** assuming that magnetic lines of force are in a direction of north to south. **The thumb points in the direction of thrust,** or movement of the conductor, **and the center finger shows the direction of the current induced into the armature.** An easy method of remembering which finger represents which quantity is.

THumb = **TH**rust

Forefinger = **F**lux

Center finger = **C**urrent

The left-hand rule clearly illustrates that if the polarity of the magnetic field is changed or if the direction of armature rotation is changed, the direction of induced current will change also.

Moving Magnetic Fields

The important factors concerning magnetic induction are a conductor, a magnetic field, and relative motion. In practice, it is often desirable to move the magnet instead of the conductor. Most alternating current generators or alternators operate on this principle. In Figure 10-6, a coil of wire is held stationary while a magnet is moved through the coil. As the magnet is moved, the lines of flux cut through the windings of the coil and induce a voltage into them.

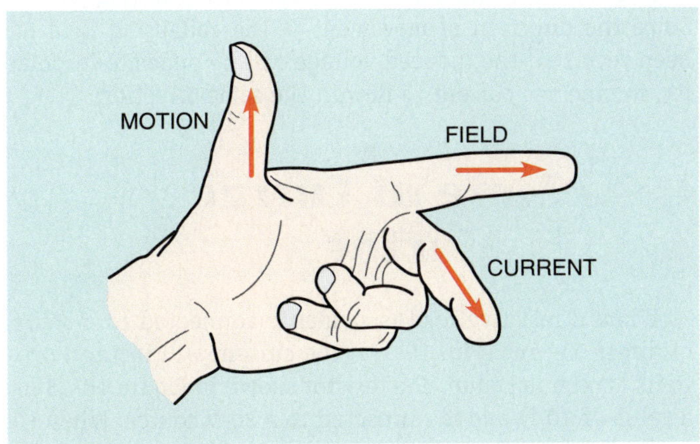

Figure 10-5 Left-hand generator rule.

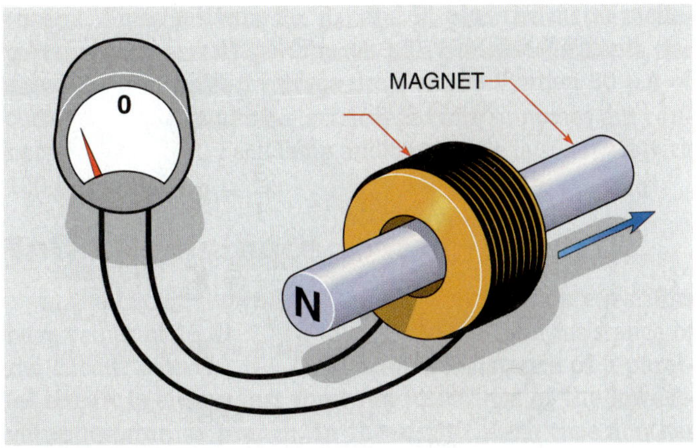

Figure 10-6 Voltage is induced by a moving magnetic field.

Determining the Amount of Induced Voltage

Three factors determine the amount of voltage that will be induced in a conductor:

1. **The number of turns of wire.**
2. **The strength of the magnetic field** (flux density).
3. **The speed of the cutting action.**

In order to induce 1 V in a conductor, the conductor must cut 100,000,000 lines of magnetic flux in 1 s. In magnetic measurement, 100,000,000 lines of flux are equal to one **weber (Wb).** Therefore, if a conductor cuts magnetic lines of flux at a rate of 1 Wb/s, a voltage of 1 V will be induced. A simple one-loop generator is shown in Figure 10-7. The loop is attached to a rod that is free to rotate. This assembly is suspended between the poles of two stationary magnets. If the loop is turned, the conductor cuts through magnetic lines of flux and a voltage is induced into the conductor.

If the speed of rotation is increased, the conductor cuts more lines of flux per second, and the amount of induced voltage increases. If the speed of rotation remains constant and the strength of the magnetic field is increased, there will be more lines of flux per square inch. When there are more lines of flux, the number of lines cut per second increases and the induced voltage increases. If more turns of wire are added to the loop (Fig. 10-8), more flux lines are cut per second and the amount of induced voltage increases again. Adding more turns has the effect of connecting single conductors in series, and the amount of induced voltage in each conductor adds.

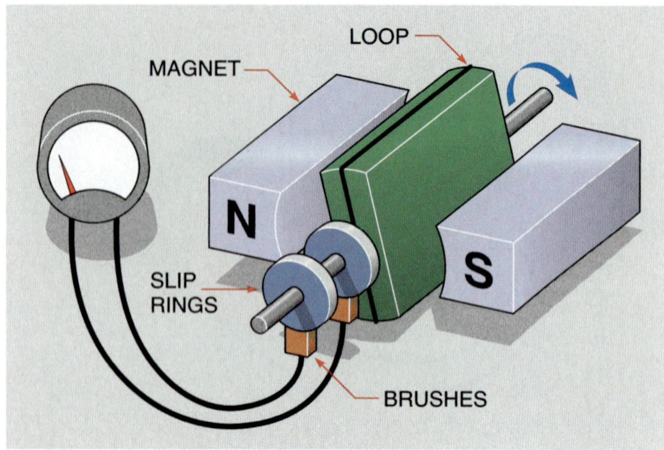

Figure 10-7 A single-loop generator.

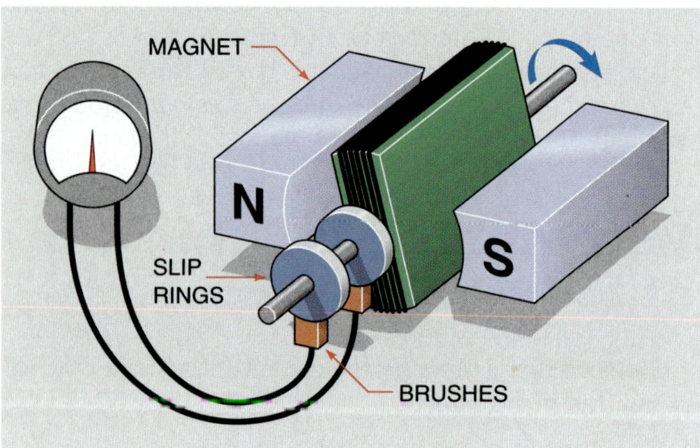

Figure 10-8 Increasing the number of turns increases the induced voltage.

Lenz's Law

When a voltage is induced in a coil and there is a complete circuit, current will flow through the coil (Fig. 10-9). When current flows through the coil, a magnetic field is created around the coil. This magnetic field develops a polarity opposite that of the moving magnet. The magnetic field developed by the induced current acts to attract the moving magnet and pull it back inside the coil.

If the direction of motion is reversed, the polarity of the induced current is reversed, and the magnetic field created by the induced current again opposes the motion of the magnet. This principle was first noticed by Heinrich Lenz many years ago and is summarized in **Lenz's law,** which states that **an induced voltage or current opposes the motion that causes it,** which is why the induced voltage is known as the counteremf (cemf), or back-emf. From this basic principle, other laws concerning inductors have been developed. One is that **inductors always oppose a change of current.** The coil in Figure 10-10, for example, has no induced voltage and therefore no induced current. If the magnet is moved toward the coil, however, magnetic lines of flux will begin to cut the conductors of the coil, and a current will be induced in the coil. The induced current causes magnetic lines of flux to expand outward around the coil (Fig. 10-11). As this expanding magnetic field cuts through the conductors of the coil, a voltage is induced in the coil. The polarity of the voltage is such that it opposes the induced current caused by the moving magnet.

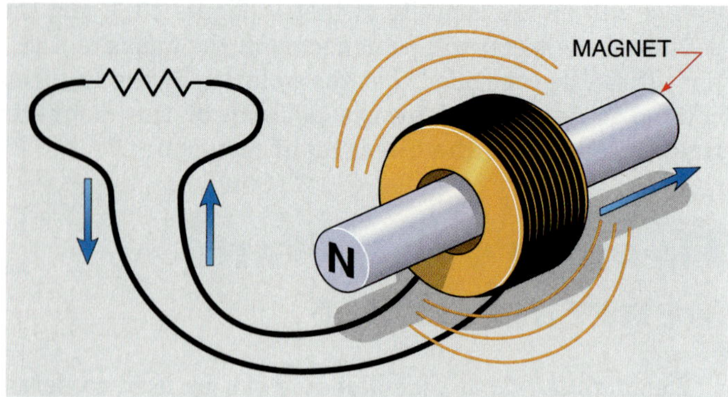

Figure 10-9 An Induced current produces a magnetic field around the coil.

Figure 10-10 No current flows through the coil.

If the magnet is moved away, the magnetic field around the coil will collapse and induce a voltage in the coil (Fig. 10-12). Since the direction of movement of the collapsing field has been reversed, the induced voltage will be opposite in polarity, forcing the current to flow in the same direction.

Rise Time of Current in an Inductor

When a resistive load is suddenly connected to a source of direct current (Fig. 10-13), the current will instantly rise to its maximum value. The resistor shown in Figure 10-13 has a value of 10 Ω and is connected to a 20-V source. When the switch is closed the current will instantly rise to a value of 2 A (20 V/10 Ω = 2 A).

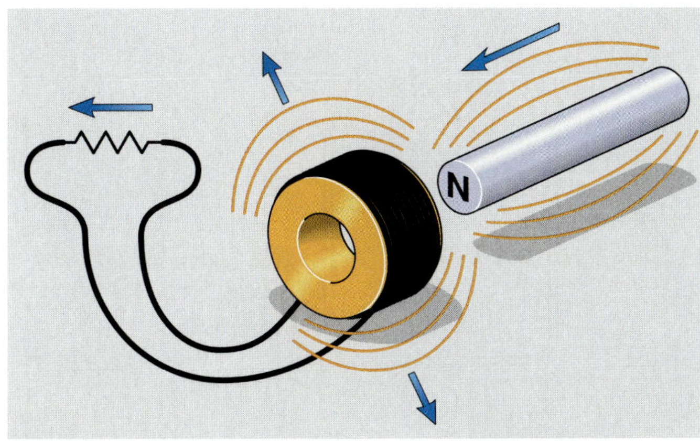

Figure 10-11 Induced current produces a magnetic field around the coil.

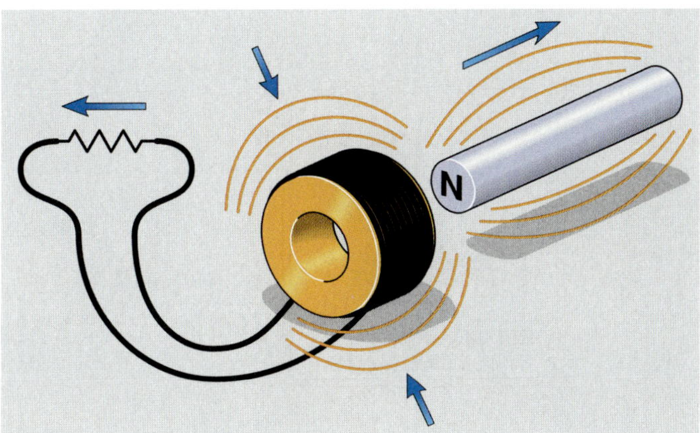

Figure 10-12 The induced voltage forces current to flow in the same direction.

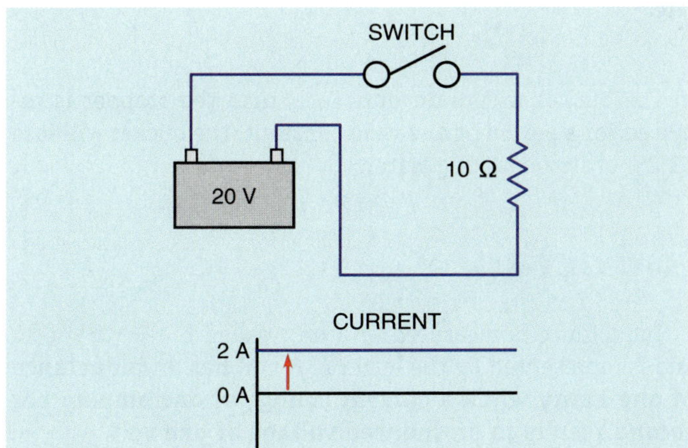

Figure 10-13 The current rises instantly in a resistive circuit.

If the resistor is replaced with an inductor that has a wire resistance of 10 Ω and the switch is closed, the current cannot instantly rise to its maximum value of 2 A (Fig. 10-14). As current begins to flow through an inductor, the expanding magnetic field cuts through the conductors, inducing a voltage into them. In accord with Lenz's law, the induced voltage is opposite in polarity to the applied voltage. The induced voltage, therefore, acts like a resistance to hinder the flow of current through the inductor (Fig. 10-15).

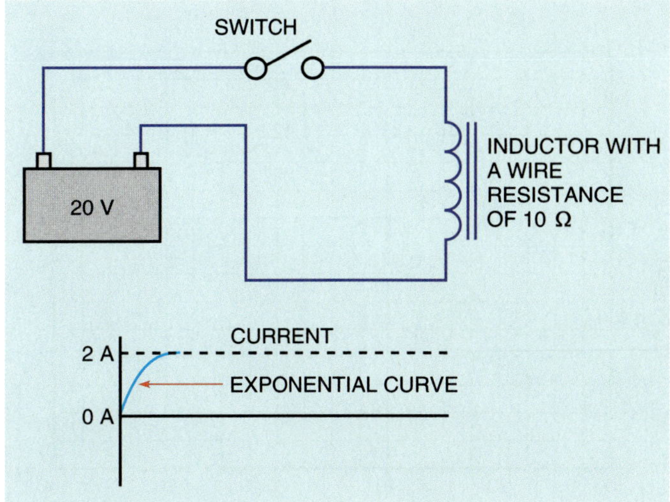

Figure 10-14 Current rises through an indicator at an exponential rate.

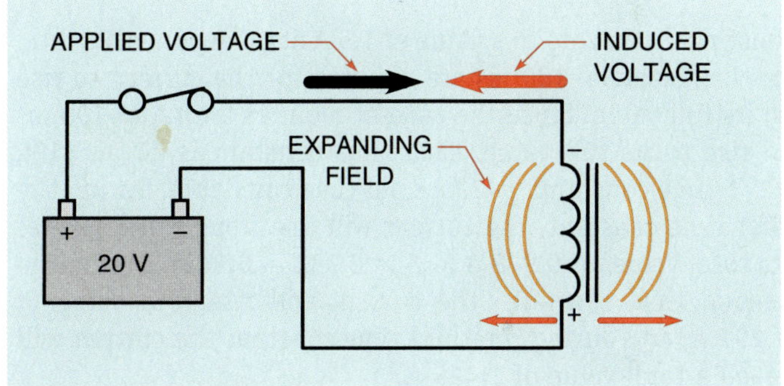

Figure 10-15 The applied voltage is opposite in polarity to the induced voltage.

The induced voltage is proportional to the rate of change of current (speed of the cutting action). When the switch is first closed, current flow through the coil tries to rise instantly. This extremely fast rate of current change induces maximum voltage in the coil. As the current flow approaches its maximum Ohm's law value, in this example 2 A, the rate of change becomes less and the amount of induced voltage decreases.

The Exponential Curve

The **exponential curve** describes a rate of certain occurrences. The curve is divided into five time constants. Each time constant is equal to 63.2% of some value. An exponential curve is shown in Figure 10-16. In this example, current

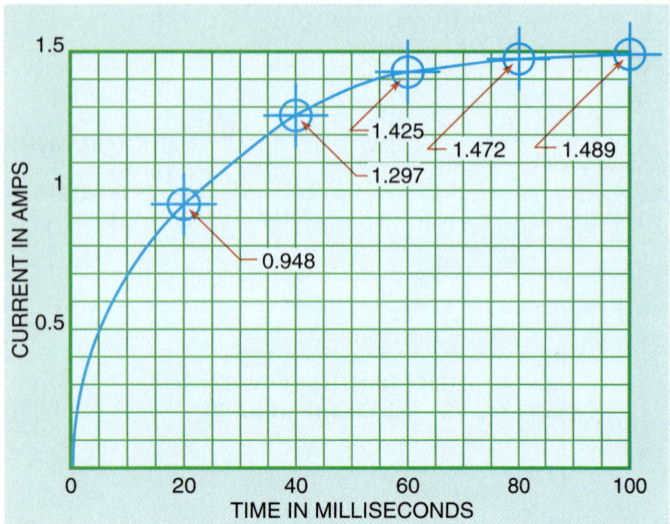

Figure 10-16 An exponential curve.

must rise from zero to a value of 1.5 A at an exponential rate. In this example, 100 ms are required for the current to rise to its full value. Since the current requires a total of 100 ms to rise to its full value, each time constant is 20 ms (100 ms/5 time constants = 20 ms per time constant). During the first time constant, the current will rise from 0 to 63.2% of its total value, or 0.948 A ($1.5 \times 0.632 = 0.948$). During the second time constant, the current will rise to a value of 1.297 A, and during the third time constant the current will reach a total value of 1.425 A.

Because the current increases at a rate of 63.2% during each time constant, it is theoretically impossible to reach the total value of 1.5 A. After five time constants, however, the current has reached approximately 99.3% of the maximum value and for all practical purposes is considered to be complete.

The exponential curve can often be found in nature. If clothes are hung on a line to dry, they will dry at an exponential rate. Another example of the exponential curve can be seen in Figure 10-17. In this example, a bucket has been filled to a certain mark with water. A hole has been cut at the bottom of the bucket and a stopper placed in the hole. When the stopper is removed from the bucket, water will flow out at an exponential rate. Assume, for example, it takes 5 min for the water to flow out of the bucket. Exponential curves are always divided into five time constants so in this case each time constant has a value of 1 min. If the stopper is removed and water is permitted to drain from the bucket for a period of 1 min before the stopper is replaced, during that first time constant 63.2% of the water in the bucket will drain out. (see Fig. 10-18.) If the stopper is again removed for a period of 1 min, 63.2% of the water remaining in the bucket will drain out. Each time the stopper is removed for a period of one time constant, the bucket will lose 63.2% of its remaining water.

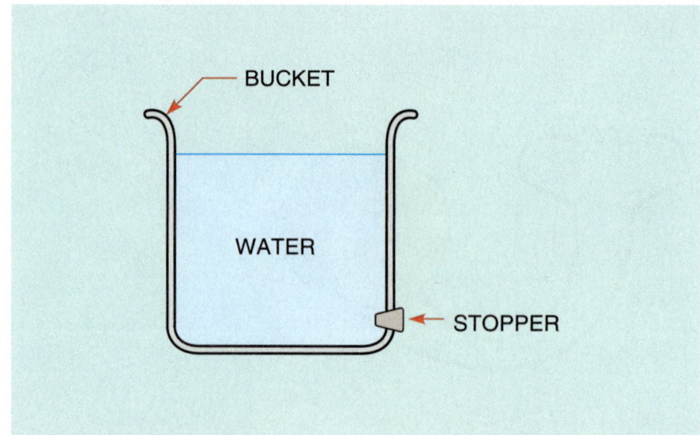

Figure 10-17 Exponential curves can be found in nature.

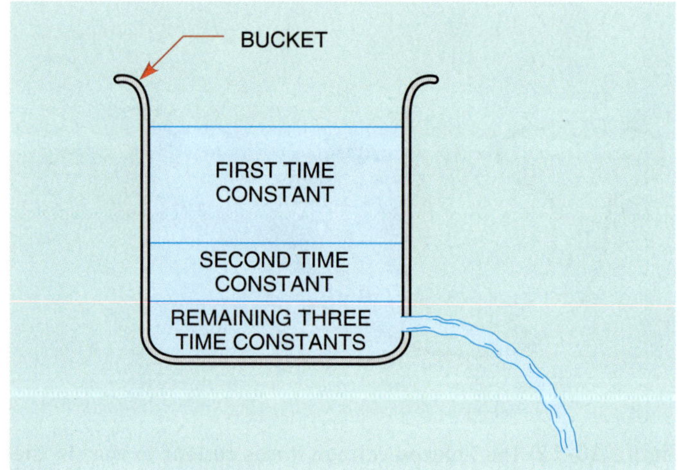

Figure 10-18 Water flows from a bucket at an exponential rate.

Inductance

Inductance is measured in a unit called the **henry (H)** and is represented by the letter *L*. **A coil has an inductance of one henry when a current change of one ampere per second results in an induced voltage of one volt.**

The amount of inductance a coil will have is determined by its physical properties and construction. A coil wound on a nonmagnetic core material such as wood or plastic is referred to as an *air core* inductor. If the coil is wound on a core made of magnetic material such as silicon steel or soft iron, it is called an *iron core* inductor. Iron core inductors

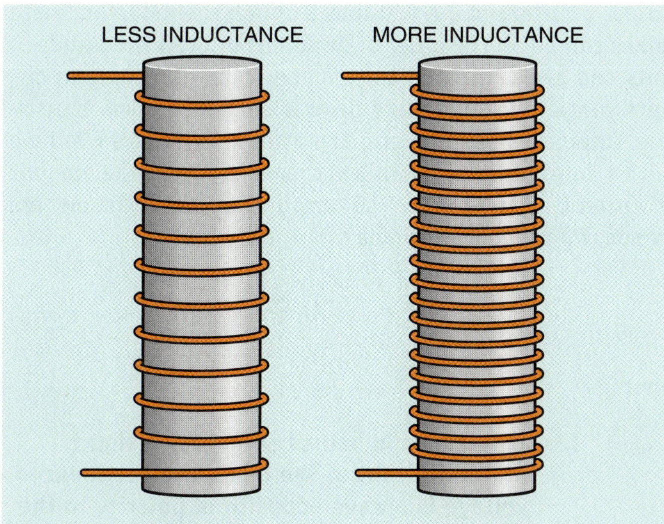

Figure 10-19 Inductance is determined by the physical construction of the coil.

produce more inductance with fewer turns than air core inductors because of the good magnetic path provided by the core material. Iron core inductors cannot be used for high-frequency applications, however, because of **eddy current** loss and **hysteresis loss** in the core material.

Another factor that determines inductance is how far the windings are separated from each other. If the turns of wire are far apart they will have less inductance than turns wound closer together (Fig. 10-19).

The inductance of a coil can be determined using the formula

$$L = \frac{0.4\pi N^2 \mu A}{l}$$

where

L = **inductance in henrys**

π = **3.1416**

N = **number of turns of wire**

μ = **permeability of the core material** -DIFFERENT MAGNETS

A = **cross sectional area of the core**

l = **length of the core**

The formula indicates that the inductance is proportional to the number of turns of wire, the type of core material used, and the cross sectional area of the core, but inversely proportional to core length. An inductor is basically an electromagnet that changes its polarity at regular intervals. Because the inductor is an electromagnet, the same factors that affect magnets affect inductors. The permeability of the core material is just as important to an inductor as it is to any other electromagnet. Flux lines pass through a core material with a high permeability (such as silicon, steel, or soft iron) better than through a material with a low permeability (such as brass, copper, or aluminum). Once the core material has become saturated, however, the permeability value becomes approximately 1 and an increase in turns of wire has only a small effect on the value of inductance.

R-L Time Constants

The time necessary for current in an inductor to reach its full Ohm's law value, called the **R-L time constant,** can be computed using the formula

$$T = \frac{L}{R}$$

where

T = **time in seconds**

L = **inductance in henrys**

R = **resistance in ohms**

This formula computes the time of one time constant.

Example 1

A coil has an inductance of 1.5 H and a wire resistance of 6 Ω. If the coil is connected to a battery of 3 V, how long will it take the current to reach its full Ohm's law value of 0.5 A (3 V/6 Ω = 0.5 A)?

SOLUTION

To find the time of one time constant, use the formula

$$T = \frac{L}{R}$$

$$T = \frac{1.5}{6}$$

$$T = 0.25 \text{ s}$$

The time for one time constant is 0.25 s. Since five time constants are required for the current to reach its full value of 0.5 A, 0.25 s will be multiplied by 5.

$$0.25 \times 5 = 1.25 \text{ s}$$

Induced Voltage Spikes

A **voltage spike** occurs when the current flow through an inductor stops, and the current decreases at an exponential rate also (Fig. 10-20). As long as a complete circuit exists when the power is interrupted, there is little or no problem. In the circuit shown in Figure 10-21, a resistor and inductor are connected in parallel. When the switch is closed, the battery will supply current to both. When the switch is opened, the magnetic field surrounding the inductor will collapse and induce a voltage into the inductor. The induced voltage will attempt to keep current flowing in the same direction. Recall that inductors oppose a change of current. The amount of current flow and the time necessary for the flow to stop will be determined by the resistor and the properties of the inductor. The amount of voltage produced by the collapsing magnetic field is determined by the maximum current in the circuit and the total resistance in the circuit. In the circuit shown in Figure 10-21, assume that the inductor has a wire resistance of 6 Ω, and the resistor has a resistance of 100 Ω. Also assume that when the switch is closed, a current of 2 A will flow through the inductor. These spikes can be on the order of hundreds or even thousands of volts and can damage circuit components, especially in circuits containing solid state devices such as diodes, transistors, integrated circuits, etc. The amount of induced voltage can be determined if the inductance of the coil, the amount of current change, and the amount of time change are known, by using the formula

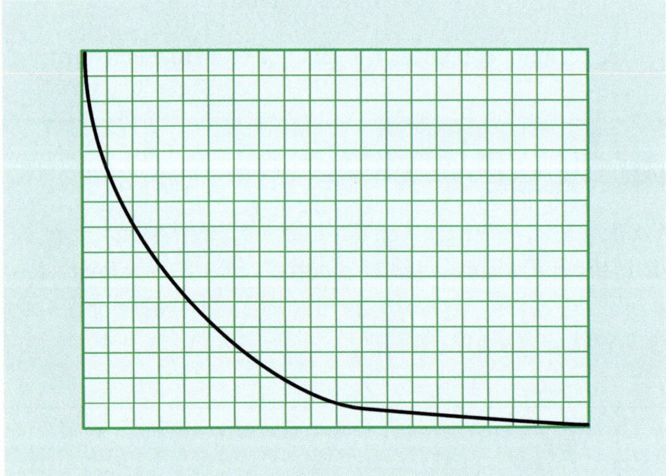

Figure 10-20 **Current flow through an inductor decreases at an exponential rate.**

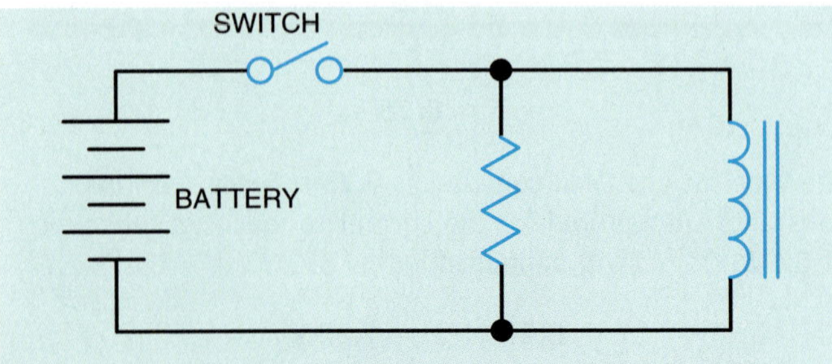

Figure 10-21 **The resistor helps prevent voltage spikes caused by the inductor.**

$$EMF = -L\left(\frac{\Delta I}{\Delta t}\right)$$

where

L = inductance in henrys (A negative sign is placed in front of the L because the induced voltage is always opposite in polarity to the voltage that produces it.)

Δ I = change of current

Δ t = change of time

Assume a 1.5-henry inductor has a current flow of 2.5 amperes. When a switch is opened, the current changes from 2.5 amperes to 0.5 amperes in 0.005 second. How much voltage is induced into the inductor?

$$EMF = -L\left(\frac{\Delta I}{\Delta t}\right)$$

$$EMF = -1.5\left(\frac{2}{0.005}\right)$$

$$EMF = -1.5 \times 400$$

$$EMF = -600 \text{ volts}$$

When the switch is opened, a series circuit exists composed of the resistor and inductor (Fig. 10-22). The maximum voltage developed in this circuit would be 212 V (2 A × 106 Ω = 212 V). If the circuit resistance were increased, the induced voltage would become greater. If the circuit resistance were decreased, the induced voltage would become less.

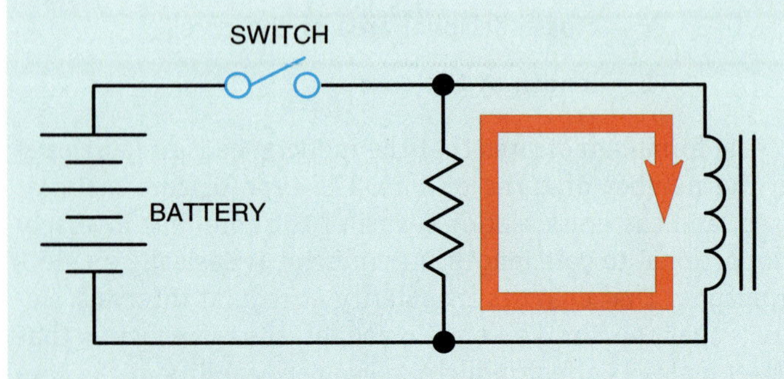

Figure 10-22 **When the switch is opened, a series path is formed by the resistor and inductor.**

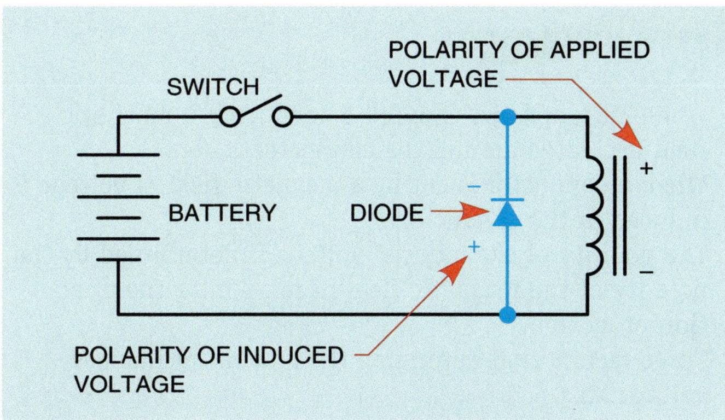

Figure 10.23 A diode is used to prevent induced voltage spikes.

Another device often used to prevent induced voltage spikes when the current flow through an inductor is stopped is the diode (Fig. 10-23). The diode is an electronic component that operates like an electrical check valve. The diode will permit current to flow through it in only one direction. The diode is connected in parallel with the inductor in such a manner that when voltage is applied to the circuit, the diode is reverse-biased and acts like an open switch. When the diode is reverse-biased no current will flow through it.

When the switch is opened, the induced voltage produced by the collapsing magnetic field will be opposite in polarity to the applied voltage. The diode then becomes forward-biased and acts like a closed switch. Current can now flow through the diode and complete a circuit back to the inductor. A silicon diode has a forward voltage drop of approximately 0.7 V regardless of the current flowing through it. Since the diode is connected in parallel with the inductor, and voltage drops of devices connected in parallel must be the same, the induced voltage is limited to approximately 0.7 V. The diode can be used to eliminate inductive voltage spikes in direct current circuits only; it cannot be used for this purpose in alternating current circuits.

A device that can be used for spike suppression in either direct or alternating current circuits is the **metal oxide varistor (MOV).** The MOV is a bidirectional device, which means that it will conduct current in either direction and therefore, can be used in alternating current circuits. The metal oxide varistor is an extremely fast-acting solid state component that will exhibit a change of resistance when the voltage reaches a certain point. Assume that the MOV shown in Figure 10-24 has a voltage rating of 140 V, and that the voltage applied to the circuit is 120 V. When the switch is closed and current flows through the circuit, a magnetic field will be established around the inductor (Fig. 10-25). As long as the voltage applied to the MOV is less than 140 V, it will exhibit an extremely high resistance, in the range of several hundred thousand ohms.

When the switch is opened, current flow through the coil suddenly stops, and the magnetic field collapses. This sudden collapse of the magnetic field will cause an extremely high voltage to be induced in the coil. When this induced voltage reaches 140 V, however, the MOV will suddenly change from a high resistance to a low resistance, preventing the voltage from becoming greater than 140 V (Fig. 10-26).

Metal oxide varistors are extremely fast-acting. They can typically change resistance values in less than 20 ns (nanoseconds). They are often found connected across the coils of relays and motor starters in control systems to prevent voltage spikes from being induced back into the line. They are also found in the surge protectors used to protect many home appliances such as televisions, stereos, and computers.

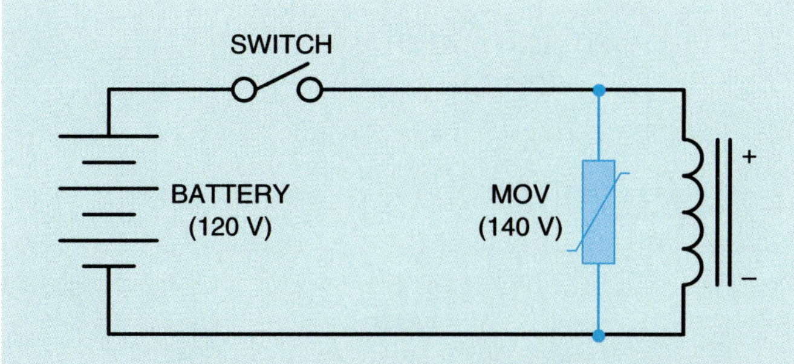

Figure 10-24 Metal oxide varistor used to suppress a voltage spike.

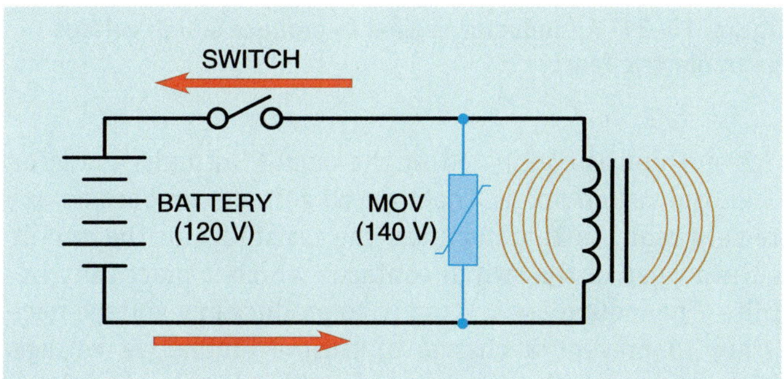

Figure 10-25 When the switch is closed, a magnetic field is established around the inductor.

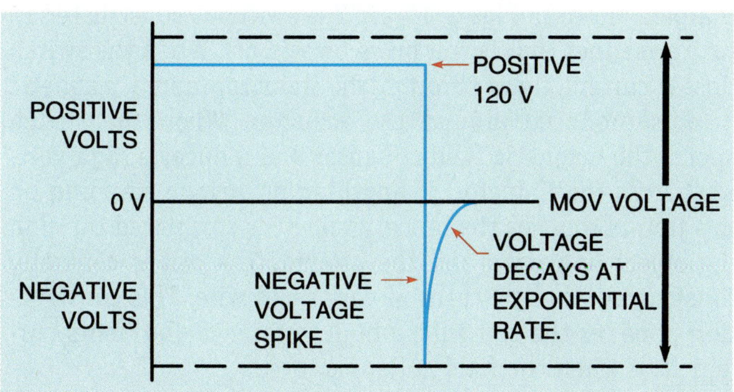

Figure 10-26 The MOV prevents the spike from becoming too high.

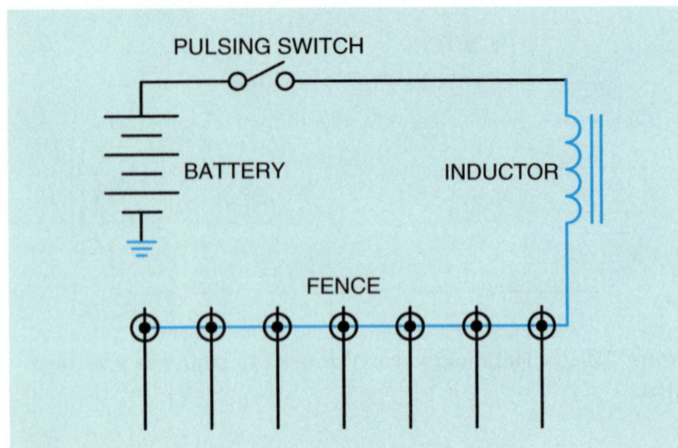

Figure 10-27 An inductor is used to produce a high voltage for an electric fence.

If nothing is connected in the circuit with the inductor when the switch opens, the induced voltage can become extremely high. In this instance, the resistance of the circuit is the air gap of the switch contacts, which is practically infinite. The inductor will attempt to produce any voltage necessary to prevent a change of current. Inductive voltage spikes can reach thousands of volts. This is the principle of operation of many high-voltage devices such as the ignition systems of many automobiles.

Another device that uses the collapsing magnetic field of an inductor to produce a high voltage is the electric fence charger, shown in Figure 10-27. The switch is constructed in such a manner that it will pulse on and off. When the switch closes, current flows through the inductor, and a magnetic field is produced around the inductor. When the switch opens, the magnetic field collapses and induces a high voltage across the inductor. If anything or anyone standing on the ground touches the fence, a circuit is completed through the object or person and the ground. The coil is generally constructed of many turns of very small wire. This construction provides the coil with a high resistance and limits current flow when the field collapses.

Summary

- When current flows through a conductor, a magnetic field is created around the conductor.
- When a conductor is cut by a magnetic field, a voltage is induced in the conductor.
- The polarity of the induced voltage is determined by the polarity of the magnetic field in relation to the direction of motion.
- Three factors that determine the amount of induced voltage are
  - A. the number of turns of wire
  - B. the strength of the magnetic field
  - C. the speed of the cutting action
- One volt is induced in a conductor when magnetic lines of flux are cut at a rate of one weber per second.
- Induced voltage is always opposite in polarity to the applied voltage.
- Inductors oppose a change of current.
- Current rises in an inductor at an exponential rate.
- An exponential curve is divided into five time constants.
- Each time constant is equal to 63.2% of some value.
- Inductance is measured in units called henrys (H).
- A coil has an inductance of 1 H when a current change of 1 A per second results in an induced voltage of 1 V.
- Air core are wound on cores of nonmagnetic material.
- Iron core inductors are wound on cores of magnetic material.
- The amount of inductance an inductor will have is determined by the number of turns of wire and the physical construction of the coil.
- Inductors can produce extremely high voltages when the current flowing through them is stopped.
- Two devices used to help prevent large spike voltages are the resistor and the diode.

Review Questions

1. What determines the polarity of magnetism when current flows through a conductor?
2. What determines the strength of the magnetic field when current flows through a conductor?
3. Name three factors that determine the amount of induced voltage in a coil.
4. How many lines of magnetic flux must be cut in 1 s to induce a voltage of 1 V?
5. What is the effect on induced voltage of adding more turns of wire to a coil?
6. Into how many time constants is an exponential curve divided?
7. Each time constant of an exponential curve is equal to what percentage of the whole?
8. An inductor has an inductance of 0.025 H and a wire resistance of 3 Ω. How long will it take the current to reach its full Ohm's law value?
9. Refer to the circuit shown in Figure 10-21. Assume that the inductor has a wire resistance of 0.2 Ω and the resistor has a value of 250 Ω. If a current of 3 A is flowing through the inductor, what will be the maximum induced voltage when the switch is opened?
10. What electronic component is often used to prevent large voltage spikes from being produced when the current flow through an inductor is suddenly terminated?

SECTION TWO

Alternating Current

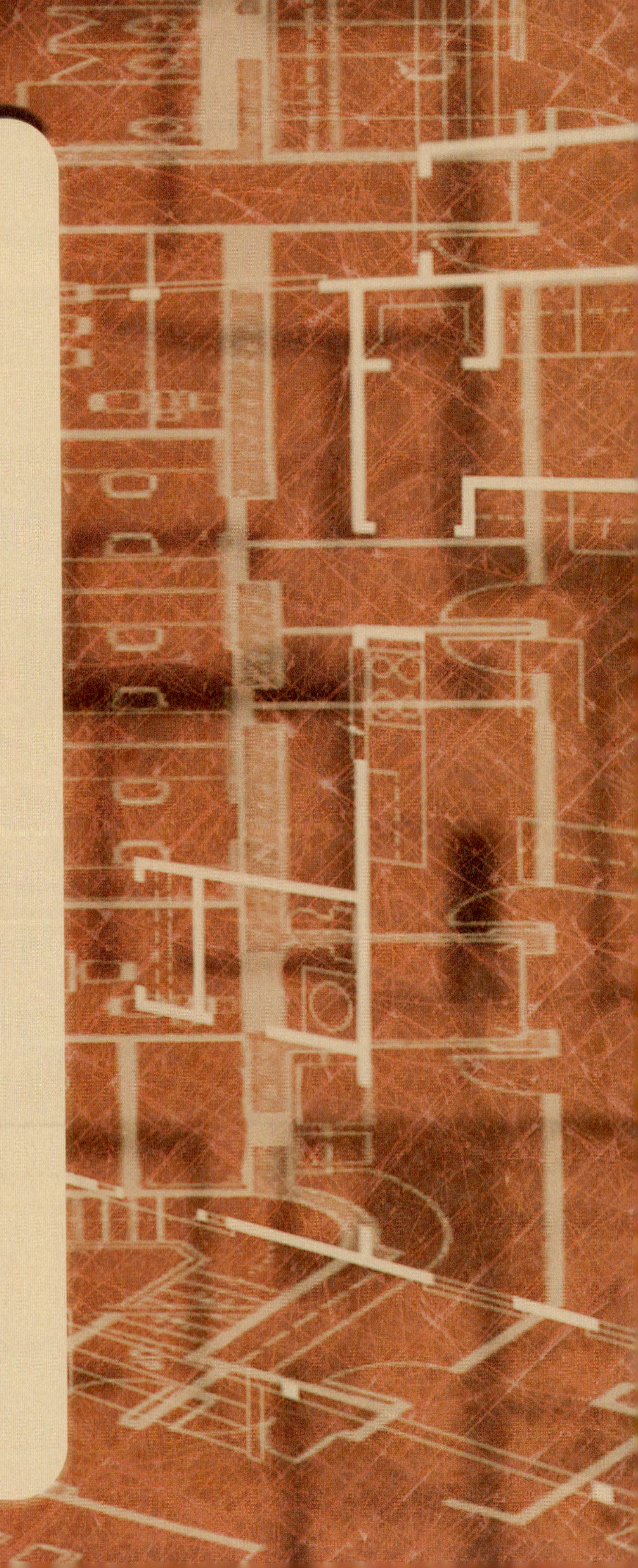

SECTION TWO

ALTERNATING CURRENT

Chapter 11 Alternating Current

Most of the electrical power produced in the world is alternating current. It is used to operate everything from home appliances such as television sets, computers, microwave ovens, and electric toasters, to the largest motors found in industry. Alternating current has several advantages over direct current that make it a better choice for the large-scale production of electrical power.

OBJECTIVES

After studying this unit, you should be able to:

- **discuss differences between direct and alternating current.**
- **be able to compute instantaneous values of voltage and current for a sine wave.**
- **be able to compute peak, RMS, and average values of voltage and current.**
- **discuss the phase relationship of voltage and current in a pure resistive circuit.**

Glossary of Alternating Current Terms

amplitude the maximum height reached by a waveform

average the value of voltage when an AC sine wave voltage is changed to DC

cycle one cycle is completed when an AC voltage completes 360 degrees

frequency the number of complete cycles that occur in one second

hertz a measure of frequency

in phase a condition that occurs when the voltage and current reach peak values and zero crossing at the same time

linear wave a waveform that increases in the same proportion as the angle of rotation

oscillators devices used to convert direct current into alternating current

peak the maximum value obtained by a waverform

resistive loads loads that employ resistance to limit circuit current

ripple the turning on and off effect that occurs when alternating current is converted into direct current

root mean square (RMS) the amount of AC voltage that will produce the same heating effect as a like amount of DC voltage

sine wave a waveform produced by any rotating machine. The voltage at any point on the waveform can be determined by multiplying the peak value by the sine of the angle of rotation

skin effect the effect that causes electrons to move toward the outside of a conductor in an alternating current circuit

triangle wave a linear waveform in which the voltage is proportional to the peak value and percent of angle of rotation

watts a measure of electric power

Advantages of Alternating Current

Probably the single greatest advantage of alternating current is that AC current can be transformed and DC current cannot. A transformer permits voltage to be stepped up or down. Voltage can be stepped up for transmission and then stepped back down when it is to be used by some device. Transmission voltages of 69 kV, 138 kV, and 345 kV are common. The advantage of high-voltage transmission is that less current is required to produce the same amount of power. The reduction of current permits smaller wires to be used, which results in a savings of material.

In the very early days of electric power generation, Thomas Edison, an American inventor, proposed powering the country with low-voltage direct current. He reasoned that low-voltage direct current was safer for people to use than higher-voltage alternating current. A Serbian immigrant named Nikola Tesla, however, argued that direct current was impractical for large-scale applications. The disagreement was finally settled at the 1904 World's Fair held in St. Louis, Missouri. The 1904 World's Fair not only introduced the first ice cream cone and the first iced tea, it was also the first World's Fair to be lighted with "electric candles." At that time, the only two companies capable of providing electric lighting for the World's Fair were the Edison Company, headed by Thomas Edison, and the Westinghouse Company, headed by George Westinghouse, a close friend of Nikola Tesla. The Edison Company submitted a bid of over one dollar per lamp to light the fair with low-voltage direct current. The Westinghouse Company submitted a bid of less than 25 cents per lamp to light the fair using higher-voltage alternating current. This set the precedent for how electric power would be supplied throughout the world.

AC Wave Forms

Square Waves

Alternating current differs from direct current in that AC current reverses its direction of flow at periodic intervals (Fig. 11-1). Alternating current wave forms can vary depending on how the current is produced. One wave form frequently encountered is the square wave (Fig. 11-2). It is assumed that the oscilloscope in Figure 11-2 has been adjusted so that 0 V is represented by the center horizontal line. The wave form shows that the voltage is in the positive direction for some length of time and then changes polarity. The voltage remains negative for some length of time and then changes back to positive again. Each time the voltage reverses polarity, the current flow through the circuit changes direction. A square wave could be produced by a simple single-pole double-throw switch connected to two batteries as shown in Figure 11-3. Each time the switch position is changed, current flows through the resistor in a different direction. Although this circuit will produce a square wave alternating current, it is not practical. Square waves are generally produced by electronic devices called **oscillators.** The schematic diagram of a simple square wave oscillator is shown in Figure 11-4. In this circuit, two bipolar transistors are used as switches to reverse the direction of current flow through the windings of the transformer.

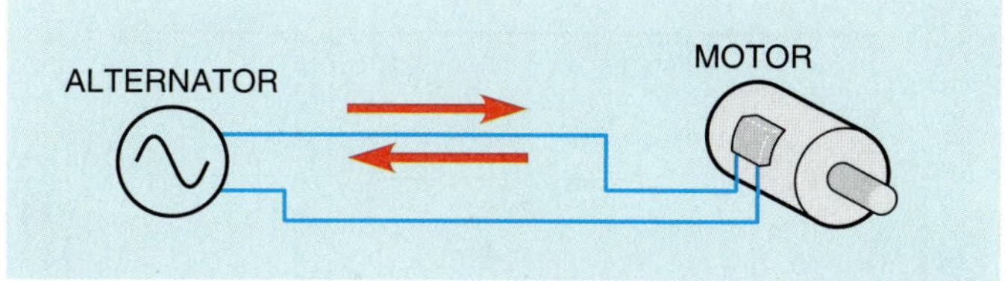

Figure 11-1 Alternating current flows first in one direction and then in the other.

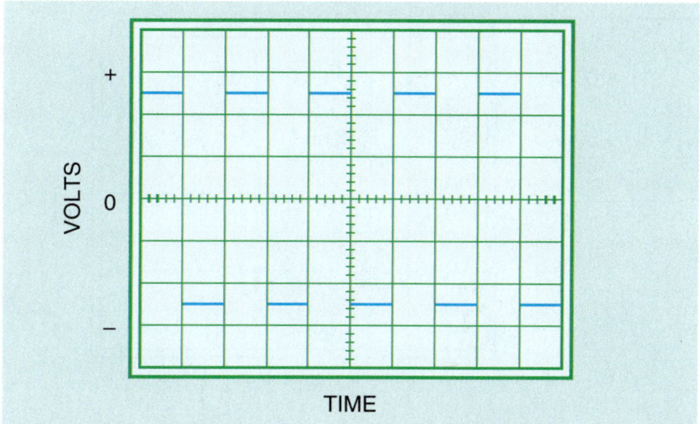

Figure 11-2 Square wave alternating current.

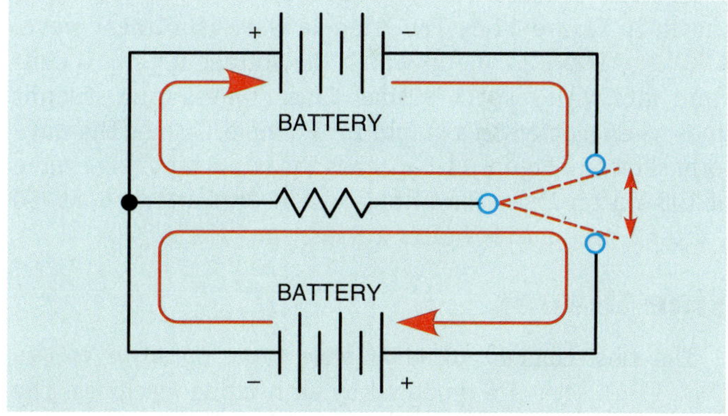

Figure 11-3 Square wave alternating current produced with a switch and two batteries.

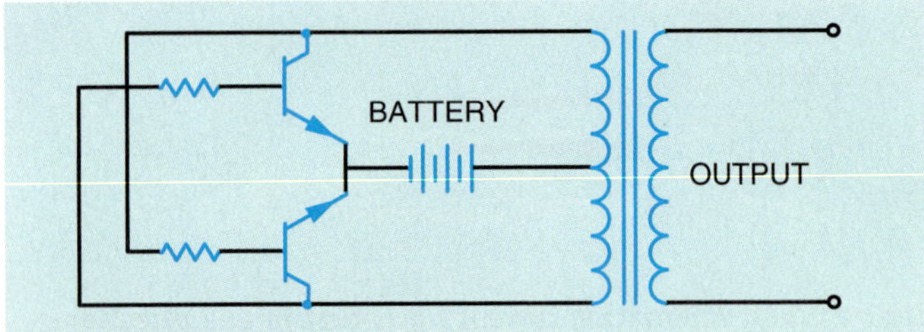

Figure 11-4 **Square wave oscillator.**

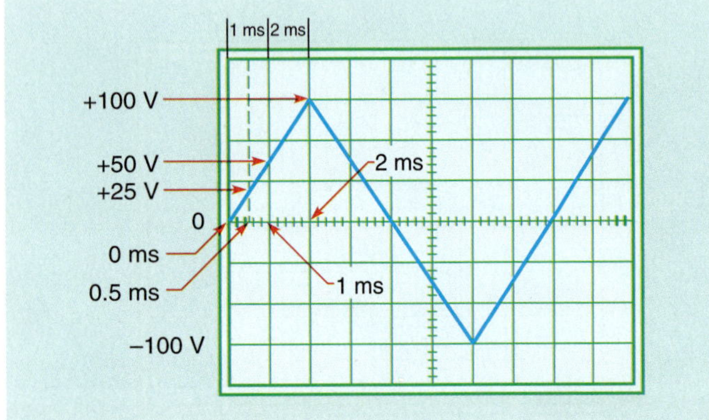

Figure 11-5 **Triangle wave.**

Triangle Waves

Another common AC wave form is the **triangle wave** shown in Figure 11-5. The triangle wave is a linear wave. A **linear wave** is one in which the voltage rises at a constant rate with respect to time. Linear waves form straight lines when plotted on a graph. For example, assume the wave form shown in Figure 11-5 reaches a maximum positive value of 100 V after 2 ms. The voltage will be 25 V after 0.5 ms, 50 V after 1 ms, and 75 V after 1.5 ms.

Sine Waves

The most common of all AC wave forms are **sine waves** (Fig. 11-6). They are produced by all rotating machines. The sine wave contains a total of 360 electrical degrees. It reaches its peak positive voltage at 90°, returns to a value of 0 V at 180°, increases to its maximum negative voltage at 270°, and returns to 0 V at 360°. Each complete wave form of 360° is called a **cycle.** The number of complete cycles that occur in one second is called the **frequency.** Frequency is measured in **hertz (Hz).** The most common frequency in the United States and Canada is 60 Hz. This means that the voltage increases from zero to its maximum value in the positive direction, returns to zero, increases to its maximum value in the negative direction, and returns to zero 60 times each second.

Sine waves are so named because the voltage at any point along the wave form is equal to the maximum, or peak, value times the sine of the angle of rotation. Figure 11-7 illustrates one-half of a loop of wire cutting through lines of magnetic flux. The flux lines are shown with

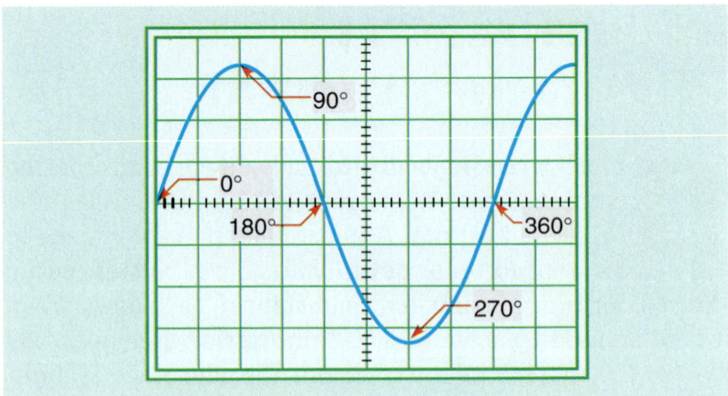

Figure 11-6 **Sine wave.**

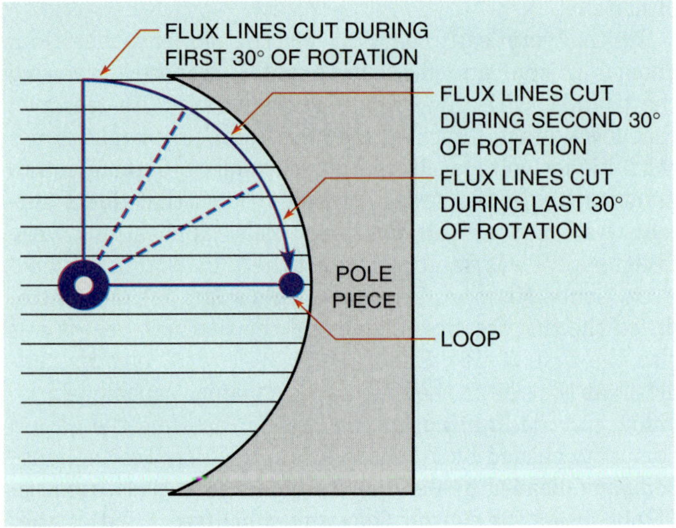

Figure 11-7 **As the loop approaches 90° of rotation, the flux lines are cut at a faster rate.**

equal spacing between each line, and the arrow denotes the arc of the loop as it cuts through the lines of flux. Notice the number of flux lines that are cut by the loop during the first 30° of rotation. Now notice the number of flux lines that are cut during the second and third 30° of rotation. Because the loop is cutting the flux lines at an angle, it must travel a greater distance between flux lines during the first degrees of rotation. Consequently, fewer flux lines are cut per second, which results in a lower induced voltage. Recall that 1 V is induced in a conductor when it cuts lines of magnetic flux at a rate of 1 Wb/s. One weber is equal to 100,000,000 lines of flux.

When the loop has rotated 90°, it is perpendicular to the flux lines and is cutting them at the maximum rate, which results in the highest, or peak, voltage being induced in the loop. The voltage at any point during the rotation is equal to the maximum induced voltage times the sine of the angle of rotation. For example, if the induced voltage after 90° of rotation is 100 V, the voltage after 30° of rotation will be 50 V because the sine of a 30° angle is 0.5 ($100 \times 0.5 = 50$ V). The induced voltage after 45° of rotation is 70.7 V because the sine of a 45° angle is 0.707 ($100 \times 0.707 = 70.7$ V). A sine wave showing

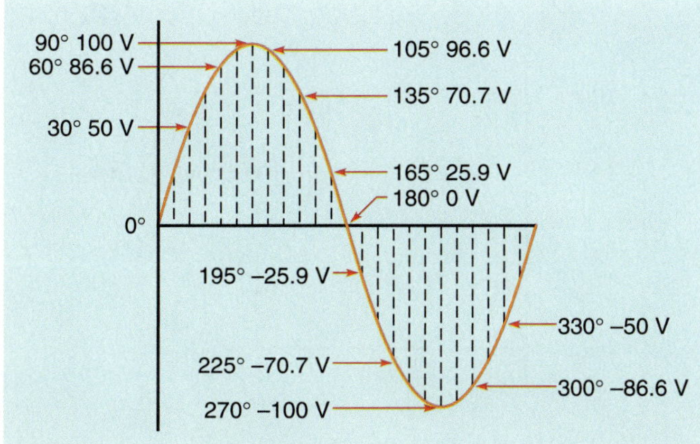

Figure 11-8 **Instantaneous values of voltage along a sine wave.**

the instantaneous voltage values after different degrees of rotation is shown in Figure 11-8. The instantaneous voltage value is the value of voltage at any instant on the wave form.

The following formula can be used to determine the instantaneous value at any point along the sine wave.

$$E_{(INST)} = E_{(MAX)} \times SIN \angle$$

where

$E_{(INST)}$ = **the voltage at any point on the wave form**

$E_{(MAX)}$ = **the maximum, or peak, voltage**

$SIN \angle$ = **the sine of the angle of rotation**

Example 1

A sine wave has a maximum voltage of 138 V. What is the voltage after 78° of rotation?

SOLUTION

$$E_{(INST)} = E_{(MAX)} \times SIN \angle$$

$$E_{(INST)} = 138 \times 0.978 \text{ (SIN of 78°)}$$

$$E_{(INST)} = 134.96 \text{ V}$$

The formula can be changed to find the maximum value if the instantaneous value and the angle of rotation are known or to find the angle if the maximum and instantaneous values are known.

$$E_{(MAX)} = \frac{E_{(INST)}}{SIN\angle}$$

$$SIN\angle = \frac{E_{(INST)}}{E_{(MAX)}}$$

Example 2

A sine wave has an instantaneous voltage of 246 V after 53° of rotation. What is the maximum value the wave form will reach?

SOLUTION

$$E_{(MAX)} = \frac{E_{(INST)}}{SIN\angle}$$

$$E_{(MAX)} = \frac{246}{0.799}$$

$$E_{(MAX)} = 307.88 \text{ V}$$

Example 3

A sine wave has a maximum voltage of 350 V. At what angle of rotation will the voltage reach 53 V?

SOLUTION

$$SIN\angle = \frac{E_{(INST)}}{E_{(MAX)}}$$

$$SIN\angle = \frac{53}{350}$$

$$SIN\angle = 0.51$$

Note: 0.151 is the *sine* of the angle, not the angle. To find the angle that corresponds to a sine of 0.151, use the trigonometric tables located in Appendices A and B or a scientific calculator.

$$\angle = 8.71°$$

Sine Wave Values

Several measurements of voltage and current are associated with sine waves. These measurements are peak-to-peak, peak, RMS, and average. A sine wave showing peak-to-peak, peak, and RMS measurements is shown in Figure 11-9.

Peak-to-Peak and Peak Values

The peak-to-peak value is measured from the maximum value in the positive direction to the maximum value in the negative direction. The peak-to-peak value is often the simplest measurement to make when using an oscilloscope.

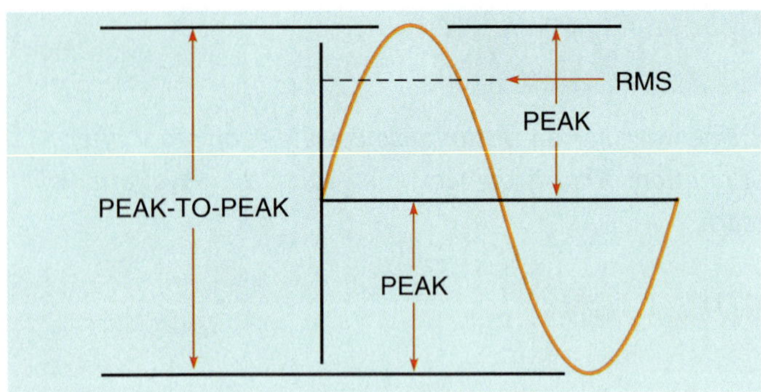

Figure 11-9 **Sine wave values.**

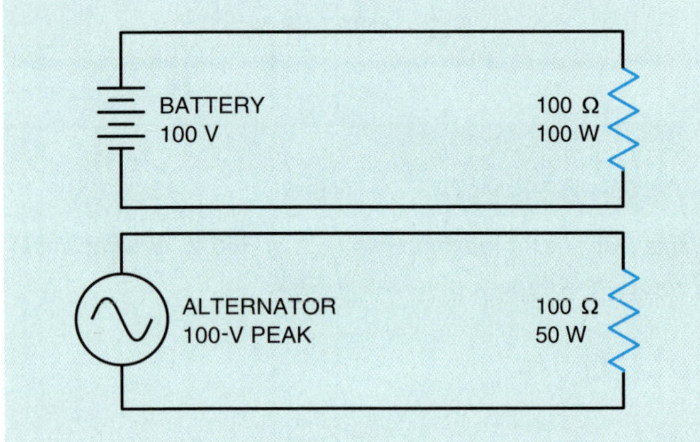

Figure 11-10 **Direct current compared with a sine wave AC current.**

The **peak** value or **amplitude** is measured from zero to the highest value obtained in either the positive or negative direction. The peak value is one-half of the peak-to-peak value.

RMS Values

In Figure 11-10, a 100-V battery is connected to a 100-Ω resistor. This connection will produce 1 A of current flow, and the resistor will dissipate 100 W of power in the form of heat. An AC alternator that produces a peak voltage of 100 V is also shown connected to a 100-Ω resistor. A peak current of 1 A will flow in the circuit, but the resistor will dissipate only 50 W in the form of heat. The reason is that the voltage produced by a pure source of direct current, such as a battery, is one continuous value (Fig. 11-11). The AC sine wave, however, begins at zero, increases to the maximum value, and decreases back to zero during an equal period of time. Since the sine wave has a value of 100 V for only a short period of time and is less than 100 V during the rest of the half cycle, it cannot produce as much power as 100 V of DC.

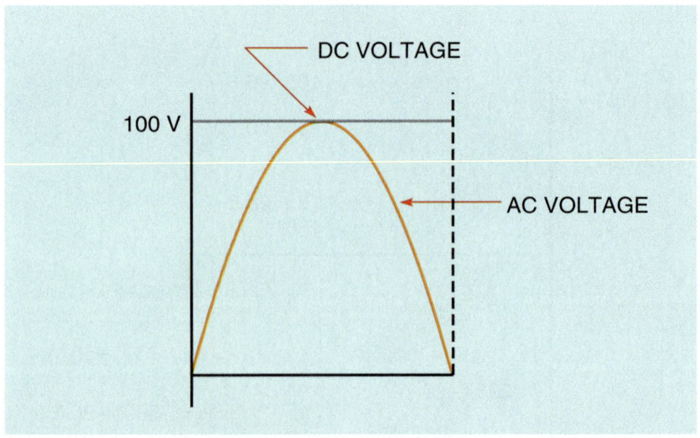

Figure 11-11 **The DC voltage remains at a constant value during a half cycle of AC voltage.**

The solution to this problem is to use a value of AC voltage that will produce the same amount of power as a like value of DC voltage. This AC value is called the **RMS,** or **effective,** value. It is the value indicated by almost all AC voltmeters and ammeters. **RMS stands for root-mean-square, which is an abbreviation for the square root of the mean of the square of the instantaneous currents.** The RMS value can be found by dividing the peak value by the square root of 2 (1.414) or by multiplying the peak value by 0.707 (the reciprocal of 1.414). The formulas for determining the RMS and peak values are

$$\text{RMS} = \text{peak} \times 0.707$$

$$\text{peak} = \text{RMS} \times 1.414$$

Example 4

A sine wave has a peak value of 354 V. What is the RMS value?

SOLUTION

$$\text{RMS} = \text{peak} \times 0.707$$

$$\text{RMS} = 354 \times 0.707$$

$$\text{RMS} = 250.3 \text{ V}$$

Example 5

An AC voltage has a value of 120 V RMS. What is the peak value of voltage?

SOLUTION

$$\text{peak} = \text{RMS} \times 1.414$$

$$\text{peak} = 120 \times 1.414$$

$$\text{peak} = 169.7 \text{ V}$$

When the RMS value of voltage and current is used, it will produce the same amount of power as a like value of DC voltage or current. If 100 V RMS is applied to a 100-Ω resistor, the resistor will produce 100 W of heat. AC voltmeters and ammeters indicate the RMS value, not the peak value. Oscilloscopes, however, display the peak-to-peak value of voltage. All values of AC voltage and current used from now on in this text will be RMS values unless otherwise stated.

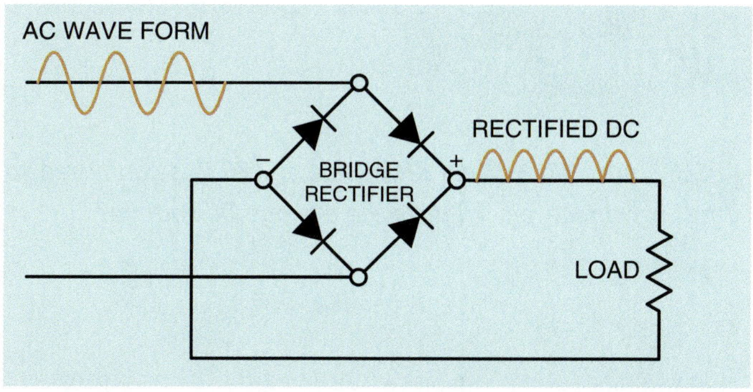

Average Values

Average values of voltage and current are actually direct current values. The average value must be found when a sine wave AC voltage is changed into DC with a rectifier (Fig. 11-12). The rectifier shown is a bridge type rectifier that produces full wave rectification. This means that both the positive and negative half of the AC wave form are changed into DC. The average value is the amount of voltage that would be indicated by a DC voltmeter if it were connected across the load resistor. The average voltage is proportional to the peak, or maximum, value of the wave form, and to the length of time it is on as compared with the length of time it is off (Fig. 11-13). Notice in Figure 11-13 that the voltage wave form turns on and off, but it never changes polarity. The current, therefore, never reverses direction. This is called pulsating direct current. The pulses are often referred to as **ripple.** The average value of voltage will produce the same amount of power as a nonpulsating source of voltage such as a battery (Fig. 11-14). For a sine wave, the average value of voltage is found by multiplying the peak value by 0.637 or by multiplying the RMS value by 0.9.

Figure 11-12 The bridge rectifier changes AC voltage into DC voltage.

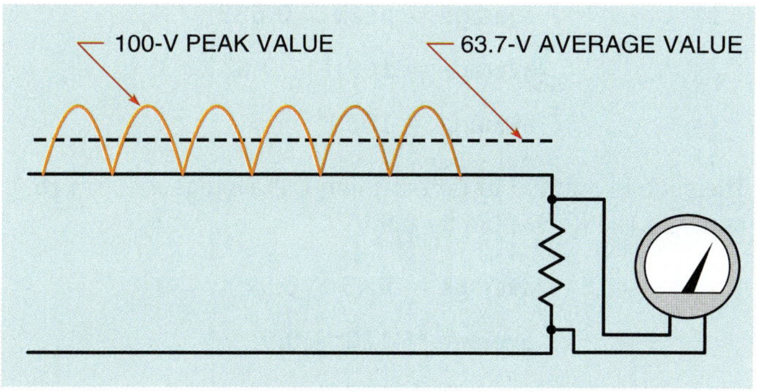

Figure 11-13 A DC voltmeter indicates the average value.

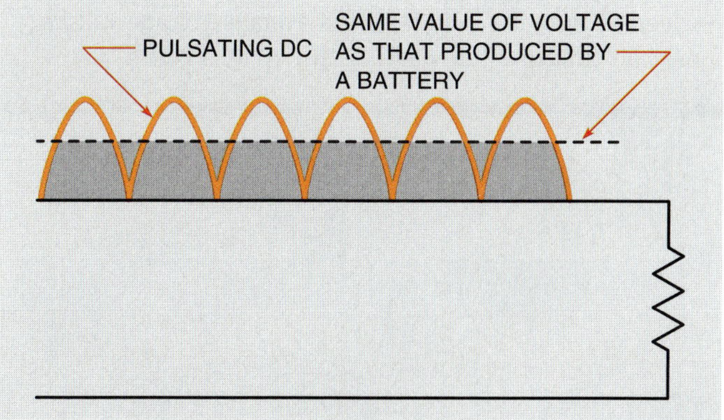

Figure 11-14 The average value produces the same amount of power as a nonpulsating source of voltage.

Example 6

An AC sine wave with an RMS value of 120 V is connected to a full wave rectifier. What is the average DC voltage?

SOLUTION

The problem can be solved in one of two ways. The RMS value can be changed into peak and then the peak value can be changed to the average value.

peak = RMS × 1.414

peak = 120 × 1.414

peak = 169.7 V

average = peak × 0.637

average = 169.7 × 0.637

average = 108 V

The second method of determining the average value is to multiply the RMS value by 0.9.

average = RMS × 0.9

average = 120 × 0.9

average = 108 V

The conversion factors given are for full wave rectification. If a half-wave rectifier is used (Fig. 11-15), only one-half of the AC wave form is converted into DC. To determine the average voltage for a half-wave rectifier, multiply the peak value by 0.637 or the RMS value by 0.9 and then divide the product by 2. Since only half of the AC wave form has been converted into direct current, the average voltage will be only half that of a full-wave rectifier (Fig. 11-16).

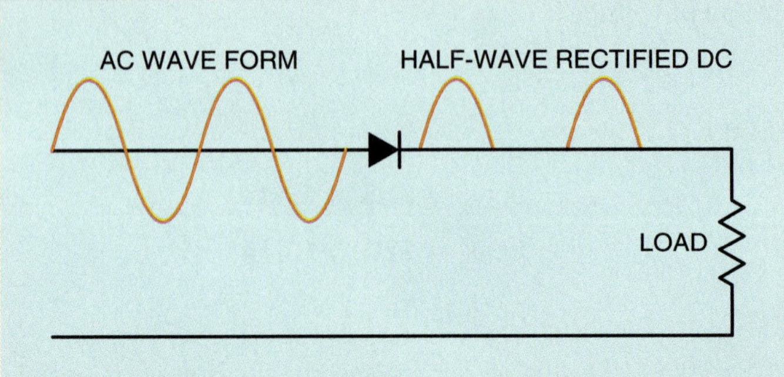

Figure 11-15 A half-wave rectifier converts only one-half of the AC wave form into DC.

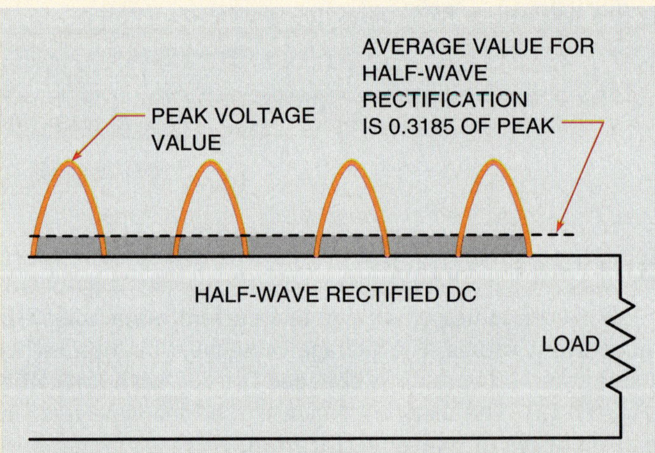

Figure 11-16 The average value for a half-wave rectifier is only half that of a full wave.

Example 7

A half-wave rectifier is connected to 277 V AC. What is the average DC voltage?

SOLUTION

$$\text{average} = \text{RMS} \times \frac{0.9}{2}$$

$$\text{average} = 277 \times \frac{0.9}{2}$$

$$\text{average} = 124.6 \text{ V}$$

Resistive Loads

In direct current circuits, there is only one basic type of load, which is resistive. Even motor loads appear to be resistive because there is a conversion of electrical energy into mechanical energy. In this type of load, the true power, or **watts,** is the product of the volts times the amps. In alternating current circuits, the type of load can vary depending on several factors. Alternating current loads are generally described as being resistive, inductive, or capacitive depending on the phase-angle relationship of voltage and current and the amount of true power produced by the circuit. Inductive and capacitive loads will be discussed in later chapters. **Resistive loads** contain pure resistance, such as electric heating equipment and incandescent lighting. Resistive loads are characterized by the following:

1. They produce heat.
2. The current and voltage are in phase with each other.

Any time a circuit contains resistance, electrical energy will be changed into heat.

When an AC voltage is applied to a resistor, the current flow through the resistor will be a copy of the voltage (Fig. 11-17). The current will rise and fall at the same rate as the voltage and will reverse the direction of flow when the voltage reverses polarity. In this condition, the current is said to be **in phase** with the voltage.

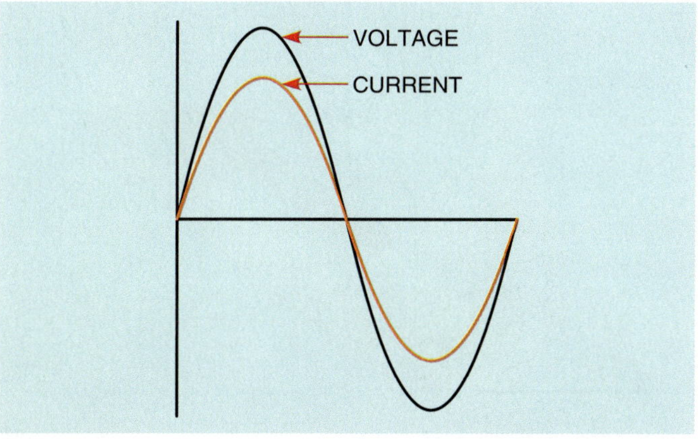

Figure 11-17 In a pure resistive circuit the voltage and current are in phase with each other.

Power in an AC Circuit

True power, or watts, can be produced only when both current and voltage are either positive or negative. When like signs are multiplied, the product is positive (+ × + = +, or − × − = +), and when unlike signs are multiplied the product is negative (+ × − = −). Since the current and voltage are either positive or negative at the same time, the product, watts, will always be positive (Fig. 11-18).

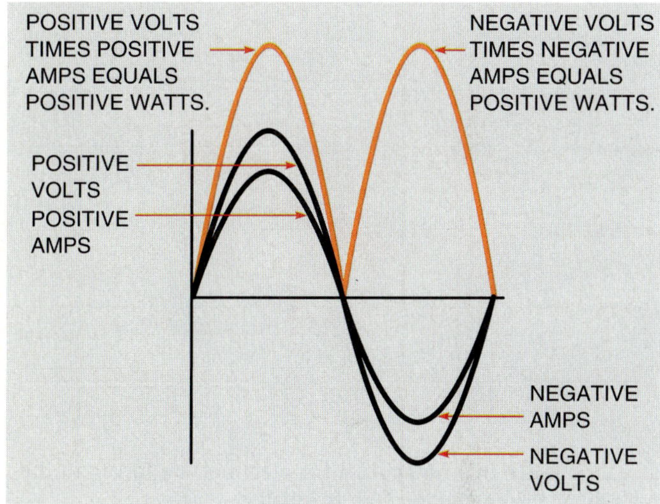

Figure 11-18 Power in a pure resistive AC circuit.

Skin Effect in AC Circuits

When current flows through a conductor connected to a source of direct current, the electrons flow through the entire conductor (Fig. 11-19). The conductor offers some amount of ohmic resistance to the flow of electrons depending on the type of material from which the conductor is made, its length, and its diameter. If that same conductor were connected to a source of alternating current, the resistance of the conductor to the flow of current would be slightly higher because of the **skin effect.** The alternating current induces eddy currents into the conductor. These eddy currents cause the electrons to be repelled toward the outer surface of the conductor (Fig. 11-20). This phenomenon is called the skin effect. Forcing the electrons toward the outer surface of the conductor has the same effect as decreasing the diameter of the conductor, which increases conductor resistance. The skin effect is proportional to the frequency. As the frequency increases,

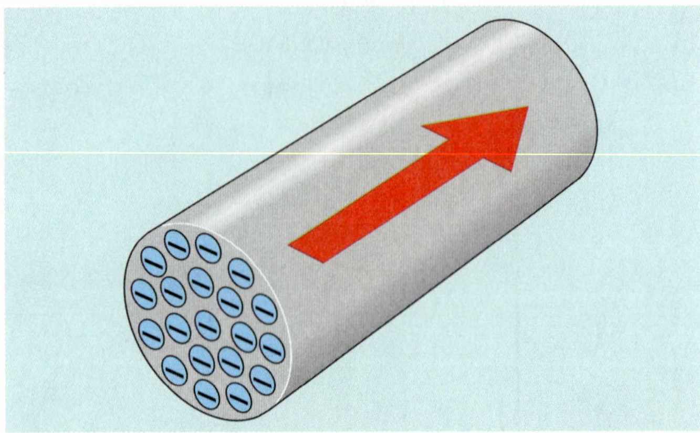

Figure 11-19 In a DC circuit, the electrons travel through the entire conductor.

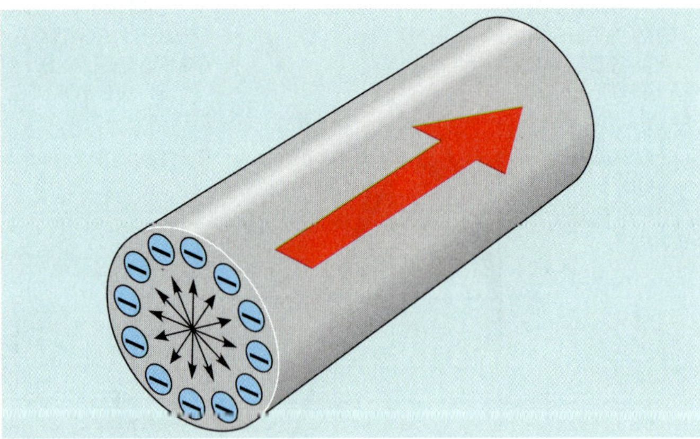

Figure 11-20 In an AC circuit the electrons are forced to the outside of the conductor. This is called skin effect.

the skin effect increases. At low frequencies, such as 60 Hz, the skin effect has very little effect on the resistance of a conductor and generally is not taken into consideration when computing the size wire needed for a particular circuit. At high frequencies, however, the skin effect can have a great effect on the operation of the circuit.

To help overcome the problem of skin effect in high-frequency circuits, conductors with a large amount of surface area must be used. Grounding a piece of equipment operating at 60 Hz, for example, may be as simple as using a grounding rod and a piece of #6 AWG copper conductor. Grounding a piece of equipment operating at 20 MHz, however, may require the use of wide copper tape or a wide, flat, braided cable. Braided cable is less affected by skin effect because it contains many small conductors, which provides a large amount of surface area.

Summary

- Most of the electrical power generated in the world is alternating current.
- Alternating current can be transformed and direct current cannot.
- Alternating current reverses its direction of flow at periodic intervals.
- The most common AC wave form is the sine wave.
- There are 360° in one complete sine wave.
- One complete wave form is called a cycle.
- The number of complete cycles that occur in one second is called the frequency.
- Sine waves are produced by rotating machines.
- Frequency is measured in hertz (Hz).
- The instantaneous voltage at any point on a sine wave is equal to the peak, or maximum, voltage times the sine of the angle of rotation.
- The peak-to-peak voltage is the amount of voltage measured from the positive-most peak to the negative-most peak.
- The peak value is the maximum amount of voltage attained by the wave form.
- The RMS value of voltage will produce as much power as a like amount of DC voltage.
- The average value of voltage is used when an AC sine wave is changed into DC.
- The current and voltage in a pure resistive circuit are in phase with each other.
- True power, or watts, can be produced only when current and voltage are both positive or both negative.
- Resistance in AC circuits is characterized by the fact that the resistive part will produce heat.
- There are three basic types of AC loads: resistive, inductive, and capacitive.
- The electrons in an alternating current circuit are forced toward the outside of the conductor by eddy current induction in the conductor itself.
- Skin effect is proportional to frequency.
- Skin effect can be reduced by using conductors with a large surface area.

Review Questions

1. What is the most common type of AC wave form?
2. How many degrees are there in one complete sine wave?
3. At what angle does the voltage reach its maximum negative value on a sine wave?
4. What is frequency?
5. A sine wave has a maximum value of 230 V. What is the voltage after 38° of rotation?
6. A sine wave has a voltage of 63 V after 22° of rotation. What is the maximum voltage reached by this wave form?
7. A sine wave has a maximum value of 560 V. At what angle of rotation will the voltage reach a value of 123 V?
8. A sine wave has a peak value of 433 V. What is the RMS value?
9. A sine wave has a peak-to-peak value of 88 V. What is the average value?
10. A DC voltage has an average value of 68 V. What is the RMS value?

Practice Problems

Refer to the alternating current formulas in the appendix to answer the following questions.

Sine Wave Values

Fill in all the missing values.

Peak Volts	Inst. Volts	Degrees
347	208	
780		43.5
	24.3	17.6
224	5.65	
48.7		64.6
	240	45
87.2	23.7	
156.9		82.3
	62.7	34.6
1256	400	
15,720		12
	72.4	34.8

$$E_{(INST)} = E_{(MAX)} \times SIN\angle$$

$$E_{(MAX)} = \frac{E_{(INST)}}{SIN\angle}$$

$$SIN\angle = \frac{E_{(INST)}}{E_{(MAX)}}$$

Peak, RMS, and Average Values

Fill in all the missing values.

Peak	RMS	Average
12.7		
	53.8	
		164.2
1235		
	240	
		16.6
339.7		
	12.6	
		9
123.7		
	74.8	
		108

Chapter 12 Inductance in Alternating Current Circuits

This unit discusses the effects of inductance on alternating current circuits. The unit explains how current is limited in an inductive circuit as well as the effect inductance has on the relationship of voltage and current.

OBJECTIVES

After studying this unit, you should be able to:

- discuss the properties of inductance in an alternating current circuit.
- discuss inductive reactance.
- compute values of inductive reactance and inductance.
- discuss the relationship of voltage and current in a pure inductive circuit.
- be able to compute values for inductors connected in series or parallel.
- discuss reactive power (VARs).
- determine the Q of a coil.

Glossary of Inductance in Alternating Current Circuits Terms

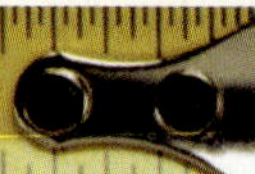

current lags voltage the relationship of current and voltage in a pure inductive circuit

impedance (Z) the total current-limiting effect in an AC circuit

induced voltage voltage that is supplied to a conductor by cutting lines of magnetic flux

inductance (L) the property of an electric circuit whereby voltage is produced by cutting magnetic flux lines

inductive reactance (X_L) the current-limiting effect of a pure inductor

quality (Q) the ratio of inductive reactance as compared with resistance

reactance the property of a circuit that limits current by means other than resistance

reactive power (VARs) Volt-Amps Reactive; often referred to as wattless power

Inductance

Inductance (L) is one of the primary types of loads in alternating current circuits. Some amount of inductance is present in all alternating current circuits because of the continually changing magnetic field (Fig. 12-1). The amount of inductance of a single conductor is extremely small, and in most instances it is not considered in circuit calculations. Circuits are generally considered to contain inductance when any type of load that contains a coil is used. For circuits that contain a coil, inductance *is* considered in circuit calculations. Loads such as motors, transformers, lighting ballast, and chokes all contain coils of wire.

In chapter 10, it was discussed that whenever current flows through a coil of wire a magnetic field is created around the wire (Fig. 12-2). If the amount of current decreases, the magnetic field will collapse (Fig. 12-3). Recall from chapter 10 several facts concerning inductance:

1. When magnetic lines of flux cut through a coil, a voltage is induced in the coil.
2. An induced voltage is always opposite in polarity to the applied voltage. This is often referred to as counter EMF (CEMF).
3. The amount of induced voltage is proportional to the rate of change of current.
4. An inductor opposes a change of current.

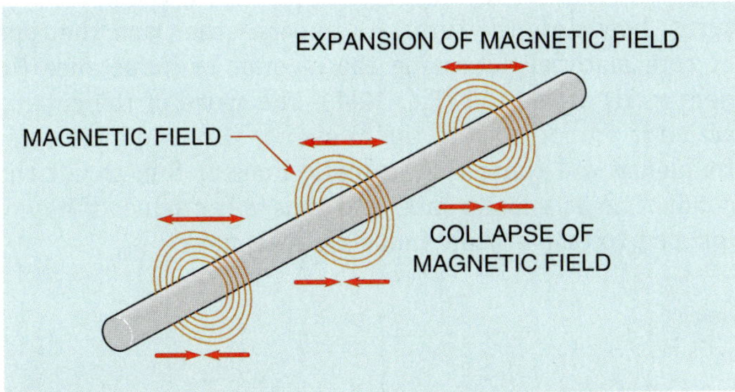

Figure 12-1 A continually changing magnetic field induces a voltage into any conductor.

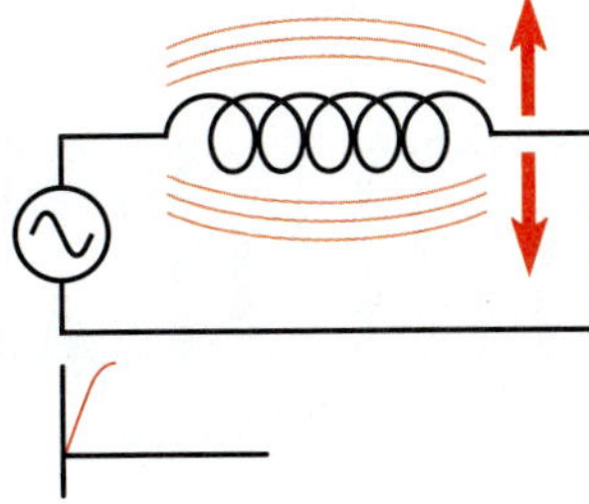

Figure 12-2 As current flows through a coil, a magnetic field is created around the coil.

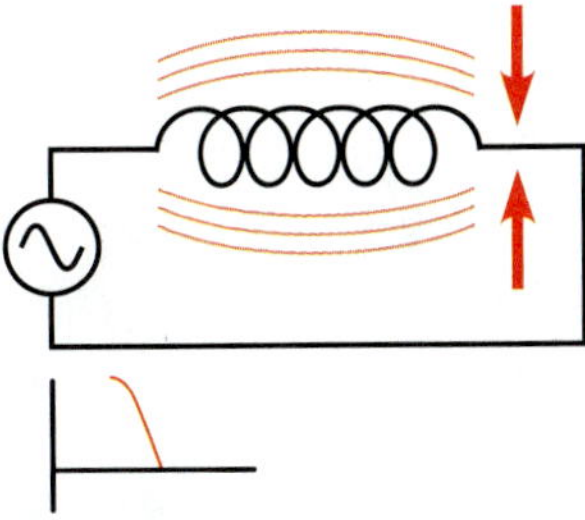

Figure 12-3 As current flow decreases, the magnetic field collapses.

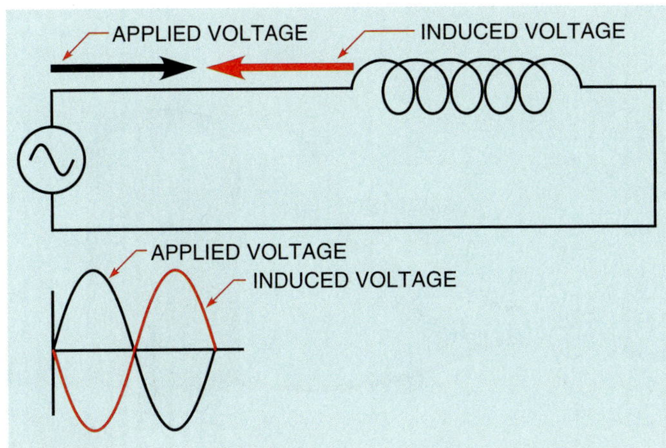

Figure 12-4 The applied voltage and induced voltage are 180° out of phase with each other.

The inductors in Figures 12-2 and 12-3 are connected to an alternating voltage. Therefore the magnetic field continually increases, decreases, and reverses polarity. Since the magnetic field continually changes magnitude and direction, a voltage is continually being induced in the coil. This **induced voltage** is 180° out of phase with the applied voltage and is always in opposition to the applied voltage (Fig. 12-4). Since the induced voltage is always in opposition to the applied voltage, the applied voltage must overcome the induced voltage before current can flow through the circuit. For example, assume an inductor is connected to a 120-V AC line. Now assume that the inductor has an induced voltage of 116 V. Since an equal amount of applied voltage must be used to overcome the induced voltage, there will be only 4 V to push current through the wire resistance of the coil (120 − 116 = 4).

Computing the Induced Voltage

The amount of induced voltage in an inductor can be computed if the resistance of the wire in the coil and the amount of circuit current are known. For example, assume that an ohmmeter is used to measure the actual amount of resistance

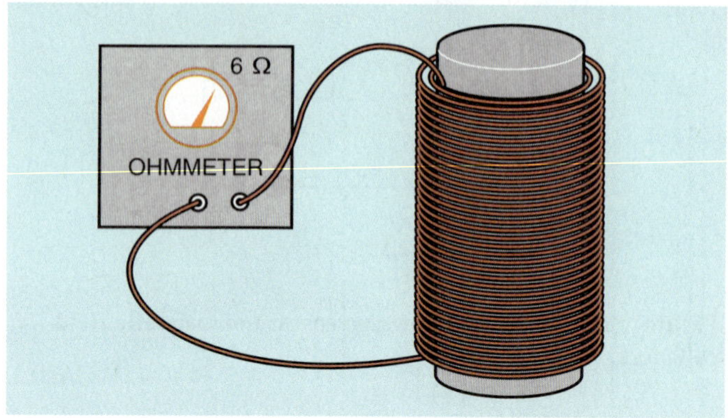

Figure 12-5 Measuring the resistance of a coil.

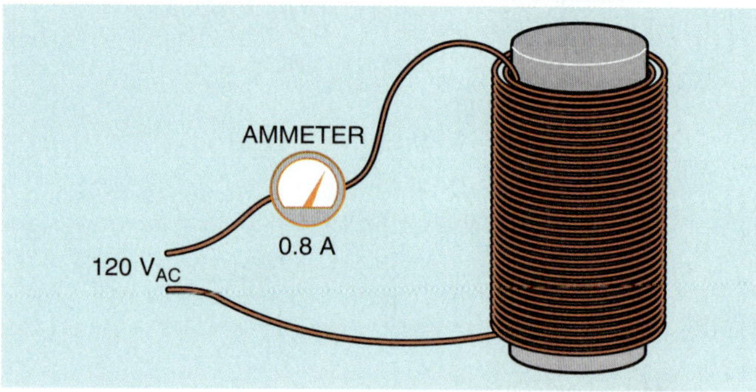

Figure 12-6 Measuring circuit current with an ammeter.

in a coil, and the coil is found to contain 6 Ω of wire resistance (Fig. 12-5). Now assume that the coil is connected to a 120-V AC circuit and an ammeter measures a current flow of 0.8 A (Fig. 12-6). Ohm's law can now be used to determine the amount of voltage necessary to push 0.8 A of current through 6 Ω of resistance.

$$E = I \times R$$

$$E = 0.8 \times 6$$

$$E = 4.8\ V$$

Since only 4.8 V is needed to push the current through the wire resistance of the inductor, the remainder of the 120 V is used to overcome the coil's induced voltage of 119.9 V ($\sqrt{120^2 - 4.8^2} = 119.9V$). Refer to vectors in Chapter 13.

Inductive Reactance

Notice that the induced voltage is able to limit the flow of current through the circuit in a manner similar to resistance. This induced voltage is *not* resistance, but it can limit the flow of current just as resistance does. This current-limiting property of the inductor is called **reactance** and is symbolized by the letter *X*. This reactance is caused by inductance, so it is called **inductive reactance** and is symbolized by $\mathbf{X_L}$, pronounced "X sub L." Inductive reactance is measured in ohms just as resistance is and can be computed when the values of inductance and frequency are known. The following formula can be used to find inductive reactance.

$$X_L = 2\pi fL$$

where

X_L = inductive reactance

2 = a constant

π = 3.1416

F = frequency in hertz (Hz)

L = inductance in henrys (H)

Inductive reactance is an induced voltage and is, therefore, proportional to the three factors that determine induced voltage:

1. The **number** of turns of wire
2. The **strength** of the magnetic field
3. The **speed** of the cutting action (relative motion between the inductor and the magnetic lines of flux)

The number of turns of wire and strength of the magnetic field are determined by the physical construction of the inductor. Factors such as the size of wire used, the number of turns, how close the turns are to each other, and the type of core material determine the amount of inductance (in henrys, H) of the coil (Fig. 12-7). The speed of the cutting action is proportional to the frequency (Hz). An increase of frequency will cause the magnetic lines of flux to cut the conductors at a faster rate, and thus will produce a higher induced voltage or more inductive reactance.

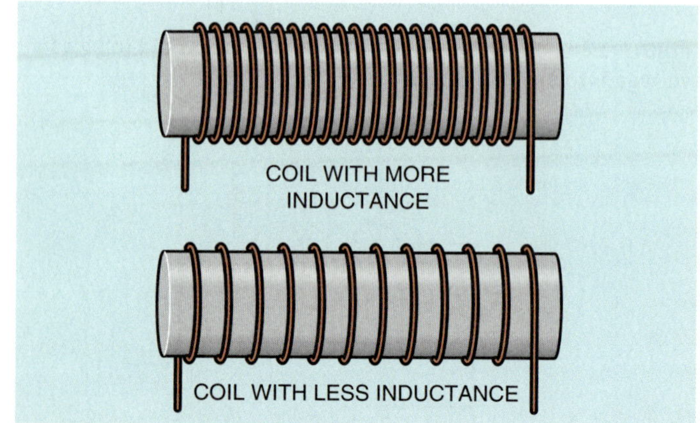

Figure 12-7 Coils with turns closer together produce more inductance than coils with turns far apart.

Example 1

The inductor shown in Figure 12-8 has an inductance of 0.8 H and is connected to a 120-V, 60-Hz line. How much current will flow in this circuit if the wire resistance of the inductor is negligible?

SOLUTION

The first step is to determine the amount of inductive reactance of the inductor.

$$X_L = 2\pi FL$$

$$X_L = 2 \times 3.1416 \times 60 \times 0.8$$

$$X_L = 301.6\ \Omega$$

Since inductive reactance is the current-limiting property of this circuit, it can be substituted for the value of R in an Ohm's law formula.

$$I = \frac{E}{X_L}$$

$$I = \frac{120}{301.6}$$

$$I = 0.398\ A$$

If the amount of inductive reactance is known, the inductance of the coil can be determined using the formula

$$L = \frac{X_L}{2\pi F}$$

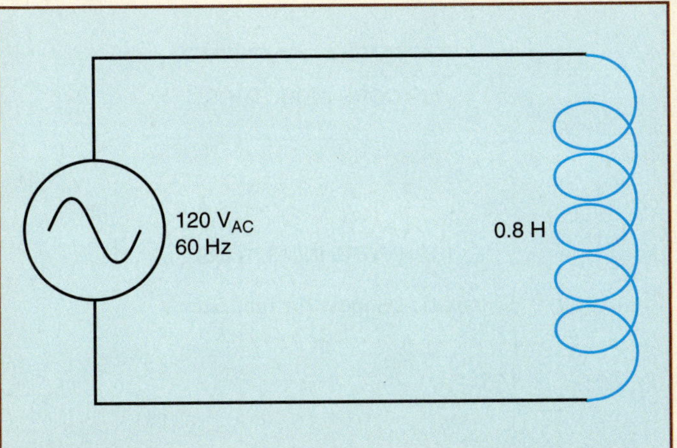

Figure 12-8 Circuit current is limited by inductive reactance.

Example 2

Assume an inductor with a negligible resistance is connected to a 36-V, 400-Hz line. If the circuit has a current flow of 0.2 A, what is the inductance of the inductor?

SOLUTION

The first step is to determine the inductive reactance of the circuit.

$$X_L = \frac{E}{I}$$

$$X_L = \frac{36}{0.2}$$

$$X_L = 180\ \Omega$$

Now that the inductive reactance of the inductor is known, the inductance can be determined.

$$L = \frac{X_L}{2\pi F}$$

$$L = \frac{180}{2 \times 3.1416 \times 400}$$

$$L = 0.0716\ H$$

Example 3

An inductor with negligible resistance is connected to a 480-V, 60-Hz line. An ammeter indicates a current flow of 24 A. How much current will flow in this circuit if the frequency is increased to 400 Hz?

SOLUTION

The first step in solving this problem is to determine the amount of inductance of the coil. Since the resistance of the wire used to make the inductor is negligible, the current is limited by inductive reactance. The inductive reactance can be found by substituting X_L for R in an Ohm's law formula.

$$X_L = \frac{E}{I}$$

$$X_L = \frac{480}{24}$$

$$X_L = 20\ \Omega$$

Now that the inductive reactance is known, the inductance of the coil can be found using the formula

$$L = \frac{X_L}{2\pi F}$$

NOTE: When using a frequency of 60 Hz, $2 \times \pi \times 60 = 377$. Since 60 Hz is the major frequency used throughout the United States and Canada, 377 should be memorized for use when necessary.

$$L = \frac{20}{377}$$

$$L = 0.053\ \text{H}$$

Since the inductance of the coil is determined by its physical construction, it will not change when connected to a different frequency. Now that the inductance of the coil is known, the inductive reactance at 400 Hz can be computed.

$$X_L = 2\pi FL$$

$$X_L = 2 \times 3.1416 \times 400 \times 0.053$$

$$X_L = 133.2\ \Omega$$

The amount of current flow can now be found by substituting the value of inductive reactance for resistance in an Ohm's law formula.

$$I = \frac{E}{X_L}$$

$$I = \frac{480}{133.2}$$

$$I = 3.6\ \text{A}$$

Schematic Symbols

The schematic symbol used to represent an inductor depicts a coil of wire. Several symbols for inductors are shown in Figure 12-9. The symbols shown with the two parallel lines represent iron core inductors, and the symbols without the parallel lines represent air core inductors.

Figure 12-9 Schematic symbols for inductors.

Inductors Connected in Series

When inductors are connected in series (Fig. 12-10), the total inductance of the circuit (L_T) equals the sum of the inductances of all the inductors.

$$L_T = L_1 + L_2 + L_3$$

The total inductive reactance (X_{LT}) of inductors connected in series equals the sum of the inductive reactances for all the inductors.

$$X_{LT} = X_{L1} + X_{L2} + X_{L3}$$

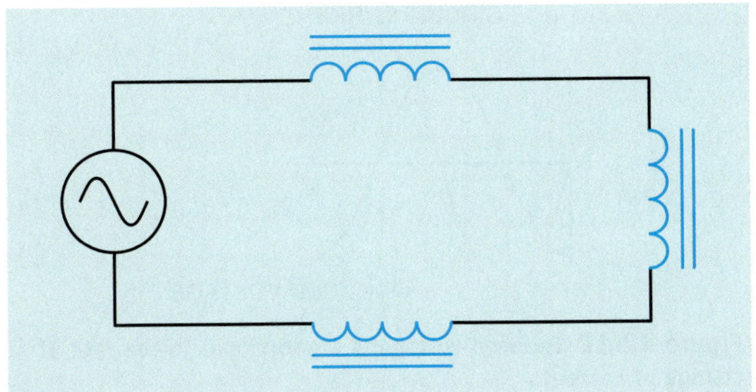

Figure 12-10 Inductors connected in series.

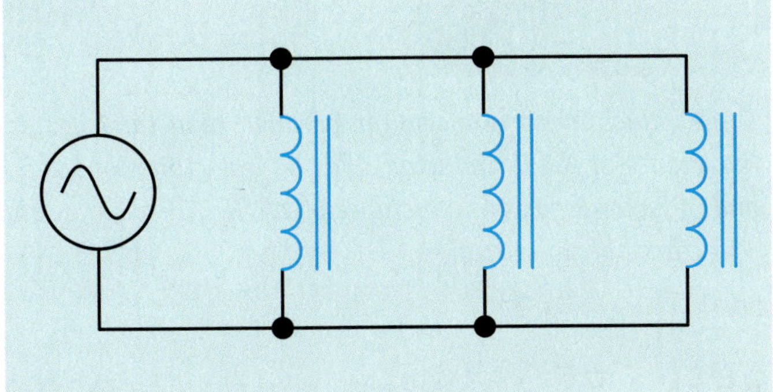

Figure 12-11 Inductors connected in parallel.

Three inductors are connected in series. Inductor 1 has an inductance of 0.6 H, inductor 2 has an inductance of 0.4 H, and inductor 3 has an inductance of 0.5 H. What is the total inductance of the circuit?

SOLUTION

$$L_T = 0.6 + 0.4 + 0.5$$

$$L_T = 1.5\ \text{H}$$

Example 5

Three inductors are connected in series. Inductor 1 has an inductive reactance of 180 Ω, inductor 2 has an inductive reactance of 240 Ω, and inductor 3 has an inductive reactance of 320 Ω. What is the total inductive reactance of the circuit?

SOLUTION

$$X_{LT} = 180\ \Omega + 240\ \Omega + 320\ \Omega$$

$$X_{LT} = 740\ \Omega$$

Inductors Connected in Parallel

When inductors are connected in parallel (Fig. 12-11), the total inductance can be found in a similar manner to finding the total resistance of a parallel circuit. The reciprocal of the total inductance is equal to the sum of the reciprocals of all the inductors.

$$\frac{1}{L_T} = \frac{1}{L_1} + \frac{1}{L_2} + \frac{1}{L_3}$$

or

$$L_T = \frac{1}{\frac{1}{L_1} + \frac{1}{L_2} + \frac{1}{L_3}}$$

The product over sum formula can be used also to find the total inductance of parallel inductors.

$$L_T = \frac{L_1 \times L_2}{L_1 \times L_2}$$

If the values of all the inductors are the same, total inductance can be found by dividing the inductance of one inductor by the total number of inductors.

$$L_T = \frac{L}{N}$$

Similar formulas can be used to find the total inductive reactance of inductors connected in parallel.

$$\frac{1}{X_{LT}} = \frac{1}{X_{L1}} + \frac{1}{X_{L2}} + \frac{1}{X_{L3}}$$

or

$$X_{LT} = \frac{1}{\frac{1}{X_{L1}} + \frac{1}{X_{L2}} + \frac{1}{X_{L3}}}$$

or

$$X_{LT} = \frac{X_{L1} \times X_{L2}}{X_{L1} \times X_{L2}}$$

or

$$X_{LT} = \frac{X_L}{N}$$

Example 6

Three inductors are connected in parallel. Inductor 1 has an inductance of 2.5 H, inductor 2 has an inductance of 1.8 H, and inductor 3 has an inductance of 1.2 H. What is the total inductance of this circuit?

SOLUTION

$$L_T = \frac{1}{\frac{1}{2.5} + \frac{1}{1.8} + \frac{1}{1.2}}$$

$$L_T = \frac{1}{1.789}$$

$$L_T = 0.559 \text{ H}$$

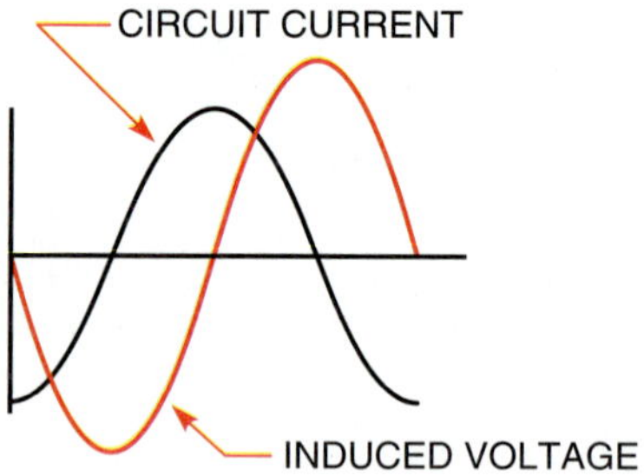

Figure 12-12 Induced voltage is proportional to the rate of change of current.

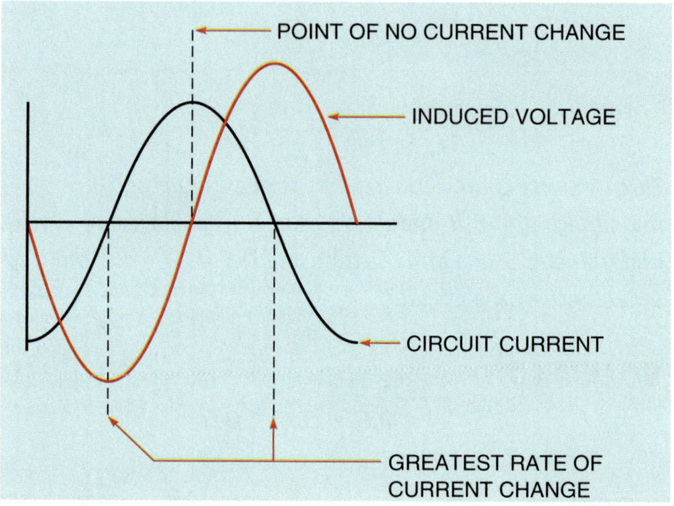

Figure 12-13 No voltage is induced when the current does not change.

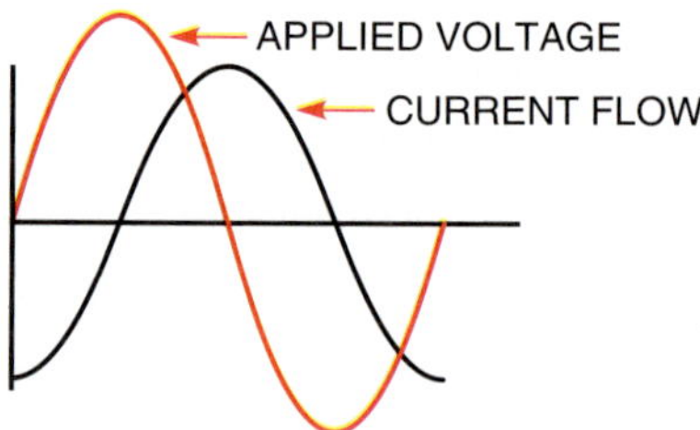

Figure 12-14 The current lags the applied voltage by 90°.

Voltage and Current Relationships in an Inductive Circuit

In Chapter 11, it was discussed that when current flows through a pure resistive circuit, the current and voltage are in phase with each other. **In a pure inductive circuit the current lags the voltage by 90°.** At first this may seem to be an impossible condition until the relationship of applied voltage and induced voltage is considered. How the current and applied voltage can become 90° out of phase with each other can best be explained by comparing the relationship of the current and induced voltage (Fig. 12-12). Recall that the induced voltage is proportional to the rate of change of the current (speed of cutting action). At the beginning of the wave form, the current is shown at its maximum value in the negative direction. At this time, the current is not changing, so induced voltage is zero. As the current begins to decrease in value, the magnetic field produced by the flow of current decreases or collapses and begins to induce a voltage into the coil as it cuts through the conductors (Fig. 12-13).

The greatest rate of current change occurs when the current passes from negative, through zero, and begins to increase in the positive direction (Fig. 12-13). Since the current is changing at the greatest rate, the induced voltage is maximum. As current approaches its peak value in the positive direction, the rate of change decreases, causing a decrease in the induced voltage. The induced voltage will again be zero when the current reaches its peak value and the magnetic field stops expanding.

It can be seen that the current flowing through the inductor is leading the induced voltage by 90°. Since the induced voltage is 180° out of phase with the applied voltage, the current will lag the applied voltage by 90° (Fig. 12-14).

Power in an Inductive Circuit

In a pure resistive circuit, the true power, or watts, is equal to the product of the voltage and current. In a pure inductive circuit, however, no true power, or watts, is pro-

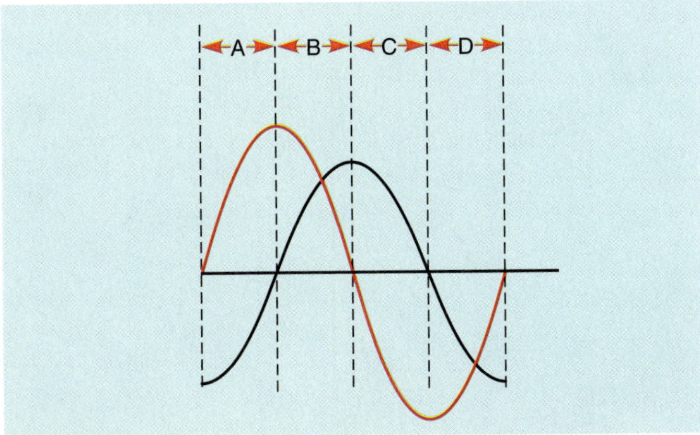

Figure 12-15 Voltage and current relationships during different parts of a cycle.

duced. Recall that voltage and current must both be either positive or negative before true power can be produced. Since the voltage and current are 90° out of phase with each other in a pure inductive circuit, the current and voltage will be at different polarities 50% of the time and at the same polarity 50% of the time. During the period of time that the current and voltage have the same polarity, power is being given to the circuit in the form of creating a magnetic field. When the current and voltage are opposite in polarity, power is being given back to the circuit as the magnetic field collapses and induces a voltage back into the circuit. Since power is stored in the form of a magnetic field and then given back, no power is used by the inductor. Any power used in an inductor is caused by losses such as the resistance of the wire used to construct the inductor, generally referred to as I^2R losses, eddy current losses, and hysteresis losses.

The current and voltage wave form in Figure 12-15 has been divided into four sections: A, B, C, and D. During the first time period, indicated by A, the current is negative and the voltage is positive. During this period, energy is being given to the circuit as the magnetic field collapses. During the second time period, section B, both the voltage and current are positive. Power is being used to produce the magnetic field. In the third time period, C, the current is positive and the voltage is negative. Power is again being given back to the circuit as the field collapses. During the fourth time period, D, both the voltage and current are negative. Power is again being used to produce the magnetic field. If the amount of power used to produce the magnetic field is subtracted from the power given back, the result will be zero.

Reactive Power

Although essentially no true power is being used, except by previously mentioned losses, an electrical measurement called **VARs** is used to measure the **reactive power** in a pure inductive circuit. VARs is an abbreviation for volt-amps-reactive. VARs can be computed like watts except that inductive values are substituted for resistive values in the formulas. VARs is equal to the amount of current flowing through an inductive circuit times the voltage applied to the inductive part of the circuit. Several formulas for computing VARs are

$$\text{VARs} = E_L \times I_L$$

$$\text{VARs} = \frac{E_L{}^2}{X_L}$$

$$\text{VARs} = I_L{}^2 \times X_L$$

where

E_L = **voltage applied to an inductor**

I_L = **current flow through an inductor**

X_L = **inductive reactance**

Q of an Inductor

So far in this chapter, it has been generally assumed that an inductor has no resistance and that inductive reactance is the only current-limiting factor. In reality, that is not true. Since inductors are actually coils of wire, they all contain some amount of internal resistance. Inductors actually appear to be a coil connected in series with some amount of resistance (Fig. 12-16). The amount of resistance compared with the inductive reactance determines the Q of the coil. The letter ***Q*** stands for **quality.** Inductors that have a higher ratio of inductive reactance to resistance are considered to be inductors of higher quality. An inductor constructed with a large wire will have a low wire resistance and, therefore, a higher Q (Fig. 12-17). Inductors constructed with many turns of small wire have a much higher resistance, and, therefore, a lower Q. To determine the Q of an inductor, divide the inductive reactance by the resistance.

$$Q = \frac{X_L}{R}$$

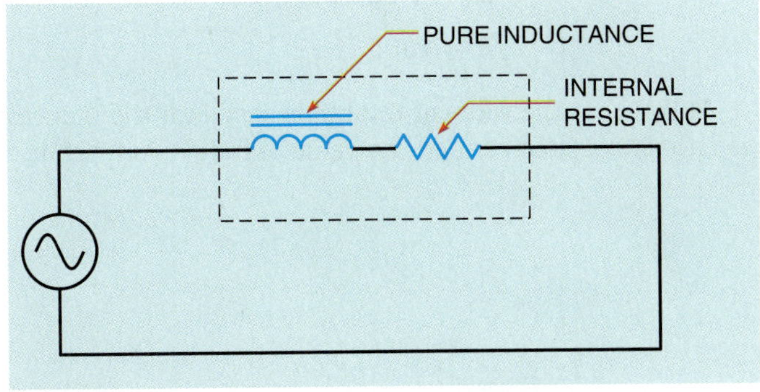

Figure 12-16 Inductors contain internal resistance.

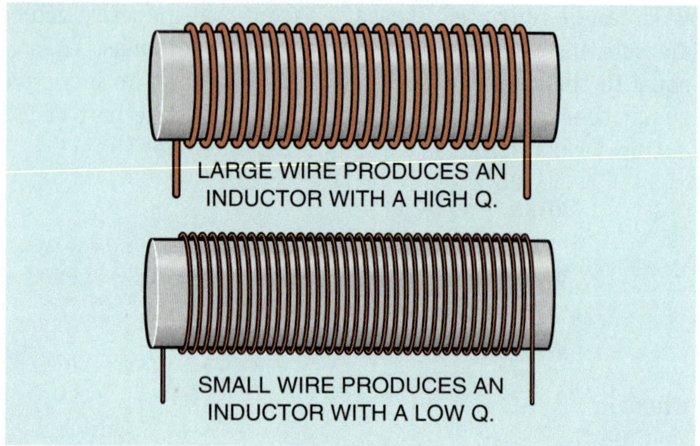

Figure 12-17 **The Q of an inductor is a ratio of inductive reactance as compared with resistance. The letter Q stands for quality.**

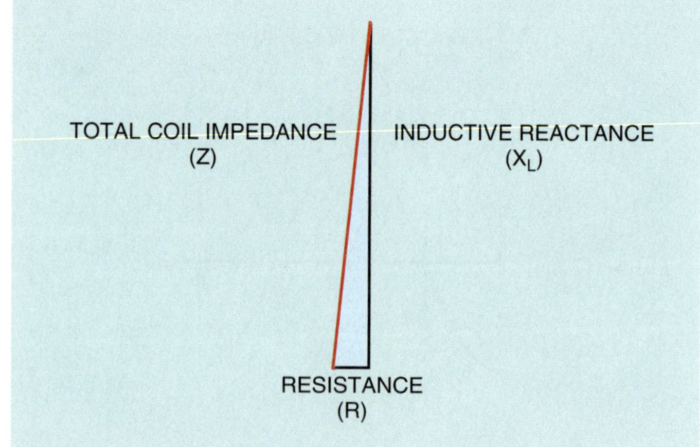

Figure 12-18 **Coil impedance is a combination of wire resistance and inductive reactance.**

Although inductors have some amount of resistance, inductors that have a Q of 10 or greater are generally considered to be pure inductors. Once the ratio of inductive reactance becomes 10 times as great as resistance, the amount of resistance is considered negligible. For example, assume an inductor has an inductive reactance of 100 Ω and a wire resistance of 10 Ω. The inductive reactive component in the circuit is 90° out of phase with the resistive component. This relationship produces a right triangle (Fig. 12-18). The total current-limiting effect of the inductor is a combination of the inductive reactance and resistance. This total current-limiting effect is called **impedance** and is symbolized by the letter ***Z***. The impedance of the circuit is represented by the hypotenuse of the right triangle formed by the inductive reactance and the resistance. To compute the value of impedance for the coil, the inductive reactance and resistance must be added. Since these two components form the legs of a right triangle and the impedance forms the hypotenuse, vector addition must be employed.

$$Z = \sqrt{R^2 + X_{L^2}}$$

$$Z = \sqrt{10^2 + 100^2}$$

$$Z = \sqrt{10{,}000}$$

$$Z = 100.5\Omega$$

Notice that the value of total impedance for the inductor is only 0.5 Ω greater than the value of inductive reactance.

Summary

- Induced voltage is proportional to the rate of change of current.
- Induced voltage is always opposite in polarity to the applied voltage.
- Inductive reactance is a countervoltage that limits the flow of current, as does resistance.
- Inductive reactance is measured in ohms.
- Inductive reactance is proportional to the inductance of the coil and the frequency of the line.
- Inductive reactance is symbolized by X_L.
- Inductance is measured in henrys (H) and is symbolized by the letter ***L***.
- When inductors are connected in series, the total inductance is equal to the sum of all the inductors.
- When inductors are connected in parallel, the reciprocal of the total inductance is equal to the sum of the reciprocals of all the inductors.
- The current lags the applied voltage by 90° in a pure inductive circuit.
- All inductors contain some amount of resistance.
- The Q of an inductor is the ratio of the inductive reactance to the resistance.
- Inductors with a Q of 10 are generally considered to be "pure" inductors.
- Pure inductive circuits contain no true power, or watts.
- Reactive power is measured in VARs.
- VARs is an abbreviation for volt-amps-reactive.

Review Questions

1. How many degrees are the current and voltage out of phase with each other in a pure resistive circuit?
2. How many degrees are the current and voltage out of phase with each other in a pure inductive circuit?
3. To what is inductive reactance proportional?
4. Four inductors, each having an inductance of 0.6 H, are connected in series. What is the total inductance of the circuit?
5. Three inductors are connected in parallel. Inductor 1 has an inductance of 0.06 H; inductor 2 has an inductance of 0.05 H; and inductor 3 has an inductance of 0.1 H. What is the total inductance of this circuit?
6. If the three inductors in question 5 were connected in series, what would be the inductive reactance of the circuit? Assume the inductors are connected to a 60-Hz line.
7. An inductor is connected to a 240-V, 1000-Hz line. The circuit current is 0.6 A. What is the inductance of the inductor?
8. An inductor with an inductance of 3.6 H is connected to a 480-V, 60-Hz line. How much current will flow in this circuit?
9. If the frequency in question 8 is reduced to 50 Hz, how much current will flow in the circuit?
10. An inductor has an inductive reactance of 250 Ω when connected to a 60-Hz line. What will be the inductive reactance if the inductor is connected to a 400-Hz line?

Practice Problems

Inductive Circuits

Fill in all the missing values. Refer to the following formulas:

$$X_L = 2\pi FL$$

$$L = \frac{X_L}{2\pi F}$$

$$F = \frac{X_L}{2\pi L}$$

Inductance (H)	Frequency (Hz)	Induct. Rct. (Ω)
1.2	60	
0.085		213.628
	1000	4712.389
0.65	600	
3.6		678.584
	25	411.459
0.5	60	
0.85		6408.849
	20	201.062
0.45	400	
4.8		2412.743
	1000	40.841

Chapter 13 Resistive-Inductive Series Circuits

This unit will cover the relationship of resistance and inductance used in the same circuit. The resistors and inductors will be connected in series. Concepts such as circuit impedance, power factor, and vector addition will be introduced. Although it is true that some circuits are basically purely resistive or purely inductive, many circuits contain a combination of both resistive and inductive elements.

OBJECTIVES

After studying this unit, you should be able to:

- discuss the relationship of resistance and inductance in an alternating current series circuit.
- define power factor.
- calculate values of voltage, current, apparent power, true power, reactive power, impedance, resistance, inductive reactance, and power factor in an R-L series circuit.
- compute the phase angle for current and voltage in an R-L series circuit.
- connect an R-L series circuit and make measurements using test instruments.
- discuss vectors and be able to plot electrical quantities using vectors.

Glossary of Resistive-Inductive Series Circuit Terms

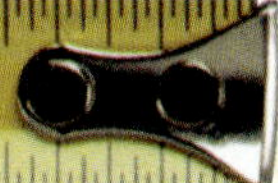

angle theta ($\angle\theta$) the phase angle difference between voltage and current in a circuit containing a reactive component such as an inductor or capacitor

apparent power volt amps (VA) the product of the total applied voltage and the total current flow in an AC circuit

parallelogram method a method of obtaining values of voltage, current, impedance, and power with vectors

power factor (PF) the ratio of true power to apparent power

quadrature power often referred to as VARs, reactive power, or wattless power

total current (I_T) the total current flow in an electric circuit

total impedance the total current-limiting effect in an AC circuit

true power the amount of electrical energy that is converted into some other form of energy such as thermal or kinetic; also known as watts

vector a line that indicates both magnitude and direction

vector addition the process of using vectors to add like electrical values

wattless power VARs; computed by using values that pertain only to reactive components such as inductors or capacitors

RL Series Circuits

When a pure resistive load is connected to an alternating current circuit, the voltage and current are in phase with each other. When a pure inductive load is connected to an alternating current circuit, the voltage and current are 90° out of phase with each other (Fig. 13-1). When a circuit containing both resistance, R, and inductance, L, is connected to an alternating current circuit, the voltage and current will be out of phase with each other by an amount between 0° and 90°. **The exact amount of phase angle difference is determined by the ratio of resistance as compared to inductance.** In the following example, a series circuit containing 30 Ω of resistance (R) and 40 Ω of inductive reactance (X_L) is connected to a 240-V 60-Hz line (Fig. 13-2). It is assumed that the inductor has negligible resistance. The following unknown values will be computed:

Z — total circuit impedance
I — current flow
E_R — voltage drop across the resistor
P — watts (true power)
L — inductance of the inductor
E_L — voltage drop across the inductor
VARs — reactive power
VA — apparent power
PF — power factor
$\angle\theta$ — the angle the voltage and current are out of phase with each other

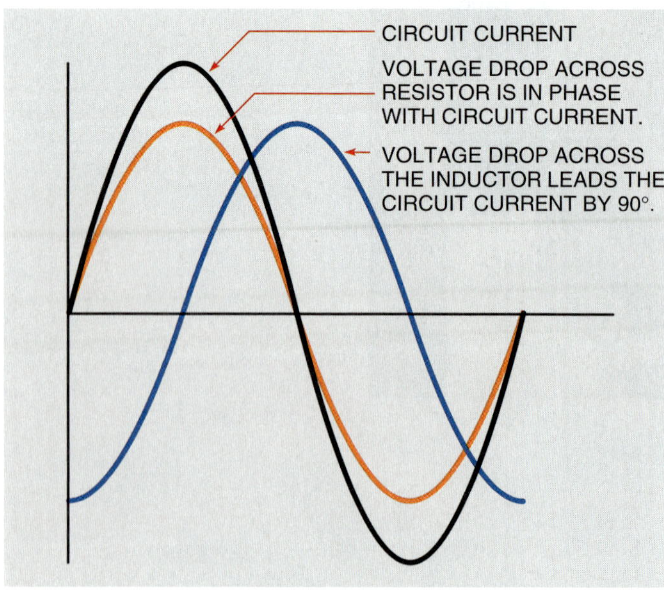

Figure 13-1 **Relationship of resistive and inductive current with voltage.**

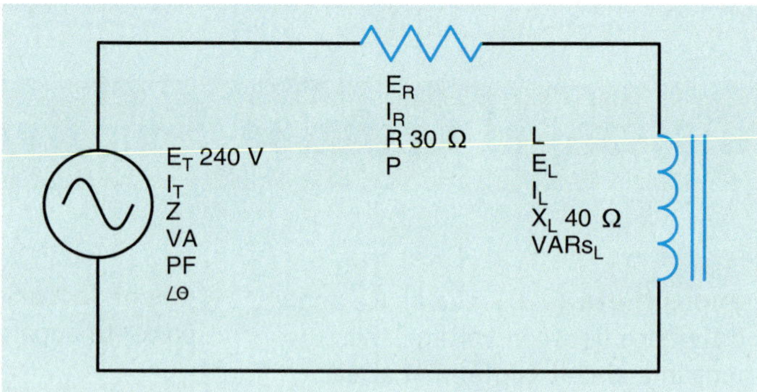

Figure 13-2 **R-L series circuit.**

Impedance

In Chapter 12, impedance was defined as a measure of the part of the circuit that impedes, or hinders, the flow of current. It is measured in ohms and symbolized by the letter *Z*. In this circuit, impedance will be a combination of resistance and inductive reactance.

In a series circuit, the total resistance is equal to the sum of the individual resistors. In this instance, however, the **total impedance (Z)** will be the sum of the resistance and the inductive reactance. It would first appear that the sum of these two quantities should be 70 Ω (30 Ω + 40 Ω = 70 Ω). In practice, however, the resistive part of the circuit and the reactive part of the circuit are out of phase with each other by 90°. To find the sum of these two quantities, vector addition must be used. Because the quantities are 90° out of phase with each other, the resistive and inductive reactance form the two legs of a right triangle, and the impedance is the hypotenuse (Fig. 13-3).

The total impedance (Z) can be computed using the formula

$$Z = \sqrt{R^2 + X_L^2}$$
$$Z = \sqrt{30^2 + 40^2}$$
$$Z = \sqrt{900 + 1600}$$
$$Z = \sqrt{2500}$$
$$Z = 50\ \Omega$$

Vectors

Using the right triangle is just one method of graphically showing how out-of-phase quantities can be added. Another way is to use vectors. A **vector** is a line that indicates both magnitude and direction. The magnitude is indicated by its length, and the direction is indicated by its angle of rotation from 0°. Vectors should not be confused with *scalars*, which are used to represent magnitude only and do not take direction into consideration. Imagine, for example, that you are

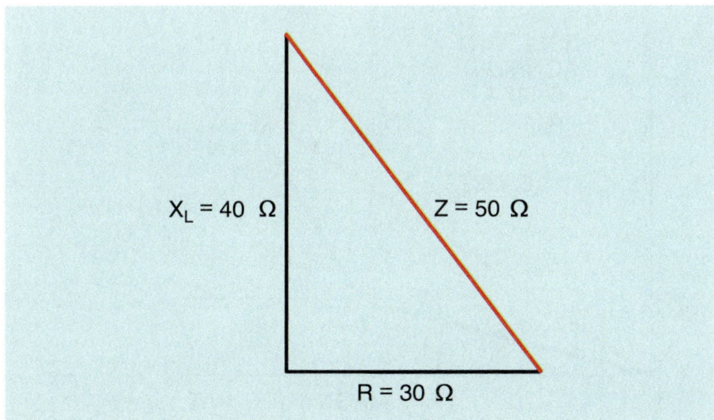

Figure 13-3 Impedance is the combination of resistance and inductive reactance.

Figure 13-4 A vector with a magnitude of 5 and an angle of 0°.

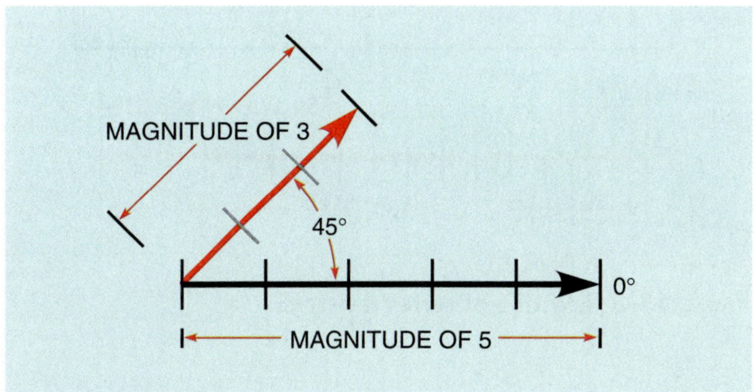

Figure 13-5 A second vector with a magnitude of 3 and a direction of 45°.

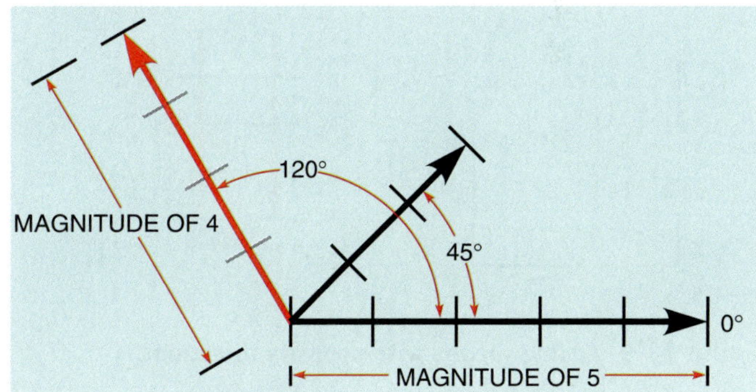

Figure 13-6 A third vector with a magnitude of 4 and a direction of 120° is added.

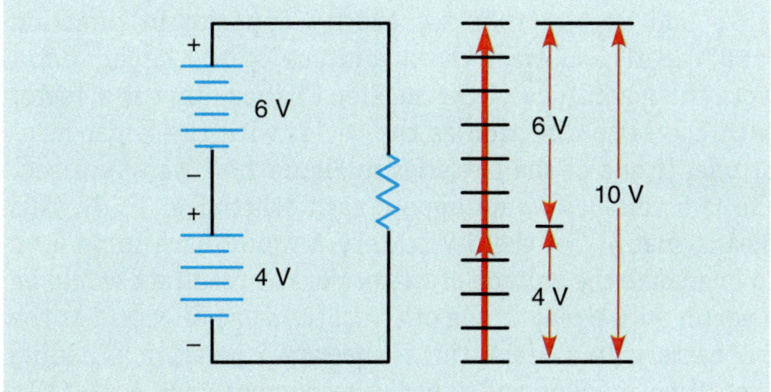

Figure 13-7 Adding vectors in the same direction.

in a strange city and you ask someone for directions to a certain building. If the person said to walk three blocks, that would be a scalar, because it contains only the magnitude, three blocks. If the person said to walk three blocks south, it would be a vector because it contains both the magnitude, three blocks, and the direction, south.

Zero degrees is indicated by a horizontal line. An arrow is placed at one end of the line to indicate direction. The magnitude can represent any quantity such as inches, meters, miles, volts, amps, ohms, or power. A vector with a magnitude of 5 at an angle of 0° is shown in Figure 13-4. Vectors rotate in a counterclockwise direction. Assume that a vector with a magnitude of 3 is to be drawn at a 45° angle from the first vector (Fig. 13-5). Now assume that a third vector with a magnitude of 4 is to be drawn in a direction of 120° (Fig. 13-6). Notice that the direction of the third vector is referenced from the horizontal 0° line and not from the second vector line, which was drawn at an angle of 45°.

Adding Vectors

Because vectors are used to represent quantities such as volts, amps, ohms, and power, they can be added, subtracted, multiplied, and divided. In electrical work, however, addition is the only function needed, so it will be the only one discussed. Several methods can be used to add vectors. Regardless of which is used, because vectors contain both magnitude and direction, they must be added with a combination of geometric and algebraic addition. This is often referred to as **vector addition.**

One method is to connect the starting point of one vector to the end point of another. This works especially well when all vectors are in the same direction. Consider the circuit shown in Figure 13-7 in which two batteries, one rated at 6 V and the other at 4 V, are connected so that their voltages add. If vector addition is used, the starting point of one vector will be placed at the end point of the other. Notice that the sum of the two vector quantities is equal to the sum of the two voltages, 10 V. Another example of this type of vector addition is shown in Figure 13-8. In this example three resistors are connected in series. The first resistor has a resistance of 80 Ω, the second a resistance of 50 Ω, and the third a resistance of 30 Ω. Because there is no phase angle shift of voltage or current, the impedance will be 160 Ω, which is the sum of the resistances of the three resistors.

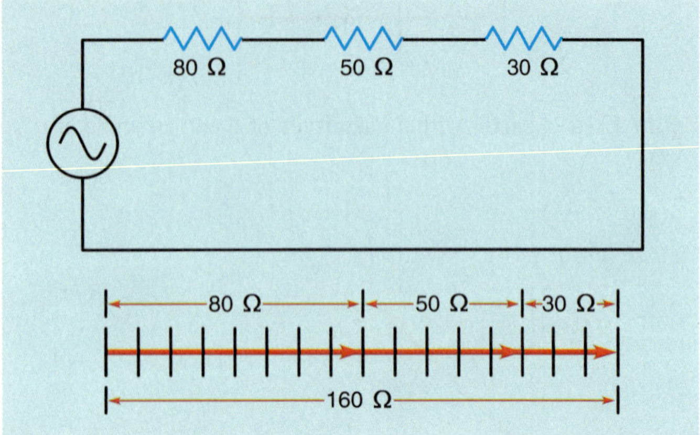

Figure 13-8 Addition of series resistors.

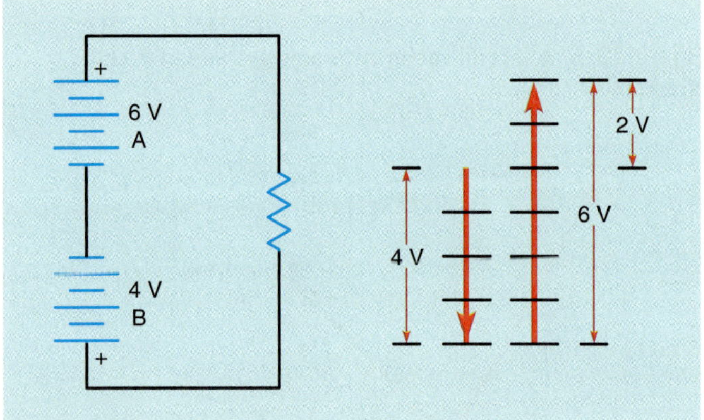

Figure 13-9 Adding vectors with opposite directions.

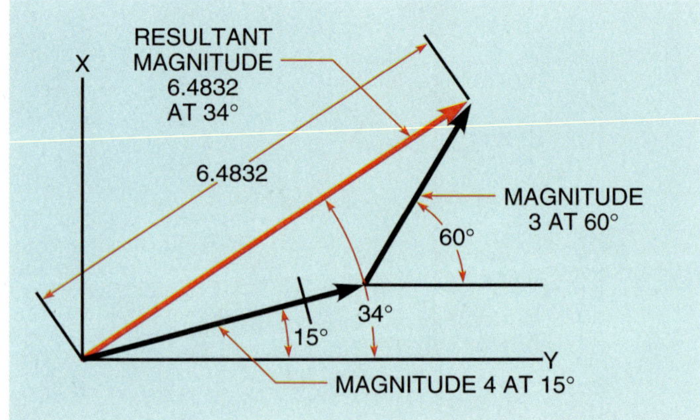

Figure 13-10 Adding two vectors with different directions.

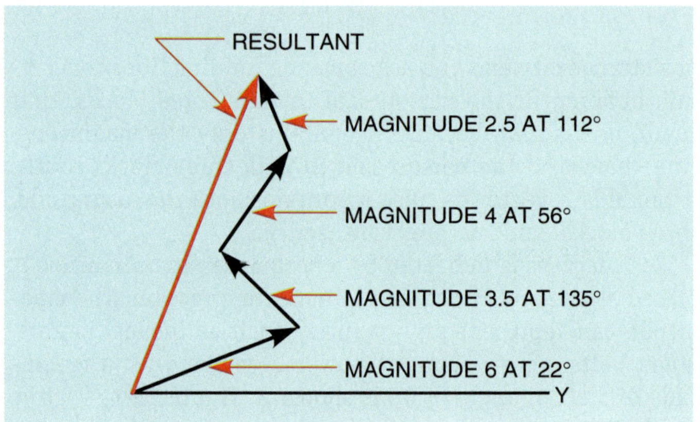

Figure 13-11 Adding vectors with different magnitude and direction.

Adding Vectors with Opposite Directions

To add vectors that are exactly opposite in direction (180° apart), subtract the magnitude of the larger vector from the magnitude of the smaller. The resultant is a vector with the same direction as the vector with the larger magnitude. If one of the batteries in Figure 13-7 were reversed, the two voltages would oppose each other (Fig. 13-9). This means that 4 V of the 6-V battery A would have to be used to overcome the voltage of battery B. The resultant would be a vector with a magnitude of 2 V in the same direction as the 6-V battery. In algebra, this is the same operation as adding a positive number and a negative number ($+6 + [-4] = +2$). When the -4 is brought out of brackets, the equation becomes $6 - 4 = 2$.

Adding Vectors of Different Directions

Vectors that have directions other than 180° from each other can also be added. Figure 13-10 illustrates the addition of a vector with a magnitude of 4 and a direction of 15° to a vector with a magnitude of 3 and a direction of 60°. The addition is made by connecting the starting point of the second vector to the ending point of the first vector. The resultant is drawn from the starting point of the first vector to the ending point of the second. It is possible to add several different vectors using this method. Figure 13-11 illustrates the addition of several different vectors and the resultant.

To find the impedance of the circuit in Figure 13-3 using this method of vector addition, connect the starting point of one vector to the ending point of the other. Since resistance and inductive reactance are 90° out of phase with each other, the two vectors must be placed at a 90° angle. If the resistive vector has a magnitude of 30 Ω and the inductive vector has a magnitude of 40 Ω, the resultant (impedance) will have a magnitude of 50 Ω (Fig. 13-12). Notice that the result is the same as that found using the right triangle.

The Parallelogram Method of Vector Addition

The **parallelogram method** can be used to find the resultant of two vectors that originate at the same point. A parallelogram is a four-sided figure whose opposite sides

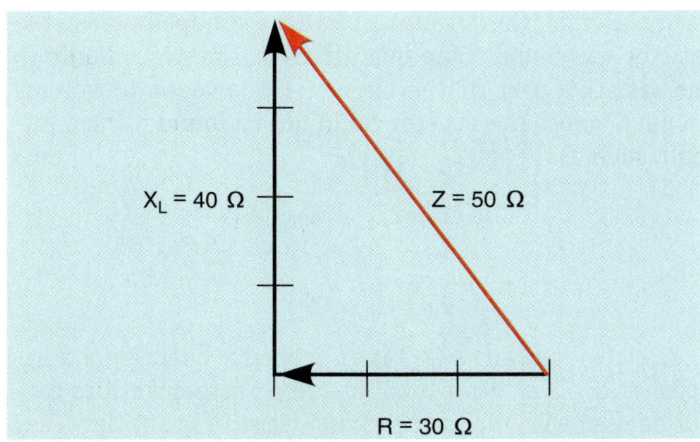

Figure 13-12 Determining impedance using vector addition.

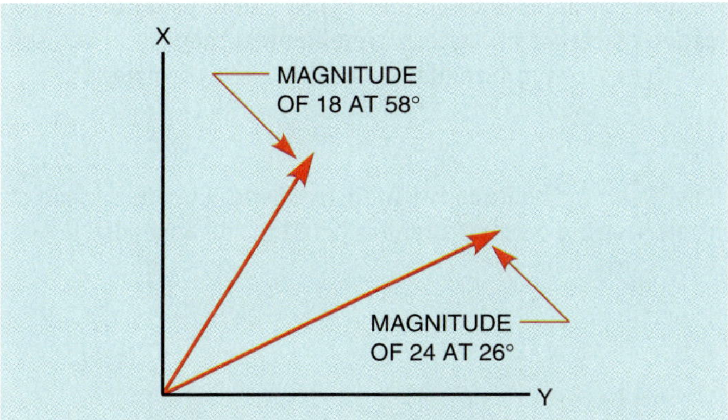

Figure 13-13 Vectors that originate from the same point.

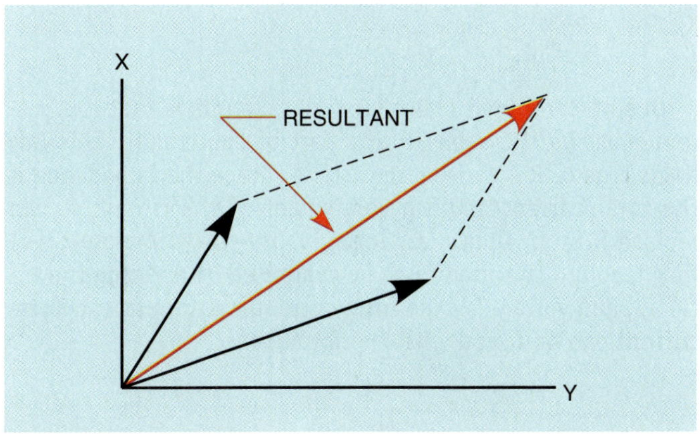

Figure 13-14 The parallelogram method of vector addition.

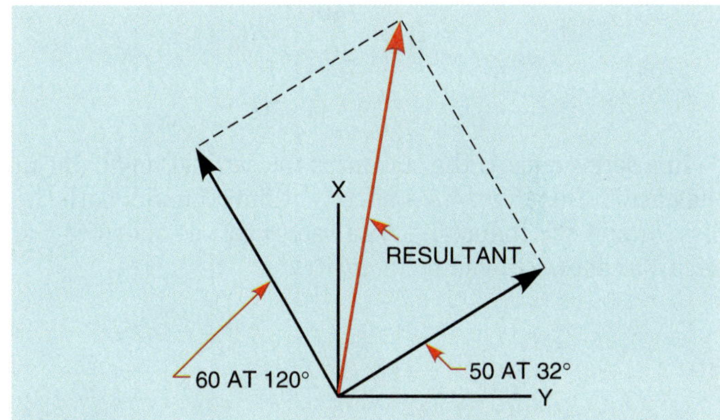

Figure 13-15 The parallelogram method of vector addition, Example 2.

form parallel lines. A rectangle, for example, is a parallelogram with 90° angles. Assume that a vector with a magnitude of 24 and a direction of 26° is to be added to a vector with a magnitude of 18 and a direction of 58°. Also assume that the two vectors originate from the same point (Fig. 13-13). To find the resultant of these two vectors, form a parallelogram using the vectors as two of the sides. The resultant is drawn from the corner of the parallelogram where the two vectors intersect to the opposite corner (Fig. 13-14).

Another example of the parallelogram method of vector addition is shown in Figure 13-15. In this example, one vector has a magnitude of 50 and a direction of 32°. The second vector has a magnitude of 60 and a direction of 120°. The resultant is found by using the two vectors as two sides of a parallelogram.

The parallelogram method of vector addition can also be used to find the total impedance of the circuit shown in Figure 13-2. The resistance forms a vector with a magnitude of 30 Ω and a direction of 0°. The inductive reactance forms a vector with a magnitude of 40 Ω and a direction of 90°. When lines are extended to form a parallelogram and a resultant is drawn, the resultant will have a magnitude of 50 Ω, which is the impedance of the circuit (Fig. 13-16).

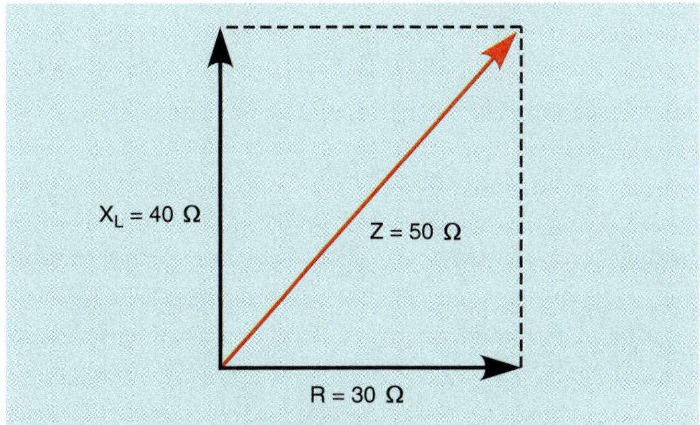

Figure 13-16 Finding the impedance of the circuit using the parallelogram method of vector addition.

Some students of electricity find the right triangle concept easier to understand, and others find vectors more helpful. For this reason, this text will use both methods to help explain the relationship of voltage, current, and power in alternating current circuits.

Total Current

One of the primary laws for series circuits is that the current must be the same in any part of the circuit. This law holds true of R-L series circuits also. Since the impedance is the total current-limiting component of the circuit, it can replace R in an Ohm's law formula. The **total current (I)** flow through the circuit can be computed by dividing the total applied voltage by the total current-limiting factor. Total current can be found with the formula

$$I_T = \frac{E_T}{Z}$$

The total current for the circuit in Figure 13-2, then, is

$$I_T = \frac{240}{50}$$

$$I_T = 4.8 \text{ A}$$

In a series circuit, the current is the same at any point in the circuit. Therefore, 4.8 A of current flow through both the resistor and the inductor. These values can be added to the circuit as shown in Figure 13-17.

Voltage Drop across the Resistor

Now that the amount of current flow through the resistor is known, the *voltage drop across the resistor (E_R)* can be computed using the formula

$$E_R = I_R \times R$$

The voltage drop across the resistor in our circuit is

$$E_R = 4.8 \times 30$$

$$E_R = 144 \text{ V}$$

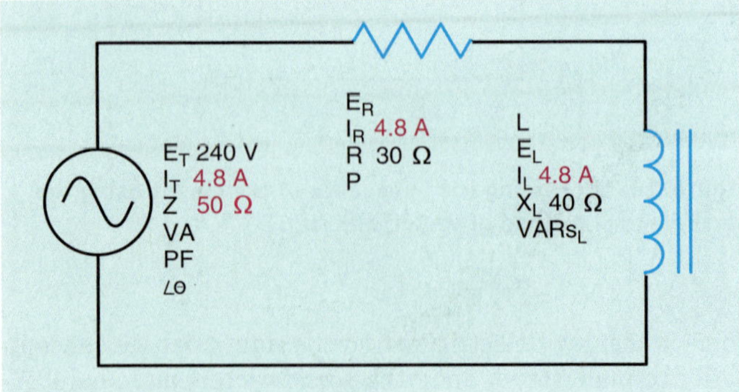

Figure 13-17 Total voltage divided by impedance equals total current.

Notice that the amount of voltage dropped across the resistor was found using quantities that pertained only to the resistive part of the circuit. The amount of voltage dropped across the resistor could not be found using a formula such as

$$E_R = I_R \times X_L$$

or

$$E_R = I_R \times Z$$

Inductive reactance (X_L) is an inductive quantity and impedance (Z) is a circuit total quantity. These quantities cannot be used with Ohm's law to find resistive quantities. They can, however, be used with vector addition to find like resistive quantities. For example, both inductive reactance and impedance are measured in ohms. The resistive quantity measured in ohms is resistance (R). If the impedance and inductive reactance of a circuit were known, they could be used with the following formula to find the circuit resistance.

$$R = \sqrt{Z^2 - X_L^2}$$

(*Note*: Refer to the Resistive-Inductive Series Circuits section of the alternating current formulas listed in the appendix.)

Watts

True power (P) for the circuit can be computed by using any of the watts formulas with pure resistive parts of the circuit. Watts (W) can be computed, for example, by multiplying the voltage dropped across the resistor (E_R) by the current flow through the resistor (I_R); by squaring the voltage dropped across the resistor and dividing by the resistance of the resistor; or by squaring the current flow through the resistor and multiplying by the resistance of the resistor. Watts cannot be computed by multiplying the total voltage (E_T) by the current flow through the resistor or by multiplying the square of the current by the inductive reactance. Recall that true power, or watts, can be produced only during periods of time when the voltage and current are both positive or both negative.

In an R-L series circuit, the current is the same through both the resistor and the inductor. The voltage dropped across the resistor, however, is in phase with the current, and the voltage dropped across the inductor is 90° out of phase with the current (Fig. 13-18). Since true power, or watts, can be produced only when the current and voltage are both positive or both negative, only resistive parts of the circuit can produce watts.

The formula used in this example will be

$$P = E_R \times I_R$$

$$P = 144 \times 4.8$$

$$P = 691.2 \text{ W}$$

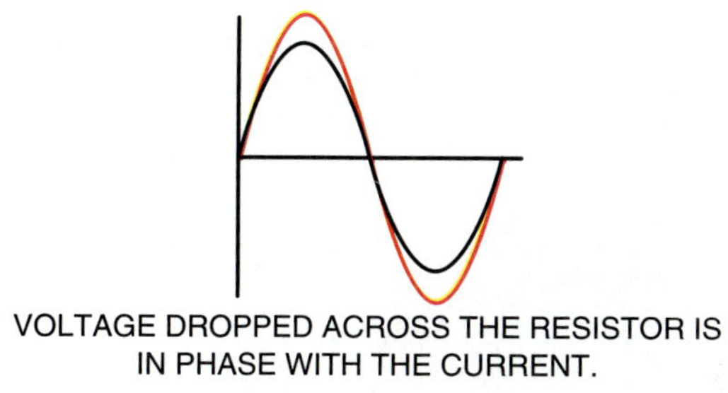

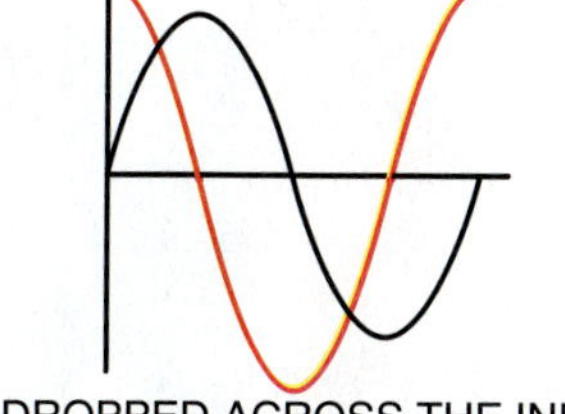

Figure 13-18 Relationship of current and voltage in an R-L series circuit.

Computing the Inductance

The amount of inductance can be computed using the formula

$$L = \frac{X_L}{2\pi F}$$

$$L = \frac{40}{377}$$

$$L = 0.106 \text{ H}$$

Voltage Drop across the Inductor

The *voltage drop across the inductor (E_L)* can be computed using the formula

$$E_L = I_L \times X_L$$

$$E_L = 4.8 \times 40$$

$$E_L = 192 \text{ V}$$

Notice that only inductive quantities were used to find the voltage drop across the inductor.

Total Voltage

Although the total applied voltage in this circuit is known (240 V), the total voltage is also equal to the sum of the voltage drops, just as it is in any other series circuit. Since the voltage dropped across the resistor is in phase with the current and the voltage dropped across the inductor is 90° out of phase with the current, vector addition must be used. The total voltage will be the hypotenuse of a right triangle, and the resistive and inductive voltage drops will form the legs of the triangle (Fig. 13-19). This relationship of voltage drops can also be represented using the parallelogram method of vector addition as shown in Figure 13-20. The following formulas can be used to find total voltage or the voltage drops across the resistor or inductor if the other two voltage values are known.

$$E_T = \sqrt{E_R^2 + E_L^2}$$

$$E_R = \sqrt{E_T^2 + E_L^2}$$

$$E_L = \sqrt{E_T^2 + E_R^2}$$

Note: Refer to the Resistive-Inductive Series Circuits section of the alternating current formulas listed in the appendix.

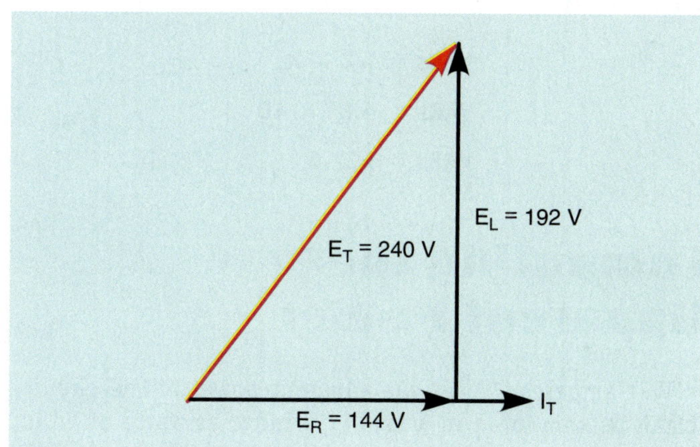

Figure 13-19 Relationship of resistive and inductive voltage drops in an R-L series circuit.

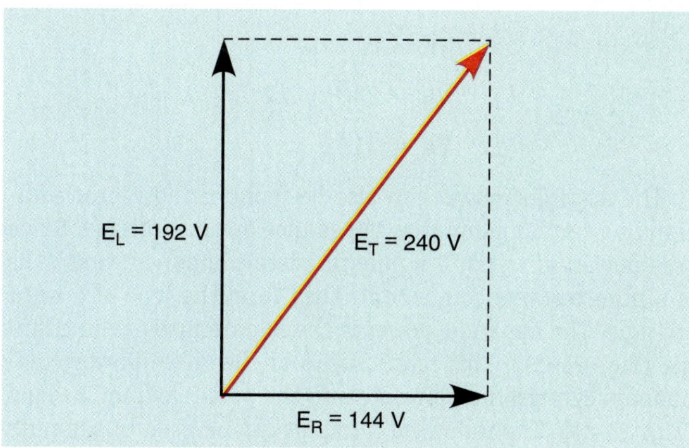

Figure 13-20 Graphic representation of voltage drops using the parallelogram method of vector addition.

Computing the Reactive Power

VARs is an abbreviation for volt-amps-reactive and is the amount of *reactive power (VARs)* in the circuit. VARs should not be confused with watts, which is true power. VARs represents the product of the volts and amps that are 90° out of phase with each other, such as the voltage dropped across the inductor and the current flowing through the inductor. Recall that true power can be produced only during periods of time that the voltage and current are both positive or both negative (Fig. 13-21). During these periods, the power is being stored in the form of a magnetic field. During the periods that voltage and current have opposite signs, the power is returned to the circuit. For this reason, VARs is often referred to as **quadrature power**, or **wattless power**. It can be computed in a similar manner as watts except that reactive values of voltage and current are used instead of resistive values. In this example the formula used is

$$\text{VARs} = I_L 2 \times X_L$$

$$\text{VARs} = 4.8^2 \times 40$$

$$\text{VARs} = 921.6$$

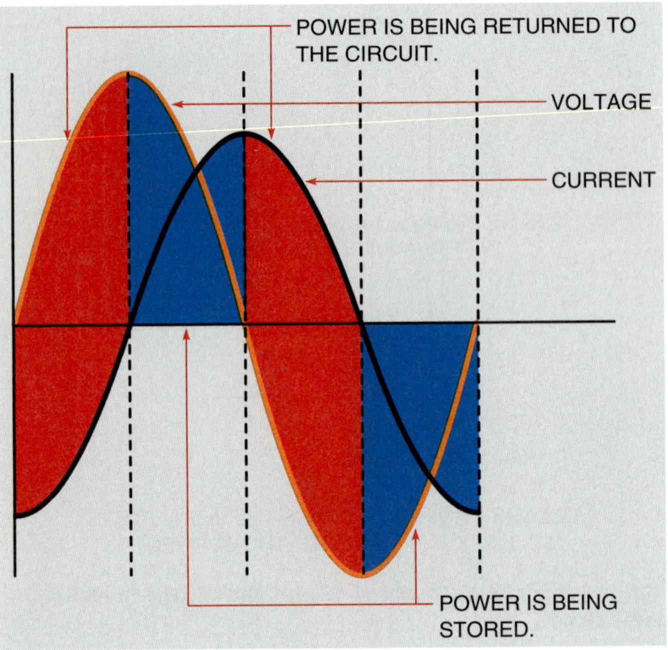

Figure 13-21 Power is stored and then returned to the circuit.

Computing the Apparent Power

Volt-amperes (VA) is the apparent power of the circuit. It can be computed in a similar manner as watts or VARs, except that total values of voltage and current are used. It is called **apparent power (VA)** because it is the value found if a voltmeter and ammeter are used to measure the circuit voltage and current and these measured values are multiplied together (Fig. 13-22). In this example the formula used is

$$\text{VA} = E_T \times I_T$$

$$\text{VA} = 240 \times 4.8$$

$$\text{VA} = 1152$$

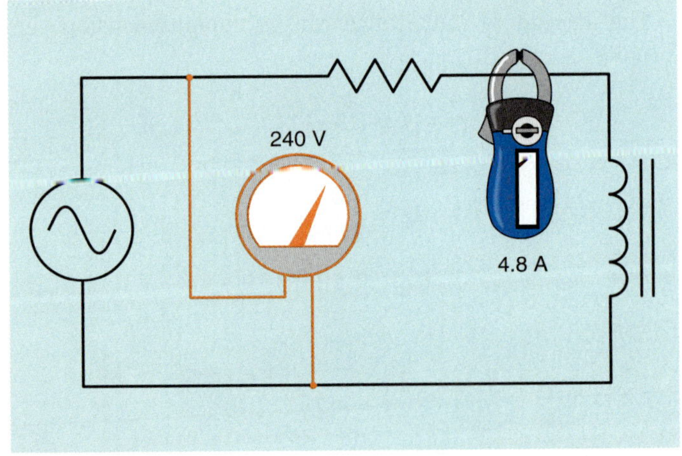

Figure 13-22 Apparent power is the product of measured values (240 V × 4.8 A = 1152 VA).

The apparent power can also be found using vector addition in a similar manner as impedance or total voltage. Since true power, or watts, is a pure resistive component and VARs is a pure reactive component, they form the legs of a right triangle. The apparent power is the hypotenuse of the triangle (Fig. 13-23). This relationship of the three power components can also be plotted using the parallelogram method (Fig. 13-24). The following formulas can be used to compute the values of apparent power, true power, and reactive power when the other two values are known.

$$\text{VA} = \sqrt{\text{P}^2 + \text{VARs}^2}$$

$$\text{P} = \sqrt{\text{VA}^2 - \text{VARs}^2}$$

$$\text{VARs} = \sqrt{\text{VA}^2 - \text{P}^2}$$

Power Factor

Power factor (PF) is a ratio of the true power to the apparent power. It can be computed by dividing any resistive value by its like total value. For example, power factor can

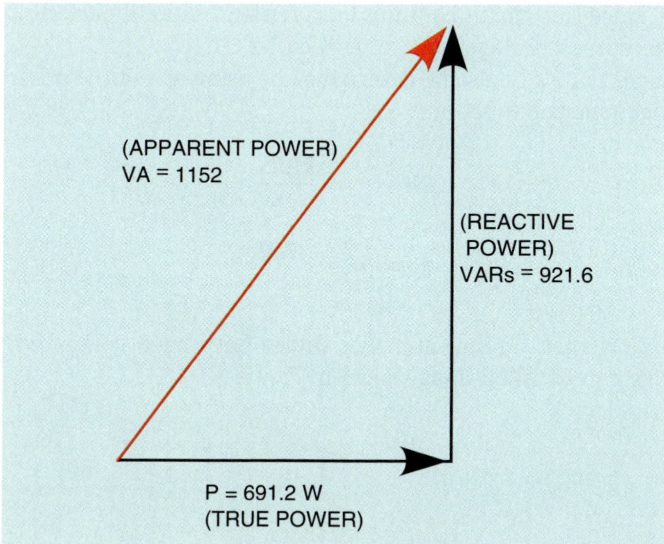

Figure 13-23 Relationship of true power (watts), reactive power (VARs), and apparent power (volt-amps) in an R-L series circuit.

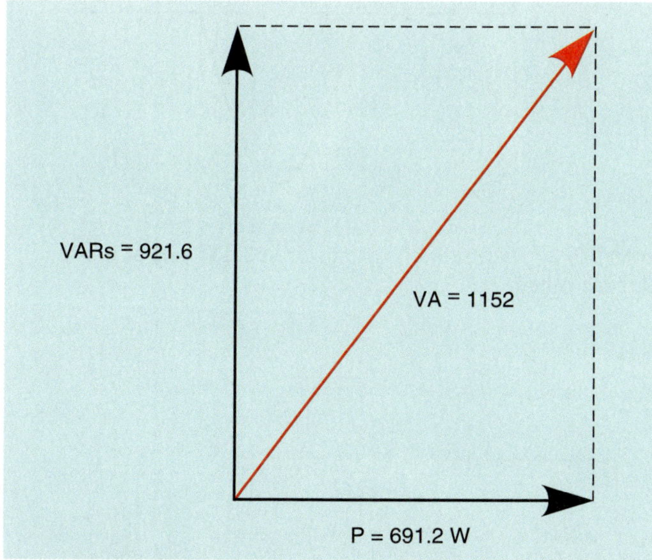

Figure 13-24 Using the parallelogram method to plot the relationship of volt-amps, watts, and VARs.

be computed by dividing the voltage drop across the resistor by the total circuit voltage; by resistance divided by impedance; or by watts divided by volt-amperes.

$$PF = \frac{E_R}{E_T}$$

$$PF = \frac{R}{Z}$$

$$PF = \frac{P}{VA}$$

Power factor is generally expressed as a percentage. The decimal fraction computed from the division will, therefore, be changed to a percent by multiplying it by 100. In this circuit the formula will be

$$PF = \frac{W}{VA}$$

$$PF = \frac{691.2}{1152}$$

$$PF = 0.6 \times 100, \text{ or } 60\%$$

Note that in a series circuit, the power factor cannot be computed using current because current is the same in all parts of the circuit.

Power factor can become very important in an industrial application. Most power companies charge a substantial surcharge when the power factor drops below a certain percent. The reason for this is that electric power is sold on the basis of true power, or watts, consumed. The power company, however, must supply the apparent power. Assume that an industrial plant has a power factor of 60% and is consuming 5 MW (megawatts) of power per hour. At a power factor of 60%, the power company must actually supply 8.33 MVA (megavolt-amps) (5 MW/0.6 = 8.33 MVA) per hour. If the power factor were to be corrected to 95%, the power company would have to supply only 5.26 MVA per hour to furnish the same amount of power to the plant.

Angle Theta

The angular displacement by which the voltage and current are out of phase with each other is called **angle theta ($\angle\theta$)**. Since the power factor is the ratio of true power to apparent power, the phase angle of voltage and current is formed between the resistive leg of the right triangle and the hypotenuse (Fig. 13-25). The resistive leg of the triangle is adjacent to the angle, and the total leg is the hypotenuse. The trigonometric function that corresponds to the adjacent side and the hypotenuse is the cosine. Angle theta will be the cosine of watts divided by volt-amps. Watts divided by volt-amps is also the power factor. Therefore, the cosine of angle theta ($\angle\theta$) is the power factor (PF).

$$\cos \angle\theta = PF$$

$$\cos \angle\theta = 0.6$$

$$\angle\theta = 53.13°$$

The vectors formed using the parallelogram method of vector addition can also be used to find angle theta as shown in Figure 13-26. Notice that the total quantity, volt-amps, and the resistive quantity, watts, again determine angle theta.

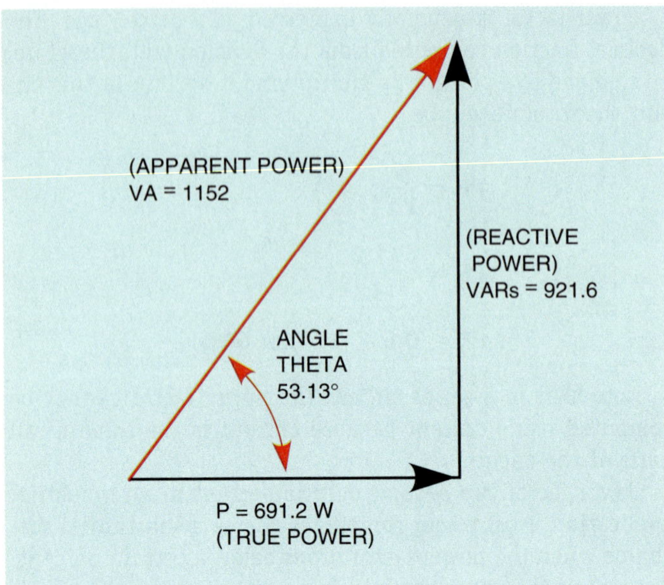

Figure 13-25 **The angle theta is the relationship of true power to apparent power.**

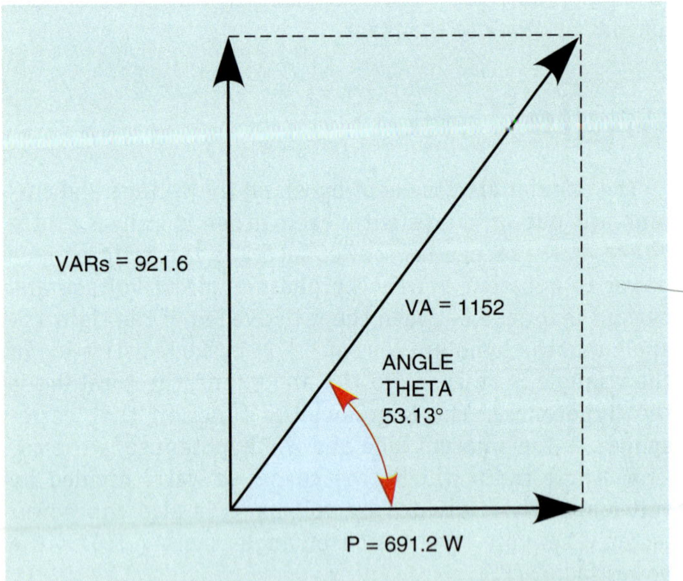

Figure 13-26 **The angle theta can be found using vectors provided by the parallelogram method.**

Since this circuit contains both resistance and inductance, the current is lagging the voltage by 53.13° (Fig. 13-27). Angle theta can also be determined by using any of the other trigonometric functions.

$$\text{SIN}\angle\theta = \frac{\text{VARS}}{\text{VA}}$$

$$\text{TAN}\angle\theta = \frac{\text{VARS}}{\text{P}}$$

Now that all the unknown values have been computed, they can be filled in as shown in Figure 13-28.

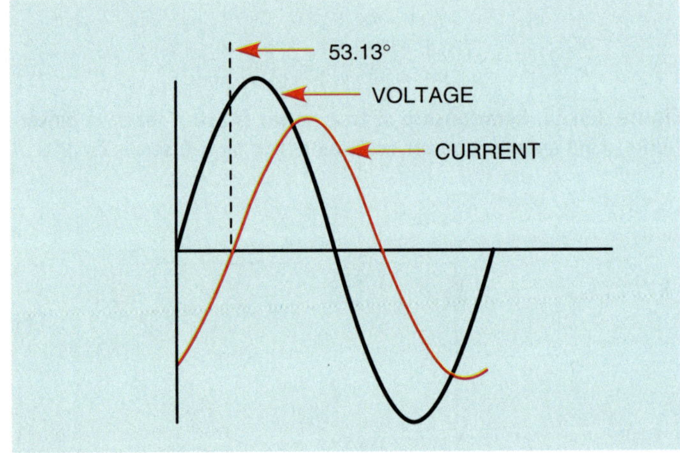

Figure 13-27 **Current and voltage are 53.13° out of phase with each other.**

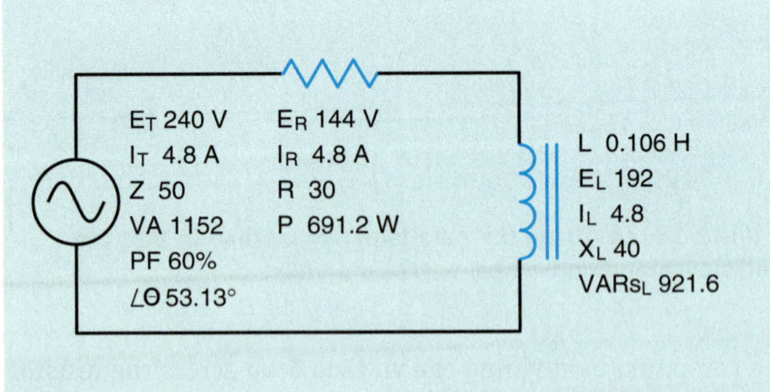

Figure 13-28 **Filling in all unknown values.**

Example 1

Two resistors and two inductors are connected in series (Fig. 13-29). The circuit is connected to a 130-V, 60-Hz line. The first resistor has a power dissipation of 56 W, and the second resistor has a power dissipation of 44 W. One inductor has a reactive power of 152 VARs and the second a reactive power of 88 VARs. Find the unknown values in this circuit.

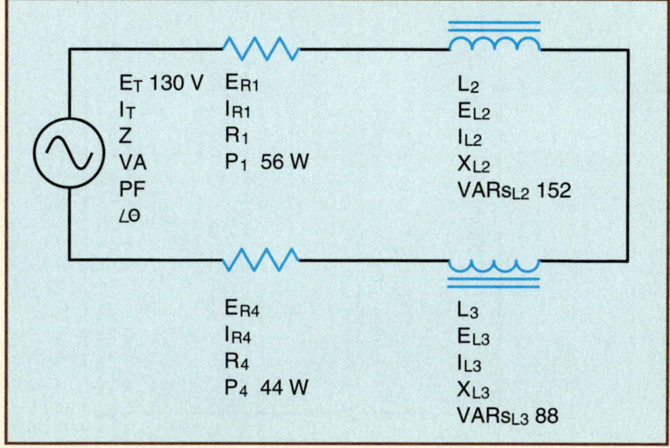

Figure 13-29 An R-L series circuit containing two resistors and two inductors.

Solution

The first step is to find the total amount of true power and the total amount of reactive power. The total amount of true power can be computed by adding the values of the resistors together. A vector diagram of these two values would reveal that they both have a direction of 0° (Fig. 13-30).

$$P_T = P_1 + P_4$$

$$P_T = 56 + 44$$

$$P_T = 100 \text{ W}$$

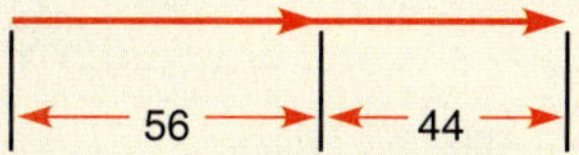

Figure 13-30 The two true power (watts) vectors are in the same direction.

The total reactive power in the circuit can be found in the same manner. Like watts, the VARs are both inductive and are, therefore, in the same direction (Fig. 13-31). The total reactive power will be the sum of the two reactive power ratings.

$$VARs_T = VARs_{L2} + VARs_{L3}$$

$$VARs_T = 152 + 88$$

$$VARs_T = 240$$

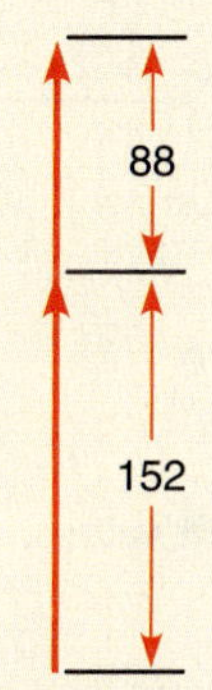

Figure 13-31 The two reactive power (VARs) vectors are in the same direction.

Apparent Power

Now that the total amount of true power and the total amount of reactive power are known, the apparent power (VA) can be computed using vector addition.

$$VA = \sqrt{P_T^2 + VARs_T^2}$$

$$VA = \sqrt{100^2 + 240^2}$$

$$VA = 260$$

The parallelogram method of vector addition is shown in Figure 13-32 for this calculation.

Total Circuit Current

Now that the apparent power is known, the total circuit current can be found using the applied voltage and Ohm's law.

$$I_T = \frac{VA}{E_T}$$

$$I_T = \frac{260}{130}$$

$$I_T = 2 \text{ A}$$

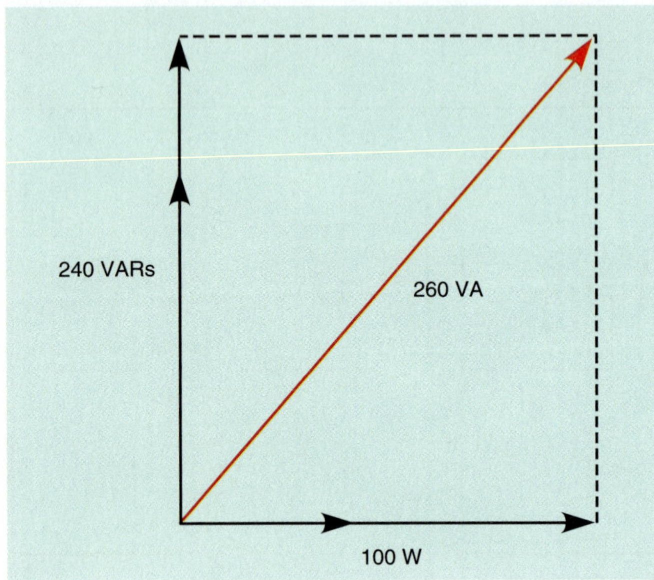

Figure 13-32 Power vector for the circuit.

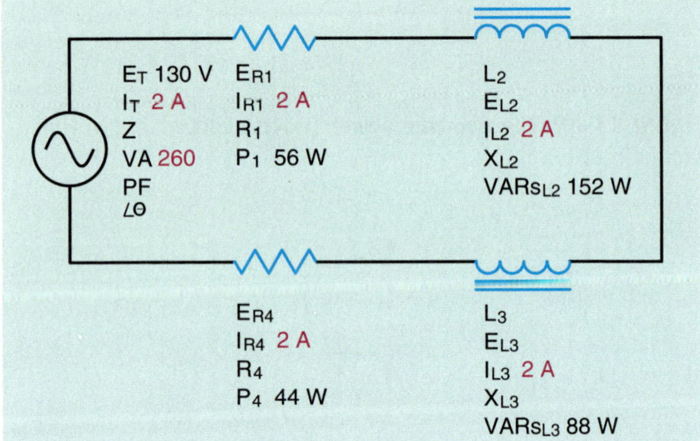

Figure 13-33 Adding circuit values.

Since this is a series circuit, the current must be the same at all points in the circuit. The known values added to the circuit are shown in Figure 13-33.

Other Circuit Values

Now that the total circuit current has been found, other values can be computed using Ohm's law.

Impedance

$$Z = \frac{E_T}{I_T}$$

$$Z = \frac{130}{2}$$

$$Z = 65\Omega$$

Power Factor

$$PF = \frac{W_T}{VA}$$

$$PF = \frac{100}{260}$$

$$PF = 38.46\%$$

Angle Theta

Angle theta is the cosine of the power factor. A vector diagram showing this relationship is shown in Figure 13-34.

$$\cos \angle \theta = PF$$

$$\cos \angle \theta = 0.3846$$

$$\angle \theta = 67.38°$$

The relationship of voltage and current for this circuit is shown in Figure 13-35.

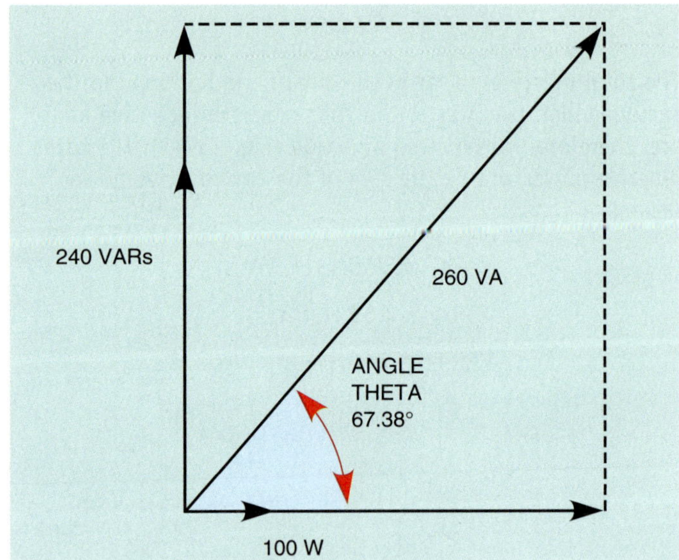

Figure 13-34 Angle theta for the circuit.

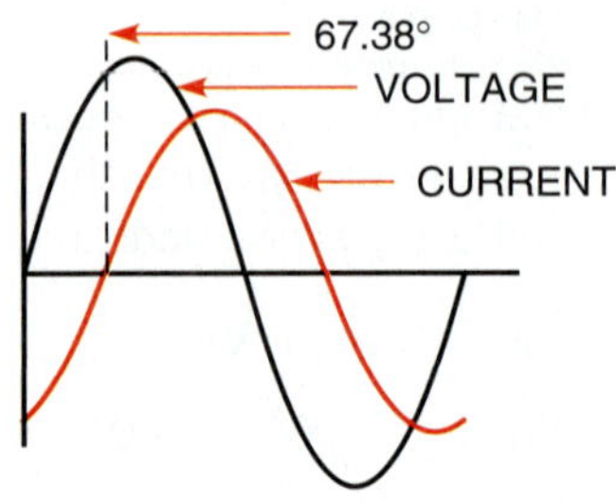

Figure 13-35 Voltage and current are 67.38° out of phase with each other.

E_{R1}

$$E_{R1} = \frac{P_1}{I_{R1}}$$

$$E_{R1} = \frac{56}{2}$$

$$E_{R1} = 28 \text{ V}$$

R_1

$$R_1 = \frac{E_{R1}}{I_{R1}}$$

$$R_1 = \frac{28}{2}$$

$$R_1 = 14\Omega$$

E_{L2}

$$E_{L2} = \frac{VARs_{L2}}{I_{L2}}$$

$$E_{L2} = \frac{152}{2}$$

$$E_{L2} = 76 \text{ V}$$

X_{L2}

$$X_{L2} = \frac{E_{L2}}{I_{L2}}$$

$$X_{L2} = \frac{76}{2}$$

$$X_{L2} = 38\Omega$$

L_2

$$L_2 = \frac{X_{L2}}{2\pi F}$$

$$L_2 = \frac{38}{377}$$

$$L_2 = 0.101 \text{ H}$$

E_{L3}

$$E_{L3} = \frac{VARs_{L3}}{I_{L3}}$$

$$E_{L3} = \frac{88}{2}$$

$$E_{L3} = 44 \text{ V}$$

X_{L3}

$$X_{L3} = \frac{E_{L3}}{I_{L3}}$$

$$X_{L3} = \frac{44}{2}$$

$$X_{L3} = 22\Omega$$

L_3

$$L_3 = \frac{X_{L3}}{2\pi F}$$

$$L_3 = \frac{22}{377}$$

$$L_3 = 0.058 \text{ H}$$

E_{R4}

$$E_{R4} = \frac{P_4}{I_{R4}}$$

$$E_{R4} = \frac{44}{2}$$

$$E_{R4} = 22 \text{ V}$$

R_4

$$R_4 = \frac{E_{R4}}{I_{R4}}$$

$$R_4 = \frac{22}{2}$$

$$R_4 = 11\Omega$$

The complete circuit with all values is shown in Figure 13-36.

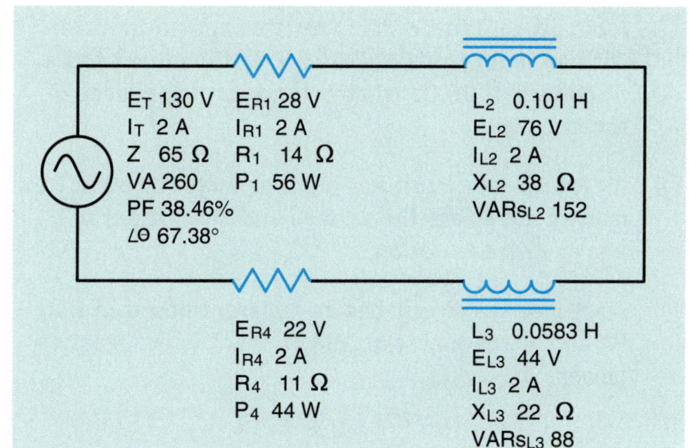

Figure 13-36 All values for the circuit.

Summary

- In a pure resistive circuit the voltage and current are in phase with each other.
- In a pure inductive circuit the voltage and current are 90° out of phase with each other.
- In an R-L series circuit the voltage and current will be out of phase with each other by some value between 0° and 90°.
- The amount the voltage and current are out of phase with each other is determined by the ratio of resistance to inductance.
- Total circuit values include total voltage, E_T; total current, I_T; volt-amps, VA; and impedance, Z.
- Pure resistive values include voltage drop across the resistor, E_R; current flow through the resistor, I_R; resistance, R; and watts, P.
- Pure inductive values include inductance of the inductor, L; voltage drop across the inductor, E_L; current through the inductor, I_L; inductive reactance, X_L; and inductive VARs, $VARs_L$.
- Angle theta measures the phase angle difference between the applied voltage and total circuit current.
- The cosine of angle theta is equal to the power factor.
- Power factor is a ratio of true power to apparent power.
- Vectors are lines that indicate both magnitude and direction.

Review Questions

1. **What is the relationship of voltage and current (concerning phase angle) in a pure resistive circuit?**
2. **What is the relationship of voltage and current (concerning phase angle) in a pure inductive circuit?**
3. **What is power factor?**
4. **A circuit contains a 20-Ω resistor and an inductor with an inductance of 0.093 H. If the circuit has a frequency of 60 Hz what is the total impedance of the circuit?**
5. **An R-L series circuit has a power factor of 86%. How many degrees are the voltage and current out of phase with each other?**
6. **An R-L series circuit has an apparent power of 230 VA and a true power of 180 W. What is the reactive power?**
7. **The resistor in an R-L series circuit has a voltage drop of 53 V, and the inductor has a voltage drop of 28 V. What is the applied voltage of the circuit?**
8. **An R-L series circuit has a reactive power of 1234 VARs and an apparent power of 4329 VA. How many degrees are voltage and current out of phase with each other?**
9. **An R-L series circuit contains a resistor and an inductor. The resistor has a value of 6.5 Ω. The circuit is connected to 120 V and has a current flow of 12 A. What is the inductive reactance of this circuit?**
10. **What is the voltage drop across the resistor in the circuit in question 9?**

Practice Problems

Refer to the circuit shown in Figure 13-2 and the Resistive-Inductive Series Circuits section of the alternating current formulas listed in the appendix.

1. Assume that the circuit shown in Figure 13-2 is connected to a 480-V, 60-Hz line. The inductor has an inductance of 0.053 H, and the resistor has a resistance of 12 Ω.

E_T 480______	E_R ______	E_L ______
I_T ______	I_R ______	I_L ______
Z ______	R 12 ______	X_L ______
VA ______	P ______	VARs ______
PF ______	$\angle\theta$ ______	L 0.053 H ____

2. Assume that the voltage drop across the resistor, E_R, is 78 V; that the voltage drop across the inductor, E_L, is 104 V; and the circuit has a total impedance, Z, of 20 Ω. The frequency of the AC voltage is 60 Hz.

E_T ______	E_R 78 ______	E_L 104 V ____
I_T ______	I_R ______	I_L ______
Z 20 ______	R ______	X_L ______
VA ______	P ______	$VARs_L$ ______
PF ______	$\angle\theta$ ______	L ______

3. Assume the above circuit has an apparent power of 144 VA and a true power of 115.2 W. The inductor has an inductance of 0.15915 H, and the frequency is 60 Hz.

E_T ______	E_R ______	E_L ______
I_T ______	I_R ______	I_L ______
Z ______	R ______	X_L ______
VA 144	P 115.2 W	$VARs_L$ ______
PF ______	$\angle\theta$ ______	L 0.15915 H

4. Assume the above circuit has a power factor of 78%, an apparent power of 374.817 VA, and a frequency of 400 Hz. The inductor has an inductance of 0.0382 H.

E_T ______	E_R ______	E_L ______
I_T ______	I_R ______	I_L ______
Z ______	R ______	X_L ______
VA 374.817	P ______	$VARs_L$ ______
PF 78%	$\angle\theta$ ______	L 0.0382 H

Chapter 14 Resistive-Inductive Parallel Circuits

This unit discusses circuits that contain resistance and inductance connected in parallel with each other. Mathematical calculations will be used to show the relationship of current and voltage on the entire circuit, and the relationship of current through different branches of the circuit.

OBJECTIVES

After studying this unit, you should be able to:

- **discuss the operation of a parallel circuit containing resistance and inductance.**
- **compute circuit values of an R-L parallel circuit.**
- **connect an R-L parallel circuit and measure circuit values with test instruments.**

Glossary of Resistive-Inductive Parallel Circuit Terms

angle theta ($\angle\theta$) the phase angle difference between voltage and current in a circuit containing a reactive component such as an inductor or capacitor

apparent power (VA) the value found by multiplying the applied voltage by the total current of an AC circuit. Apparent power is measured in volt-amps (VA) and should not be confused with true power, measured in watts.

current flow through the inductor (I_L) the amount of current flowing through the inductor

current flow through the resistor (I_R) the amount of current flowing through the resistor

power factor (PF) the ratio of true power to apparent power

reactive power VARs

total current (I_T) the total current flow in an electric circuit; in an RL parallel circuit it is determined by vector addition of the resistive current and inductive current

total impedance the total current-limiting effect in an AC circuit

true power the amount of electrical energy that is converted into some other form of energy such as thermal or kinetic; also known as watts

watts the true power in a circuit; indicates the amount of electrical energy converted to some other form

Resistive-Inductive Parallel Circuits

A circuit containing a resistor and an inductor connected in parallel is shown in Figure 14-1. Since the voltage applied to any device in parallel must be the same, the voltage applied to the resistor and inductor must be in phase and have the same value. The current flow through the inductor will be 90° out of phase with the voltage, and the current flow through the resistor will be in phase with the voltage (Fig. 14-2). This configuration produces a phase angle difference of 90° between the current flow through a pure inductive load and a pure resistive load (Fig. 14-3).

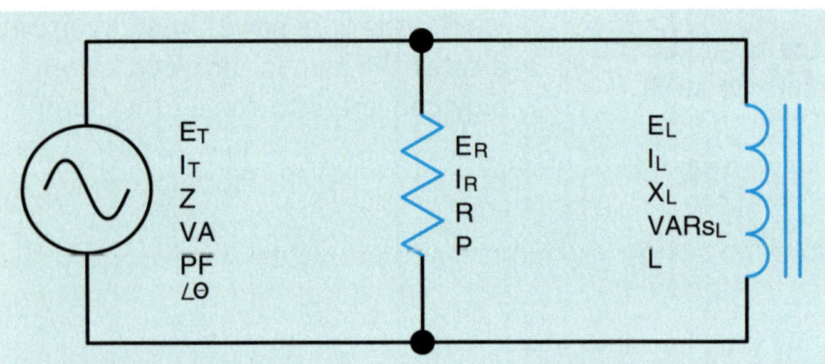

Figure 14-1 **A resistive-inductive parallel circuit.**

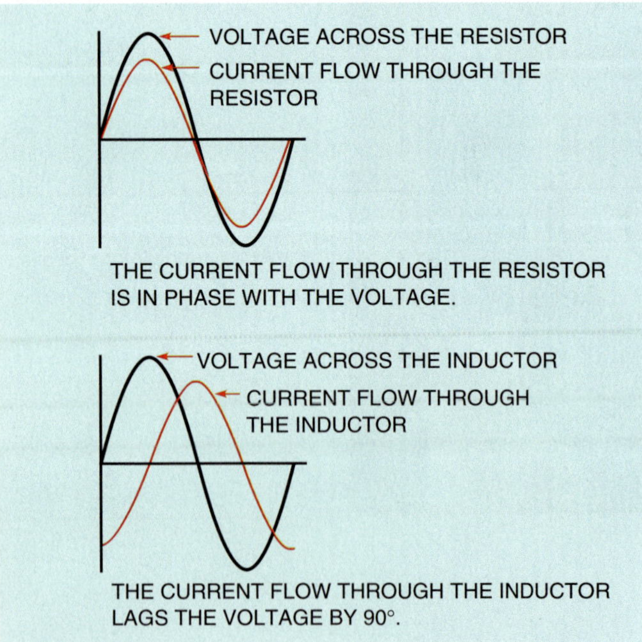

Figure 14-2 **Relationship of voltage and current in an R-L parallel circuit.**

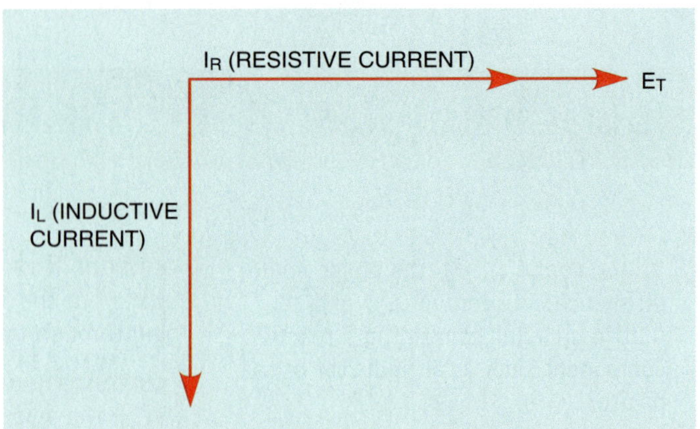

Figure 14-3 **Resistive and inductive currents are 90° out of phase with each other in an R-L parallel circuit.**

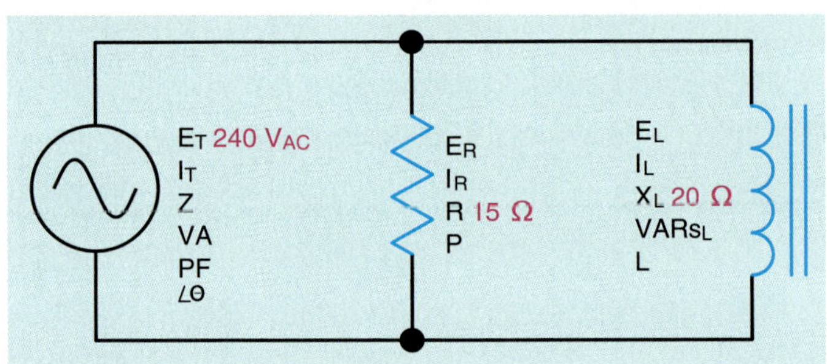

Figure 14-4 **Typical R-L parallel circuit.**

The amount of phase angle shift between the total circuit current and voltage is determined by the ratio of the amount of resistance to the amount of inductance. The circuit power factor is still determined by the ratio of apparent power to true power.

Computing Circuit Values

In the circuit shown in Figure 14-4 a resistance of 15 Ω is connected in parallel with an inductive reactance of 20 Ω. The circuit is connected to a voltage of 240 V AC and a frequency of 60 Hz. In this example problem, the following circuit values will be computed:

I_R	—	**current flow through the resistor**
P	—	**watts (true power)**
I_L	—	**current flow through the inductor**
VARs	—	**reactive power**
I_T	—	**total circuit current**
Z	—	**total circuit impedance**

VA — **apparent power**
PF — **power factor**
∠θ — **the angle the voltage and current are out of phase with each other.**

Resistive Current

In any parallel circuit, the voltage is the same across each component in the circuit. Therefore, 240 V are applied across both the resistor and the inductor. Because the amount of voltage applied to the resistor is known, the amount of **current flow through the resistor (I_R)** can be computed by using the formula

$$I_R = \frac{E}{R}$$

$$I_R = \frac{240}{15}$$

$$I_R = 16A$$

Watts

True power (P), or **watts**, can be computed using any of the watts formulas and pure resistive values. The amount of true power in this circuit will be computed using the formula

$$P = E_R \times I_R$$

$$P = 240 \times 16$$

$$P = 3840 \text{ W}$$

Inductive Current

Because the voltage applied to the inductor is known, the current flow can be found by dividing the voltage by the inductive reactance. The amount of **current flow through the inductor (I_L)** will be computed using the formula

$$I_L = \frac{E}{X_L}$$

$$I_L = \frac{240}{20}$$

$$I_L = 12A$$

VARs

The amount of **reactive power, VARs**, will be computed using the formula

$$VARs = E_L \times I_L$$

$$VARs = 240 \times 12$$

$$VARs = 2880$$

Inductance

The frequency and the inductive reactance are known, so the inductance of the coil can be found using the formula

$$L = \frac{X_L}{2\pi F}$$

$$L = \frac{20}{377}$$

$$L = 0.053H$$

Total Current

The **total current (I_T)** flow through the circuit can be computed by adding the current flow through the resistor and the inductor. Since these two currents are 90° out of phase with each other, vector addition will be used. If these current values were plotted, they would form a right triangle similar to the one shown in Figure 14-5. Notice that the current flow through the resistor and inductor form the legs of a right triangle, and the total current is the hypotenuse. Therefore, the Pythagorean theorem can be used to add these currents together.

$$I_T = \sqrt{I_R^2 + I_L^2}$$

$$I_T = \sqrt{16^2 + 12^2}$$

$$I_T = \sqrt{256 + 144}$$

$$I_T = \sqrt{400}$$

$$I_T = 20 \text{ A}$$

The parallelogram method for plotting the total current is shown in Figure 14-6.

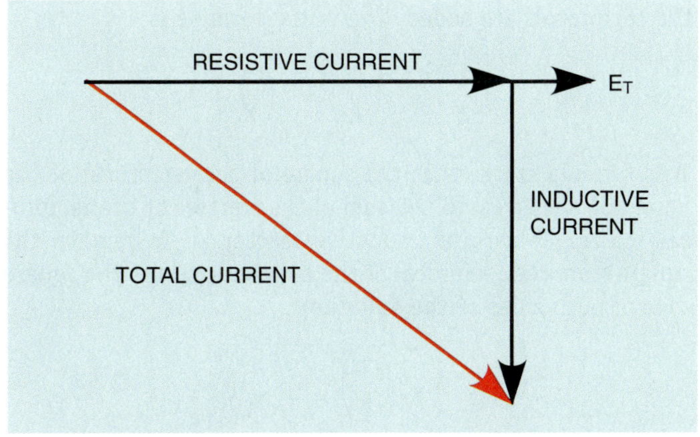

Figure 14-5 Relationship of resistive, inductive, and total current in an R-L parallel circuit.

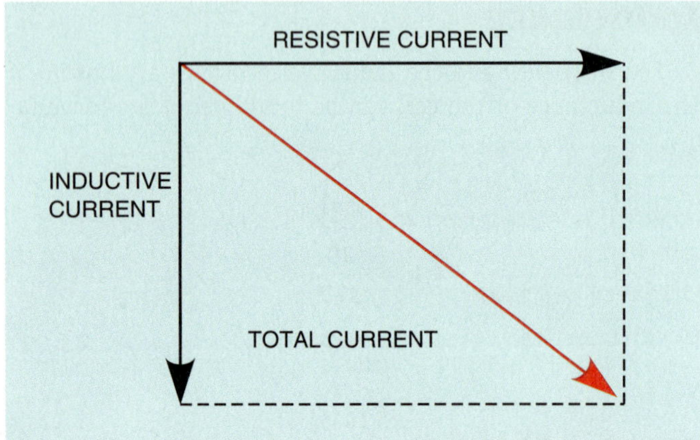

Figure 14-6 Plotting total current using the parallelogram method.

Impedance

Now that the total current and total voltage are known, the **total impedance (Z)** can be computed by substituting Z for R in an Ohm's law formula. The total impedance of the circuit can be computed using the formula

$$Z = \frac{E}{I_T}$$

$$Z = \frac{240}{20}$$

$$Z = 12\ \Omega$$

The value of impedance can also be found if total current and voltage are not known. In a parallel circuit, the reciprocal of the total resistance is equal to the sum of the reciprocals of each resistor. This same rule can be amended to permit a similar formula to be used in an R-L parallel circuit. Since resistance and inductive reactance are 90° out of phase with each other, vector addition must be used when the reciprocals are added. The initial formula is

$$\left(\frac{1}{Z}\right)^2 = \left(\frac{1}{R}\right)^2 + \left(\frac{1}{X_L}\right)^2$$

This formula states that the square of the reciprocal of the impedance is equal to the sum of the squares of the reciprocals of resistance and inductive reactance. To remove the square from the reciprocal of the impedance, take the square root of both sides of the equation.

$$\frac{1}{Z} = \sqrt{\left(\frac{1}{R}\right)^2 + \left(\frac{1}{X_L}\right)^2}$$

Notice that the formula can now be used to find the reciprocal of the impedance, not the impedance. To change the formula so that it is equal to the impedance, take the reciprocal of both sides of the equation.

$$Z = \frac{1}{\sqrt{\left(\frac{1}{R}\right)^2 + \left(\frac{1}{X_L}\right)^2}}$$

Numeric values can now be substituted in the formula to find the impedance of the circuit.

$$Z = \frac{1}{\sqrt{\left(\frac{1}{15}\right)^2 + \left(\frac{1}{20}\right)^2}}$$

$$Z = \frac{1}{\sqrt{0.004444 + 0.0025}}$$

$$Z = \frac{1}{\sqrt{0.006944}}$$

$$Z = \frac{1}{0.08333}$$

$$Z = 12\ \Omega$$

Another formula that can be used to determine the impedance of resistance and inductive reactance connected in parallel is

$$Z = \frac{R \times X_L}{\sqrt{R^2 + X_L^2}}$$

Substituting the same values for resistance and inductive reactance in this formula will result in the same answer.

$$Z = \frac{15 \times 20}{\sqrt{15^2 + 20^2}}$$

$$Z = \frac{300}{\sqrt{625}}$$

$$Z = \frac{300}{25}$$

$$Z = 12\ \Omega$$

Apparent Power

The **apparent power (VA)** can be computed by multiplying the circuit voltage by the total current flow. The relationship of volt-amps, watts, and VARs is the same for an R-L parallel circuit as it is for an R-L series circuit, because power adds in any type of circuit. Since the true power and reactive power are 90° out of phase with each other, they form a right triangle with apparent power as the hypotenuse (Fig. 14-7).

$$VA = E_T \times I_T$$

$$VA = 240 \times 20$$

$$VA = 4800$$

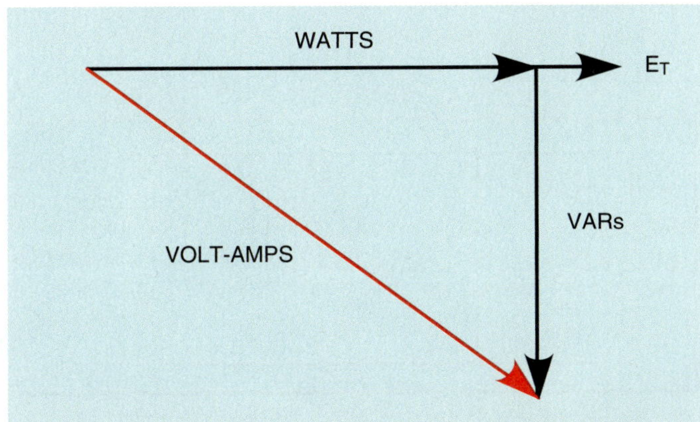

Figure 14-7 Relationship of apparent power (volt-amps), true power (watts), and reactive power (VARs) in an R-L parallel circuit.

Power Factor

Power factor (PF) in an R-L parallel circuit is the relationship of apparent power to the true power just as it was in the R-L series circuit. There are some differences in the formulas used to compute power factor in a parallel circuit, however. In an R-L series circuit, power factor could be computed by dividing the voltage dropped across the resistor by the total, or applied, voltage. In a parallel circuit, the voltage is the same, but the currents are different. Therefore, power factor can be computed by dividing the current flow through the resistive parts of the circuit by the total circuit current.

$$PF = \frac{I_R}{I_T}$$

Another formula that changes involves resistance and impedance. In a parallel circuit, the total circuit impedance will be less than the resistance. Therefore, if power factor is to be computed using impedance and resistance, the impedance must be divided by the resistance.

$$PF = \frac{Z}{R}$$

The circuit power factor in this example will be computed using the formula

$$PF = \frac{P}{VA} \times 100$$

$$PF = \frac{3840}{4800} \times 100$$

$$PF = 0.80, \text{ or } 80\%$$

Angle Theta

The cosine of **angle theta ($\angle\theta$)** is equal to the power factor.

$$\cos \angle\theta = 0.80$$

$$\angle\theta = 36.87°$$

A vector diagram using apparent power, true power, and reactive power is shown in Figure 14-8. Notice that angle theta is the angle produced by the apparent power and the true power. The relationship of current and voltage for this circuit is shown in Figure 14-9. The circuit with all values is shown in Figure 14-10.

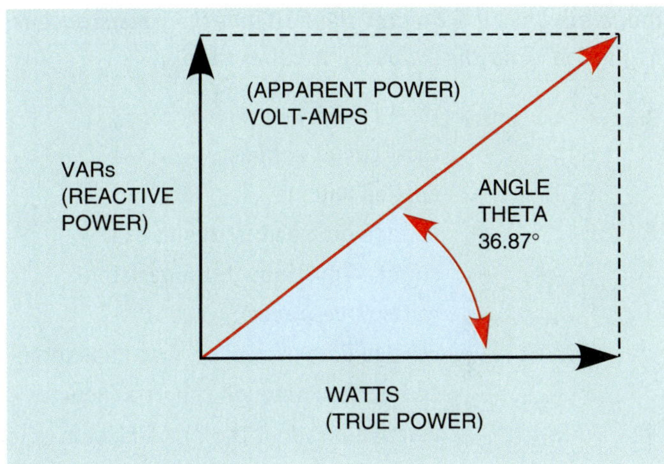

Figure 14-8 Angle theta.

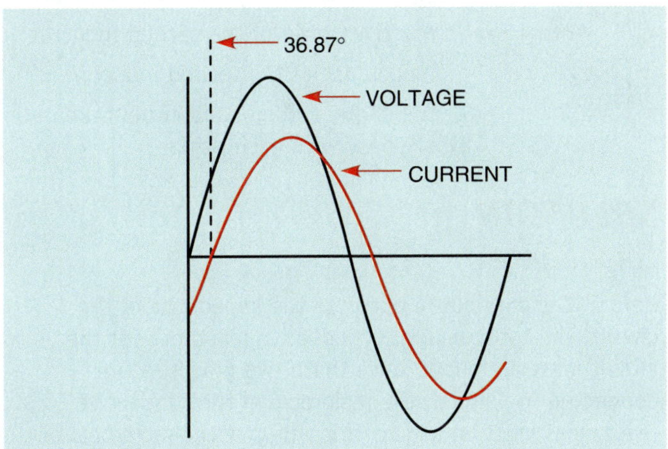

Figure 14-9 The current is 36.87° out of phase with the voltage.

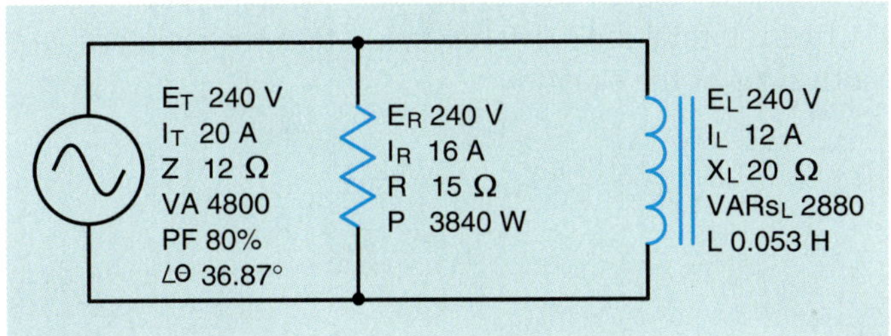

Figure 14-10 All values have been found.

Example 1

In this circuit, one resistor is connected in parallel with two inductors (Fig. 14-11). The frequency is 60 Hz. The circuit has an apparent power of 6120 VA, the resistor has a resistance of 45 Ω, the first inductor has an inductive reactance of 40 Ω, and the second inductor has an inductive reactance of 60 Ω. It is assumed that both inductors have a Q greater than 10 and their resistance is negligible. Find the following missing values.

Z	—	total circuit impedance
I_T	—	total circuit current
E_T	—	applied voltage
E_R	—	voltage drop across the resistor
I_R	—	current flow through the resistor
P	—	watts (true power)
E_{L1}	—	voltage drop across the first inductor
I_{L1}	—	current flow through the first inductor
$VARs_{L1}$	—	reactive power of the first inductor
L_1	—	inductance of the first inductor
E_{L2}	—	voltage drop across the second inductor
I_{L2}	—	current flow through the second inductor
$VARs_{L2}$	—	reactive power of the second inductor
L_2	—	inductance of the second inductor
$\angle\theta$	—	the angle, voltage, and current are out of phase with each other

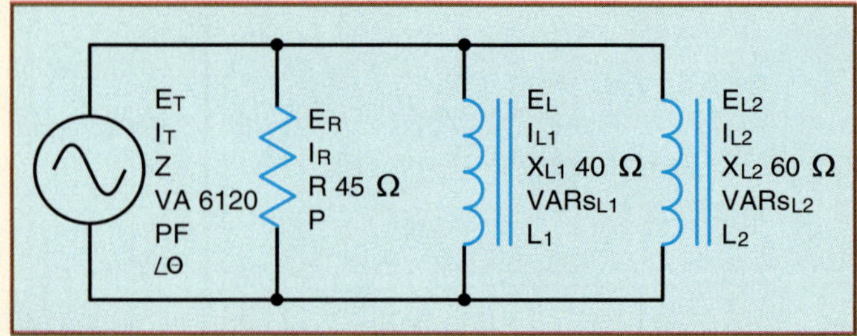

Figure 14-11 Example circuit 2.

SOLUTION

Impedance

Before it is possible to compute the impedance of the circuit, the total amount of inductive reactance for the circuit must be found. Since these two inductors are connected in parallel, the reciprocal of their inductive reactances must be added. This will give the reciprocal of the total inductive reactance:

$$\frac{1}{X_{LT}} = \frac{1}{X_{L1}} = \frac{1}{X_{L2}}$$

To find the total inductive reactance, take the reciprocal of both sides of the equation

$$X_{LT} = \frac{1}{\frac{1}{X_{L1}} + \frac{1}{X_{L2}}}$$

Refer to the formulas for pure inductive circuits shown in the Alternating Current Formulas section of the appendix. Numeric values can now be substituted in the formula to find the total inductive reactance.

$$X_{LT} = \frac{1}{\frac{1}{40} + \frac{1}{60}}$$

$$X_{LT} = \frac{1}{0.025 + 0.01667}$$

$$X_{LT} = 24\ \Omega$$

Now that the total amount of inductive reactance for the circuit is known, the impedance can be computed using the formula

$$Z = \frac{1}{\sqrt{\left(\frac{1}{R}\right)^2 + \left(\frac{1}{X_{LT}}\right)^2}}$$

$$Z = \frac{1}{\sqrt{\left(\frac{1}{45}\right)^2 + \left(\frac{1}{24}\right)^2}}$$

$$Z = 21.176\ \Omega$$

A diagram showing the relationship of resistance, inductive reactance, and impedance is shown in Figure 14-12.

E_T

Now that the circuit impedance and the apparent power are known, the applied voltage can be computed using the formula

$$E_T = \sqrt{VA \times Z}$$

$$E_T = \sqrt{6120 \times 21.176}$$

$$E_T = 360\ V$$

Example 1 Continued

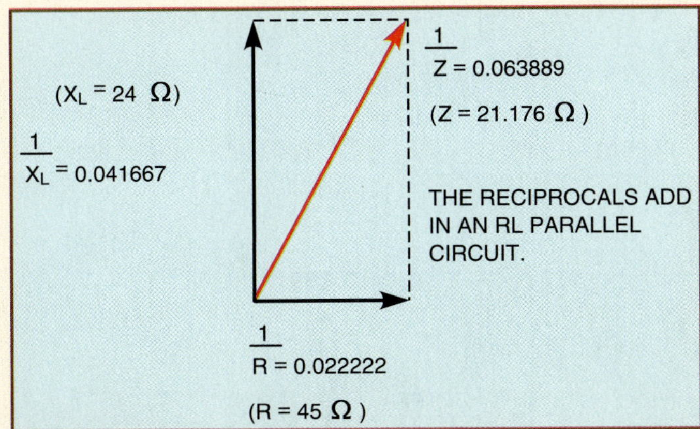

Figure 14-12 **Relationship of resistance, inductive reactance, and impedance for circuit 2.**

E_R, E_{L1}, and E_{L2}

In a parallel circuit, the voltage must be the same across any leg or branch. Therefore, 360 V is dropped across the resistor, the first inductor, and the second inductor.

$$E_R = 360\text{ V}$$

$$E_{L1} = 360\text{ V}$$

$$E_{L2} = 360\text{ V}$$

I_T

The total current of the circuit can now be computed using the formula

$$I_T = \frac{E_T}{Z}$$

$$I_T = \frac{360}{21.176}$$

$$I_T = 17\text{ A}$$

The remaining values of the circuit can be found using Ohm's law. Refer to the Resistive-Inductive Parallel Circuits listed in the alternating current formula section in the appendix.

I_R

$$I_R = \frac{E_R}{R}$$

$$I_R = \frac{360}{45}$$

$$I_R = 8\text{ A}$$

P

$$P = E_R \times I_R$$

$$P = 360 \times 8$$

$$P = 2880\text{ W}$$

I_{L1}

$$I_{L1} = \frac{E_{L1}}{X_{L1}}$$

$$I_{L1} = \frac{360}{40}$$

$$I_{L1} = 9\text{A}$$

$VARs_{L1}$

$$VARs_{L1} = E_{L1} \times I_{L1}$$

$$VARs_{L1} = 360 \times 9$$

$$VARs_{L1} = 3240$$

L_1

$$L_1 = \frac{X_{L1}}{2\pi F}$$

$$L_1 = \frac{360}{377}$$

$$L_1 = 0.106\text{ H}$$

I_{L2}

$$I_{L2} = \frac{E_{L2}}{X_{L2}}$$

$$L_1 = \frac{360}{377}$$

$$I_{L2} = 6\text{ A}$$

$VARs_{L2}$

$$VARs_{L2} = E_{L2} \times I_{L2}$$

$$VARs_{L2} = 360 \times 6$$

$$VARs_{L2} = 2160$$

Example 1 Continued

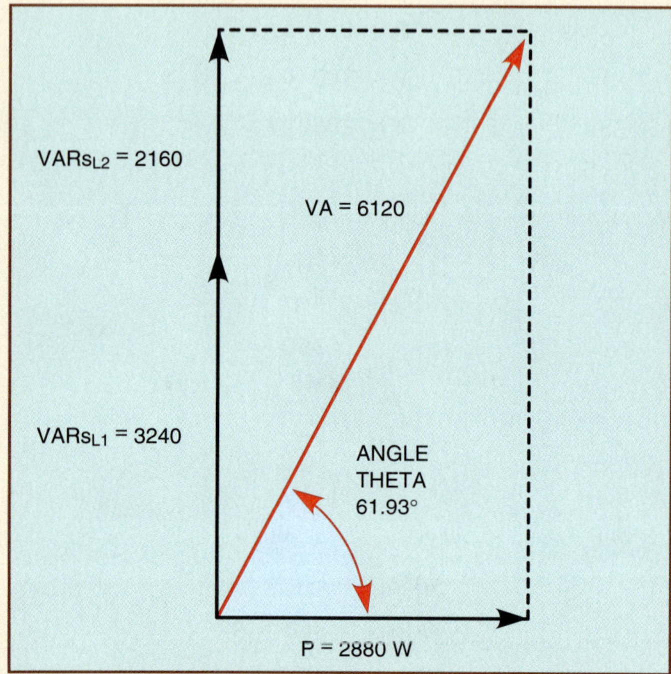

Figure 14-13 Angle theta determined by power vectors.

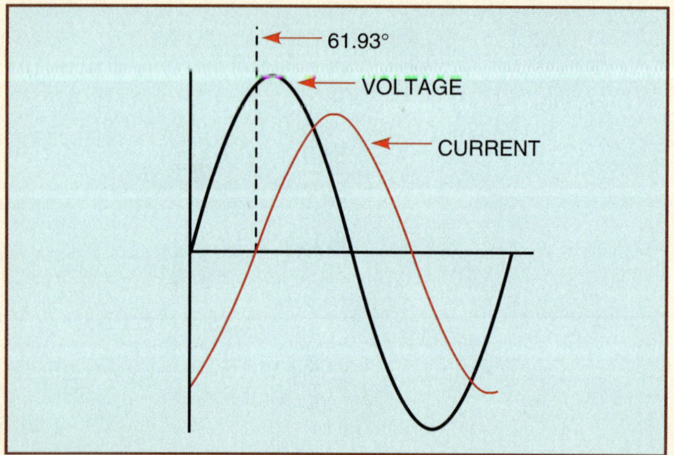

Figure 14-14 Voltage and current are 61.93° out of phase with each other.

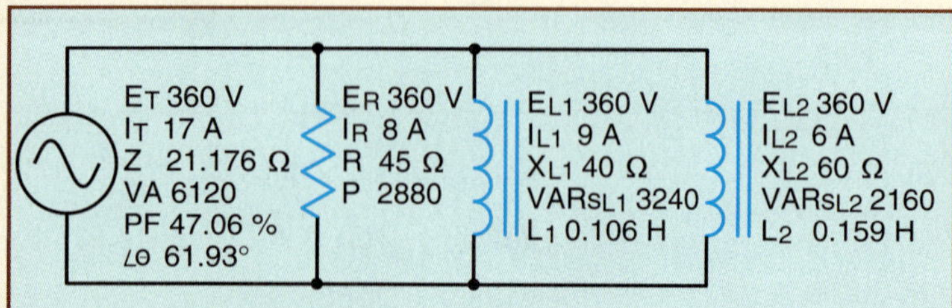

Figure 14-15 Completed values for circuit 2.

L_2

$$L_2 = \frac{X_{L2}}{2\pi F}$$

$$L_2 = \frac{60}{377}$$

$$L_2 = 0.159\ H$$

PF

$$PF = \frac{W}{VA}$$

$$PF = \frac{2880}{6120}$$

$$PF = 47.06\%$$

$\angle\theta$

$$\text{COS } \angle\theta = PF$$

$$\text{COS } \angle\theta = 0.4706$$

$$\angle\theta = 61.93°$$

A vector diagram showing angle theta is shown in Figure 14-13. The vectors used are those for apparent power, true power, and reactive power. The phase relationship of voltage and current for this circuit are shown in Figure 14-14, and the circuit with all completed values is shown in Figure 14-15.

Summary

- The voltage applied across components in a parallel circuit must be the same.
- The current flowing through resistive parts of the circuit will be in phase with the voltage.
- The current flowing through inductive parts of the circuit will lag the voltage by 90°.
- The total current in a parallel circuit is equal to the sum of the individual currents. Vector addition must be used because the current through the resistive parts of the circuit is 90° out of phase with the current flowing through the inductive parts.
- The impedance of an R-L parallel circuit can be computed using vector addition to add the reciprocals of the resistance and inductive reactance.
- Apparent power, true power, and reactive power add in any kind of circuit. Vector addition must be used, however, because true power and reactive power are 90° out of phase with each other.

Review Questions

1. **When an inductor and resistor are connected in parallel, how many degrees out of phase are the current flow through the resistor and the current flow through the inductor?**
2. **An inductor and resistor are connected in parallel to a 120-V, 60-Hz line. The resistor has a resistance of 50 Ω, and the inductor has an inductance of 0.2 H. What is the total current flow through the circuit?**
3. **What is the impedance of the circuit in question 2?**
4. **What is the power factor of the circuit in question 2?**
5. **How many degrees out of phase are the current and voltage in question 2?**
6. **In the circuit shown in Figure 14-1, the resistor has a current flow of 6.5 A, and the inductor has a current flow of 8 A. What is the total current in this circuit?**
7. **A resistor and inductor are connected in parallel. The resistor has a resistance of 24 Ω, and the inductor has an inductive reactance of 20 Ω. What is the impedance of this circuit?**
8. **The R-L parallel circuit shown in Figure 14-1 has an apparent power of 325 VA. The circuit power factor is 66%. What is the true power in this circuit?**
9. **The R-L parallel circuit shown in Figure 14-1 has an apparent power of 465 VA and a true power of 320 W. What is the reactive power?**
10. **How many degrees out of phase are the total current and voltage in question 9?**

Practice Problems

Refer to the circuit shown in Figure 14-1. Use the alternating current formulas in the Resistive-Inductive Parallel Circuits section of the appendix.

1. Assume that the circuit shown in Figure 14-1 is connected to a 60-Hz line and has a total current flow of 34.553 A. The inductor has an inductance of 0.02122 H, and the resistor has a resistance of 14 Ω.

E_T ______	E_R ______	E_L ______
I_T 34.553 ______	I_R ______	I_L ______
Z ______	R 14	X_L ______
VA ______	P ______	$VARs_L$ ______
PF ______	$\angle\theta$ ______	L 0.02122

2. Assume that the current flow through the resistor, I_R, is 15 A; the current flow through the inductor, I_L, is 36 A; and the circuit has an apparent power of 10,803 VA. The frequency of the AC voltage is 60 Hz.

E_T ______	E_R ______	E_L ______
I_T ______	I_R 15	IL 36
Z ______	R ______	X_L ______
VA 10,803	P ______	$VARs_L$ ______
PF ______	$\angle\theta$ ______	L ______

3. Assume that the circuit in Figure 14-1 has an apparent power of 144 VA and a true power of 115.2 W. The inductor has an inductance of 0.15915 H, and the frequency is 60 Hz.

E_T ______	E_R ______	E_L ______
I_T ______	I_R ______	I_L ______
Z ______	R ______	X_L ______
VA 144 ______	P 115.2 ______	$VARs_L$ ______
PF ______	$\angle\theta$ ______	L 0.15915 ______

4. Assume that the circuit in Figure 14-1 has a power factor of 78%, an apparent power of 374.817 VA, and a frequency of 400 Hz. The inductor has an inductance of 0.0382 H.

E_T ______	E_R ______	E_L ______
I_T ______	I_R ______	I_L ______
Z ______	R ______	X_L ______
VA 374.817	P ______	VARsL ______
PF 78%	$\angle\theta$ ______	L 0.0382

Chapter 15 Capacitors

Capacitors perform a variety of jobs such as power factor correction, storing an electrical charge to produce a large current pulse, timing circuits, and electronic filters. Capacitors can be nonpolarized or polarized depending on the application. Nonpolarized capacitors can be used in both AC and DC circuits, while polarized capacitors can be used in DC circuits only. Both types will be discussed in this unit.

OBJECTIVES

After studying this unit, you should be able to:

- **list the three factors that determine the capacitance of a capacitor.**
- **discuss the electrostatic charge.**
- **discuss the differences between nonpolarized and polarized capacitors.**
- **compute values for series and parallel connections of capacitors.**
- **compute an RC time constant.**
- **explain why current appears to flow through a capacitor when it is connected to an alternating current circuit.**
- **discuss capacitive reactance.**
- **compute the value of capacitive reactance in an AC circuit.**
- **compute the value of capacitance in an AC circuit.**
- **discuss the relationship of voltage and current in a pure capacitive circuit.**

Glossary of Capacitor Terms

appear to flow the charging and discharging of a capacitor will cause current to appear to flow through a capacitor when it is connected to an AC circuit

capacitive reactance (X_C) the current-limiting effect of a capacitor

dielectric the insulation between the plates of a capacitor

dielectric stress the stretching of the molecular structure of the dielectric

electrolytic a type of capacitor that is polarity sensitive

electrostatic charge the charge of a capacitor produced by the stressing of the molecules in the dielectric

exponential the rate at which a capacitor charges and discharges

farad the unit measure of capacitance

frequency the number of times per second that the current changes direction in an alternating current circuit

HIPOT an instrument that develops a high potential or high voltage for testing the dielectric of a capacitor

leakage current the amount of current that flows through the dielectric of a capacitor

nonpolarized capacitors capacitors that are not sensitive to voltage polarity

out of phase a condition that occurs between the voltage and current in a capacitive circuit

plates thin, flat metal used in the construction of a capacitor

polarized capacitors capacitors that are sensitive to voltage polarity

reactive power (VARs) volt amps reactive; often called wattless power, VARs is the product of the current and voltage in a pure reactive load

RC time constant the time required for a capacitor to charge or discharge when it is connected in series with a resistance

surface area one of the factors that determines the amount of capacitance a capacitor will have

variable capacitors capacitors that can be adjusted for different values of capacitance

voltage rating the amount of voltage that can be safely applied to a capacitor without damaging the dielectric

Capacitors

CAUTION: It is the habit of some people to charge a capacitor to high voltage and then hand it to another person. While some people think this is comical, it is an extremely dangerous practice. Capacitors have the ability to supply an almost infinite amount of current. Under some conditions, a capacitor can have enough energy to cause a person's heart to go into fibrillation.

This statement is not intended to strike fear in the heart of anyone working in the electrical field. It is intended to make you realize the danger that capacitors can pose under certain conditions.

Capacitors are devices that oppose a change of voltage. The simplest type of capacitor is constructed by separating two metal **plates** by some type of insulating material called the **dielectric** (Fig. 15-1). Three factors determine the capacitance of a capacitor:

1. The area of the plates.
2. The distance between the plates.
3. The type of dielectric used.

The greater the **surface area** of the plates, the more capacitance a capacitor will have. If a capacitor is charged by connecting it to a source of direct current (Fig. 15-2), electrons are removed from the plate connected to the positive battery terminal and are deposited on the plate connected to the negative terminal. This flow of current will continue until a voltage equal to the battery voltage is established across the plates of the capacitor (Fig. 15-3). When these two voltages become equal, the flow of electrons stops. The capacitor is now charged. If the battery is disconnected from the capacitor, the capacitor will remain charged as long as there is no path by which the electrons can move from one plate to the other (Fig. 15-4). A good rule to remember concerning a capacitor and current flow is that **current can flow only during the period of time that a capacitor is either charging or discharging.**

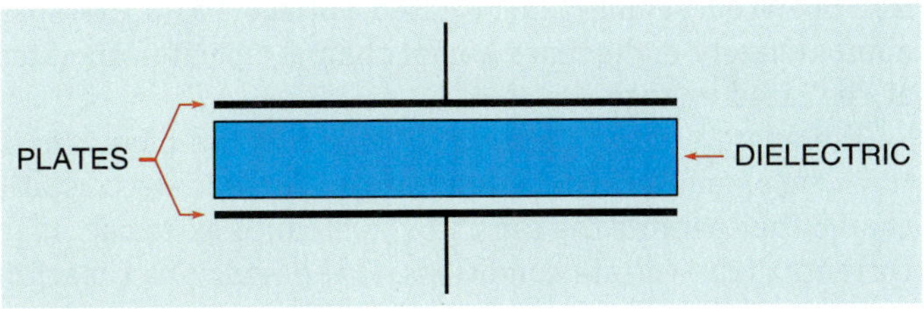

Figure 15-1 A capacitor is made by separating two metal plates by a dielectric.

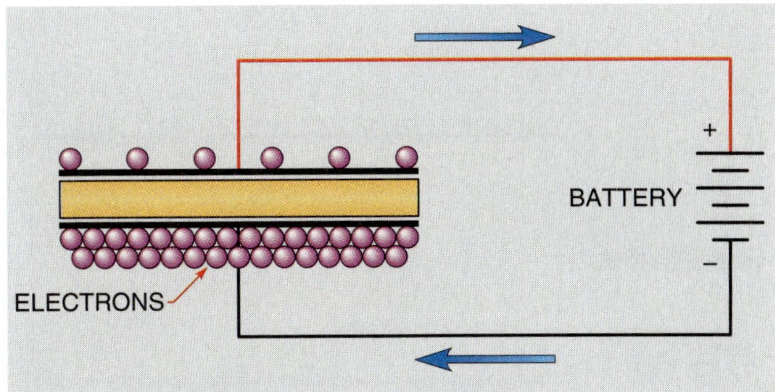

Figure 15-2 A capacitor can be charged by removing electrons from one plate and depositing them on the other plate.

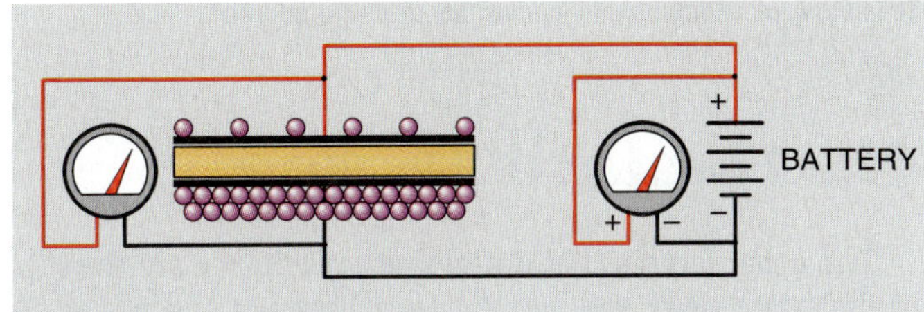

Figure 15-3 Current flows until the voltage across the capacitor is equal to the voltage of the battery.

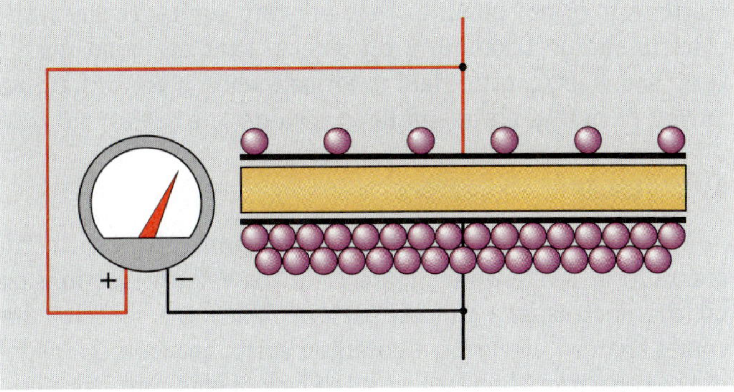

Figure 15-4 The capacitor remains charged after the battery is removed from the circuit.

In theory, it should be possible for a capacitor to remain in a charged condition forever. In actual practice, however, it cannot. No dielectric is a perfect insulator, and electrons eventually move through the dielectric from the negative plate to the positive, causing the capacitor to discharge (Fig. 15-5). This current flow through the dielectric is called **leakage current** and is proportional to the resistance of the dielectric and the charge across the plates. If the dielectric of a capacitor becomes weak, it will permit an excessive amount of leakage current to flow. A capacitor in this condition is often referred to as a leaky capacitor.

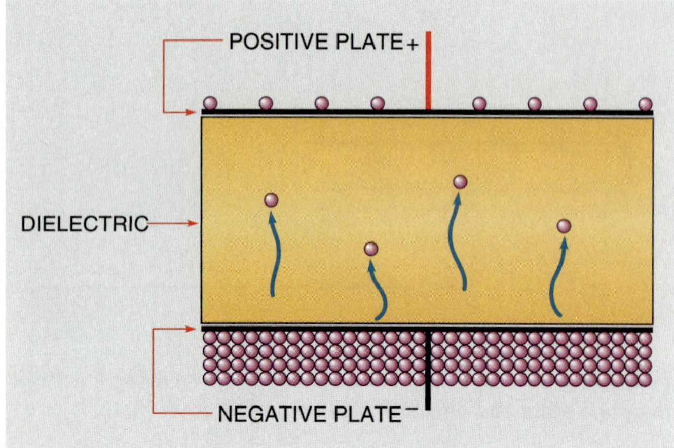

Figure 15-5 Electrons eventually leak through the dielectric. This flow of electrons is known as leakage current.

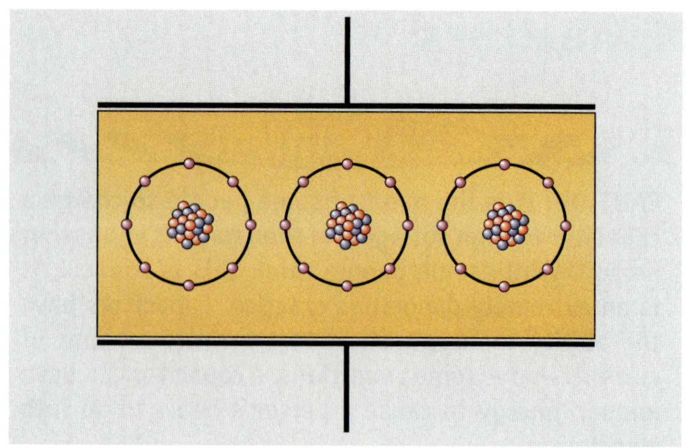

Figure 15-6 Atoms of the dielectric in an uncharged capacitor.

Electrostatic Charge

Two other factors that determine capacitance are the type of dielectric used and the distance between the plates. To understand these concepts, it is necessary to understand how a capacitor stores energy. Previous chapters stated that an inductor stores energy in the form of an electromagnetic field. A capacitor stores energy in an electrostatic field.

The term *electrostatic* refers to electrical charges that are stationary, or not moving. They are very similar to the static electric charges that form on objects that are good insulators. The electrostatic field is formed when electrons are removed from one plate and deposited on the other.

Dielectric Stress

When a capacitor is not charged, the atoms of the dielectric are uniform, as shown in Figure 15-6. The valence electrons orbit the nucleus in a circular pattern. When the capacitor becomes charged, however, a potential exists between the plates of the capacitor. The plate with the lack of electrons has a positive charge, and the plate with the excess of electrons has a negative charge. Since electrons are negative particles, they are repelled away from the negative plate and attracted to the positive plate. This attraction causes the electron orbit to become stretched, as shown in Figure 15-7. This stretching of the atoms of the dielectric is called **dielectric stress.** Placing the atoms of the dielectric under stress has the same effect as drawing back a bowstring with an arrow and holding it (Fig. 15-8); that is, it stores energy.

The amount of dielectric stress is proportional to the voltage difference between the plates. The greater the voltage, the greater the dielectric stress. If the voltage becomes too great, the dielectric will break down and permit current to flow between the plates. At this point the capacitor becomes shorted. Capacitors have a voltage rating that should not be exceeded. The voltage rating indicates the maximum amount of voltage the dielectric is intended to withstand without breaking down. The amount of voltage applied to a capacitor is critical to its life span. Capacitors operated above their voltage rating will fail relatively quickly. Many years ago, the U.S. military made a study of the voltage rating of a capacitor relative to its life span. The results showed that a capacitor operated at one-half its rated voltage has a life span approximately eight times longer than a capacitor operated at the rated voltage.

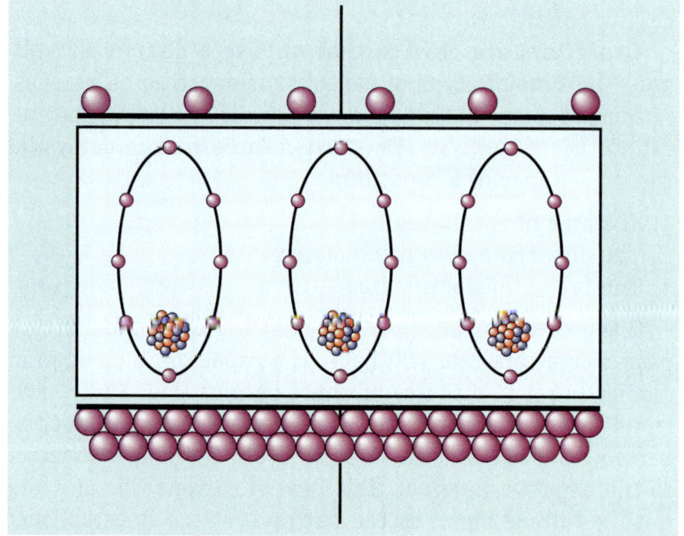

Figure 15-7 Atoms of the dielectric in a charged capacitor.

The energy of the capacitor is stored in the dielectric in the form of an **electrostatic charge.** It is this electrostatic charge that permits the capacitor to produce extremely high currents under certain conditions. If the leads of a capacitor are shorted together, it has the effect of releasing the drawn-back bowstring (Fig. 15-8). When the bowstring is released, the arrow is propelled forward at a high rate of speed. The same is true for the capacitor. When the leads are shorted, the atoms of the dielectric snap back to their normal position. Shorting causes the electrons on the negative plate to be blown off and attracted to the positive plate. Capacitors can produce currents of thousands of amperes for short periods of time.

This principle is used to operate the electronic flash of many cameras. Electronic flash attachments contain a small glass tube filled with a gas called xenon. Xenon produces a very bright white light similar to sunlight when the gas is ionized. A large amount of power is required, however, to produce a bright flash. A battery capable of directly ionizing the xenon would be very large and expensive, and would have a potential of about 500 V. The simple circuit shown in Figure 15-9 can be used to overcome the problem. In this circuit, two small 1.5-V batteries are connected to an oscillator. The oscillator changes the direct current of the batteries into square wave alternating current. The alternating current is then connected to a transformer, and the voltage is increased to about 500 V peak. A diode changes the AC voltage back into DC and charges a capacitor. The capacitor charges to the peak value of the voltage wave form. When the switch is closed, the capacitor suddenly discharges through the xenon tube and supplies the power needed to ionize the gas. It may take several seconds to store enough energy in the capacitor to ionize the gas in the tube, but the capacitor can release the stored energy in a fraction of a second.

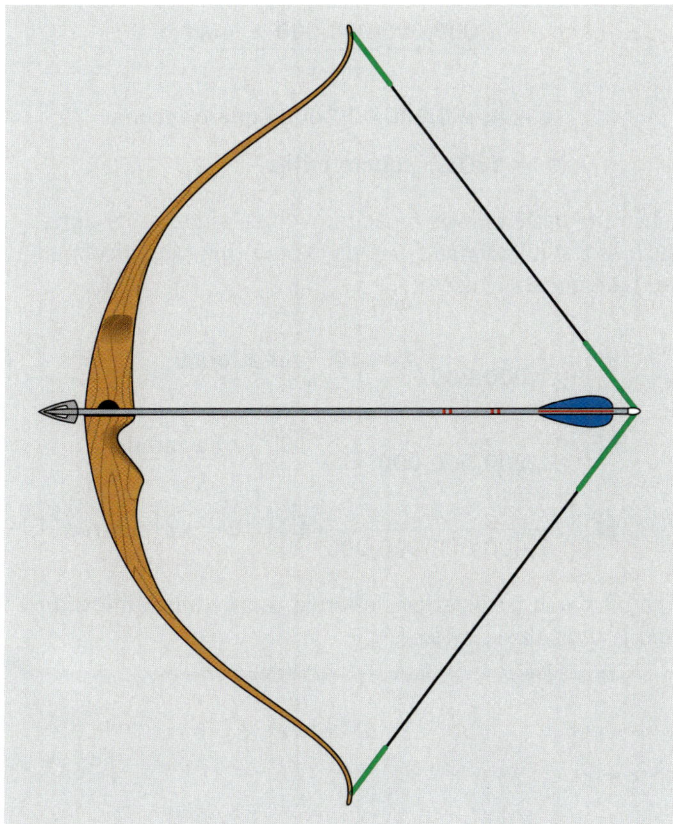

Figure 15-8 Dielectric stress is similar to drawing back a bowstring with an arrow and holding it.

To understand how the capacitor can supply the energy needed, consider the amount of gunpowder contained in a 0.357 cartridge. If the powder were to be removed from the cartridge and burned in the open air, the actual amount of energy contained in the powder would be very small. This amount of energy would not even raise the temperature by a noticeable amount in a small enclosed room. If the same amount of energy is converted into heat in a fraction of a second, however, enough force is developed to propel a heavy projectile with great force. This same principle is at work when a capacitor is charged over some period of time and then discharged in a fraction of a second.

Dielectric Constants

Since much of the capacitor's energy is stored in the dielectric, the type of dielectric used is extremely important in determining the amount of capacitance a capacitor will have. Different materials are assigned a number, called the *dielectric constant.* Air is assigned the number 1 and is used as a reference for comparison. For example, assume a capacitor uses air as the dielectric and its capacitance value is found to be 1 microfarad (μF). Now assume that some dielectric material is placed between the plates without changing the spacing, and the capacitance value becomes 5 μF. This material has a dielectric constant of 5. A chart showing the dielectric constant of different materials is shown in Figure 15-10.

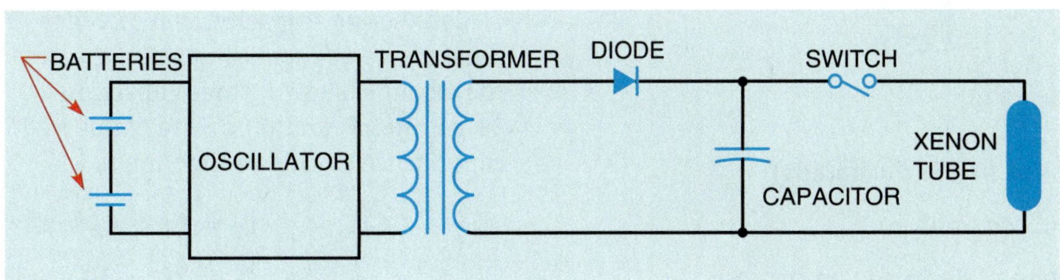

Figure 15-9 Energy is stored in a capacitor.

MATERIAL	DIELECTRIC CONSTANT
AIR	1
BAKELITE	4.0 –10.0
CASTOR OIL	4.3 – 4.7
CELLULOSE ACETATE	7.0
CERAMIC	1200
DRY PAPER	3.5
HARD RUBBER	2.8
INSULATING OILS	2.2 – 4.6
LUCITE	2.4 – 3.0
MICA	6.4 – 7.0
MYCALEX	8.0
PARAFFIN	1.9 – 2.2
PORCELAIN	5.5
PURE WATER	81
PYREX GLASS	4.1 – 4.9
RUBBER COMPOUNDS	3.0 – 7.0
TEFLON	2
TITANIUM DIOXIDE COMPOUNDS	90 – 170

Figure 15-10 **Dielectric constant of different materials.**

Capacitor Ratings

The basic unit of capacitance is the **farad,** which is symbolized by the letter *F.* It receives its name from a famous scientist named Michael Faraday. **A capacitor has a capacitance of one farad when a change of one volt across its plates results in a movement of one coulomb.**

$$Q = C \times V$$

where

Q = charge in coulombs

C = capacitance in farads

V = charging voltage

Although the farad is the basic unit of capacitance, it is seldom used because it is an extremely large amount of capacitance. The following formula can be used to determine the capacitance of a capacitor when the area of the plates, the dielectric constant, and the distance between the plates are known.

$$C = \frac{K \times A}{4.45\ D}$$

where

c = capacitance in pF (picofarads)

K = dielectric constant

A = area of one plate in square inches

D = distance between the plates in inches

Example 1

What would be the plate area of a 1-F (farad) capacitor if air is used as the dielectric and the plates are separated by a distance of 1 inch?

SOLUTION

The first step is to convert the above formula to solve for area.

$$A = \frac{C \times 4.45 \times D}{K}$$

$$A = \frac{1{,}000{,}000{,}000{,}000 \times 4.45 \times 1}{1}$$

$$A = 4{,}450{,}000{,}000{,}000 \text{ square inches}$$

$$A = 1108.5 \text{ square miles}$$

Since the basic unit of capacitance is so large, other units such as the microfarad (μF), nanofarad (nF), and picofarad (pF) are generally used.

$$\mu F = \frac{1}{1{,}000{,}000}(1 \times 10^{-6}) \text{ of a farad}$$

$$nF = \frac{1}{1{,}000{,}000{,}000}(1 \times 10^{-9}) \text{ of a farad}$$

$$pF = \frac{1}{1{,}000{,}000{,}000{,}000}(1 \times 10^{-12}) \text{ of a farad}$$

The picofarad is sometimes referred to as a micro-microfarad and is symbolized by μμF.

Capacitors Connected in Parallel

Connecting capacitors in parallel (Fig. 15-11) has the same effect as increasing the plate area of one capacitor. In the example shown, three capacitors having a capacitance of 20 μF, 30 μF, and 60 μF are connected in parallel. The total capacitance of this connection is

$$C_T = C_1 + C_2 + C_3$$

$$C_T = 20 + 30 + 60$$

$$C_T = 110\ \mu F$$

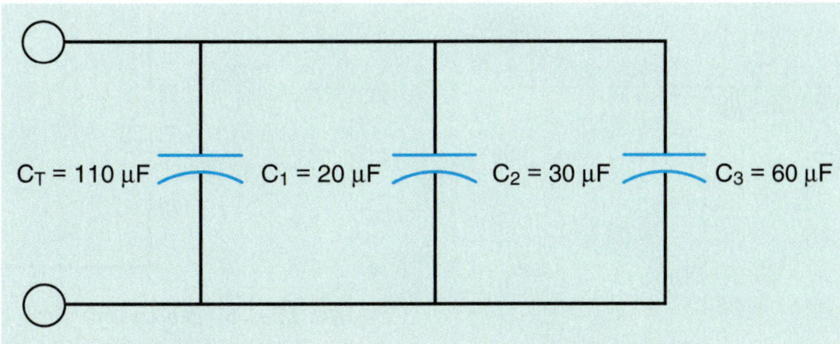

Figure 15-11 Capacitors connected in parallel.

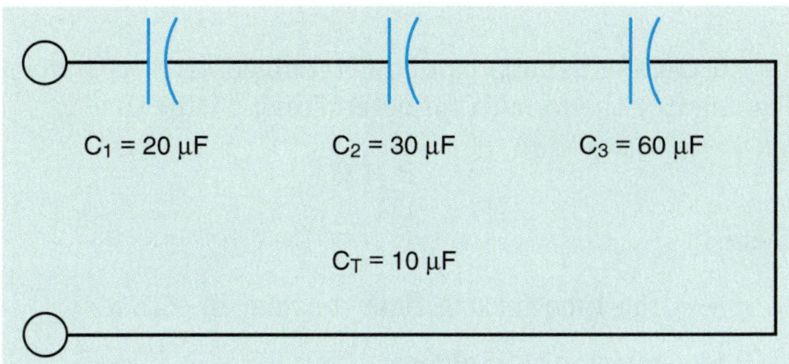

Figure 15-12 Capacitors connected in series.

Capacitors Connected in Series

Connecting capacitors in series (Fig. 15-12) has the effect of increasing the distance between the plates, thus reducing the total capacitance of the circuit. The total capacitance can be computed in a manner similar to computing parallel resistance. The following formulas can be used to find the total capacitance when capacitors are connected in series.

$$C_T = \frac{1}{\frac{1}{C_1} + \frac{1}{C_2} + \frac{1}{C_3}}$$

or

$$C_T = \frac{C_1 \times C_2}{C_1 + C_2}$$

or

$$C_T = \frac{C}{N}$$

where

C = capacitance of one capacitor

N = number of capacitors connected in series

(*Note:* The last formula can be used only when all the capacitors connected in series are of the same value.)

Example 2

What is the total capacitance of three capacitors connected in series if C_1 has a capacitance of 20 μF, C_2 has a capacitance of 30 μF, and C_3 has a capacitance of 60 μF?

SOLUTION

$$C_T = \frac{1}{\frac{1}{C_1} + \frac{1}{C_2} + \frac{1}{C_3}}$$

$$C_T = \frac{1}{\frac{1}{20} + \frac{1}{30} + \frac{1}{60}}$$

$$C_T = 10\ \mu F$$

Capacitive Charge and Discharge Rates

Capacitors charge and discharge at an **exponential** rate. A charge curve for a capacitor is shown in Figure 15-13. The curve is divided into five time constants, and each time constant is equal to 63.2% of the whole value. In Figure 15-13 it is assumed that a capacitor is to be charged to a total of 100 V. At the end of the first time constant, the voltage has reached 63.2% of 100, or 63.2 V. At the end of the second time constant, the voltage reaches 63.2% of the remaining voltage, or 86.4 V. This pattern continues until the capacitor has been charged to 100 V.

The capacitor discharges in the same manner (Fig. 15-14). At the end of the first time constant, the voltage will decrease by 63.2% of its charged value. In this example, the voltage will decrease from 100 V to 36.8 V in the first time constant. At the end of the second time constant the voltage

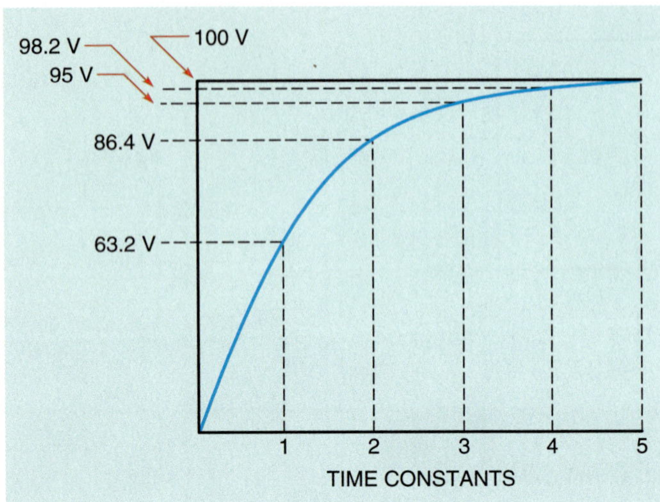

Figure 15-13 Capacitors charge at an exponential rate.

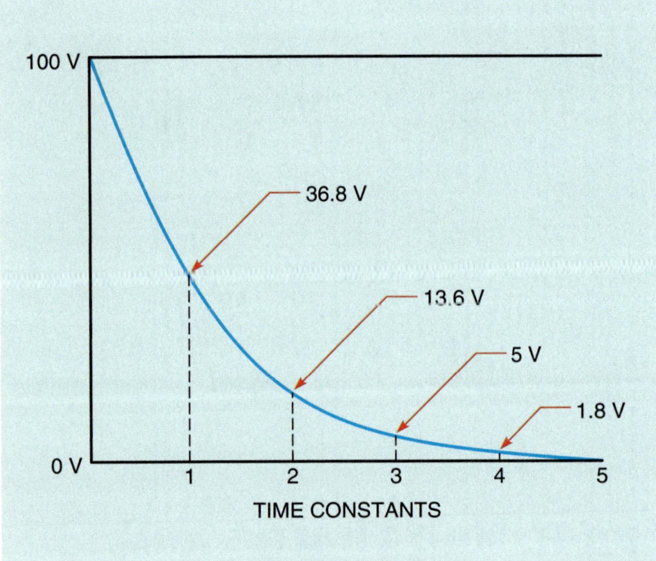

Figure 15-14 Capacitor discharge curve.

will drop to 13.6 V, and by the end of the third time constant the voltage will drop to 5 V. The voltage will continue to drop at this rate until it reaches approximately 0 after five time constants.

RC Time Constants

When a capacitor is connected in a circuit with a resistor, the amount of time needed to charge the capacitor (that is, the **RC time constant**) can be determined very accurately (Fig. 15-15). The formula for determining charge time is

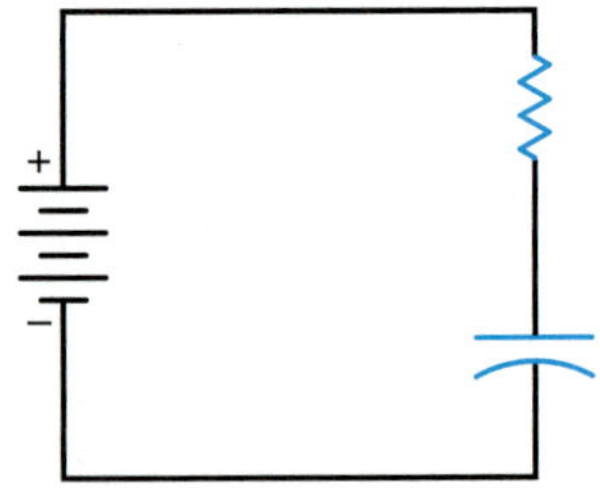

Figure 15-15 The charge time of the capacitor can be determined very accurately.

$$\tau = R \times C$$

where

τ = **the time for one time constant in seconds**

R = **resistance in ohms**

C = **capacitance in farads**

The Greek letter τ (tau) is used to represent the time for one time constant. It is not unusual, however, for the letter *T* to be used to represent time.

Example 3

How long will it take the capacitor shown in Figure 15-15 to charge if it has a value of 50 μF and the resistor has a value of 100 kΩ?

SOLUTION

$$\tau = R \times C$$

$$\tau = 0.000050\ F \times 100{,}000\ \Omega$$

$$\tau = 5\ s$$

The formula is used to find the time for one time constant. Five time constants are required to charge the capacitor.

$$\text{Total } \tau = 5\ s \times 5 \text{ time constants}$$

$$\text{Total } \tau = 25\ s$$

Example 4

How much resistance should be connected in series with a 100-pF capacitor to give it a total charge time of 0.2 s?

SOLUTION

Change the above formula to solve for resistance.

$$R = \frac{\tau}{C}$$

The total charge time is to be 0.2 s. The value of τ is therefore 0.2/5 = 0.04 s. Substitute these values in the formula.

$$R = \frac{0.04}{100 \times 10^{12}}$$

$$R = 400\ M\Omega$$

Example 5

A 500-kΩ resistor is connected in series with a capacitor. The total charge time of the capacitor is 15 s. What is the capacitance of the capacitor?

SOLUTION

Change the base formula to solve for the value of capacitance.

$$C = \frac{\tau}{R}$$

Since the total charge time is 15 s, the time of one time constant will be 3 s (15/5 = 3).

$$C = \frac{3}{500{,}000}$$

$$C = 0.000006\ F$$

or

$$C = 6\ \mu F$$

Applications for Capacitors

Capacitors are among the most used of electrical components. They are used for power factor correction in industrial applications, in the start windings of many single-phase AC motors, to produce phase shifts for SCR and Triac circuits, to filter pulsating DC, and in RC timing circuits. (SCRs and Triacs are solid state electronic devices used throughout industry to control high-current circuits.) Capacitors are used extensively in electronic circuits for control of frequency and pulse generation. The type of capacitor is dictated by the circuit application.

Nonpolarized Capacitors

Capacitors can be divided into two basic groups, nonpolarized and polarized. **Nonpolarized capacitors** are often referred to as AC capacitors, because they are not sensitive to polarity connection. Nonpolarized capacitors can be connected to either DC or AC circuits without harm to the capacitor. Nonpolarized capacitors are constructed by separating metal plates with some type of dielectric (Fig. 15-1). These capacitors come in many different styles and case types (Fig. 15-16).

A common type of AC capacitor called the paper capacitor or oil-filled capacitor is often used in motor circuits and for power factor correction (Fig. 15-17). It derives its name from the type of dielectric used. This capacitor is constructed by separating plates made of metal foil with thin sheets of paper soaked in a dielectric oil (Fig. 15-18). These capacitors are often used as the run or starting capacitor for single-phase motors. Many manufacturers of oil-filled capacitors will identify one terminal with an arrow or a painted dot, or by stamping

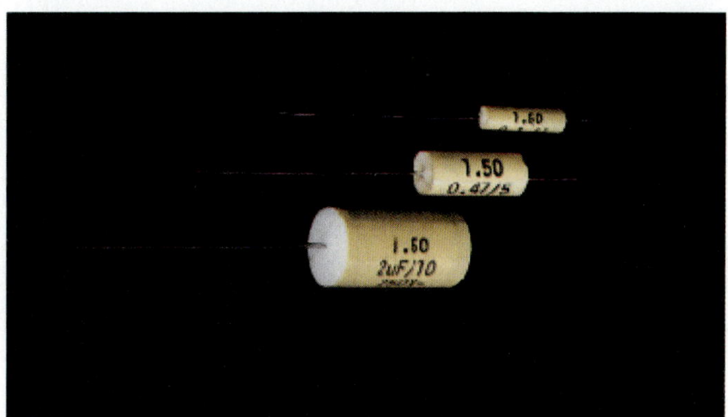

Figure 15-16 **Nonpolarized capacitors.** *Courtesy of Mallory Capacitor Co.*

Figure 15-17 Oil-filled paper capacitor.

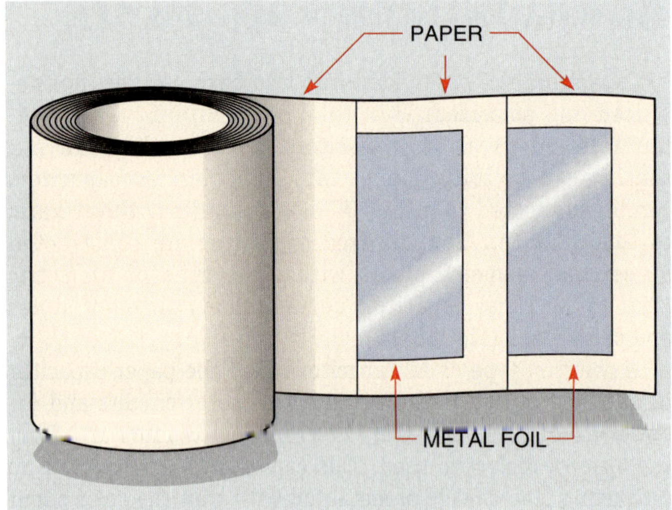

Figure 15-18 Oil-filled paper capacitor.

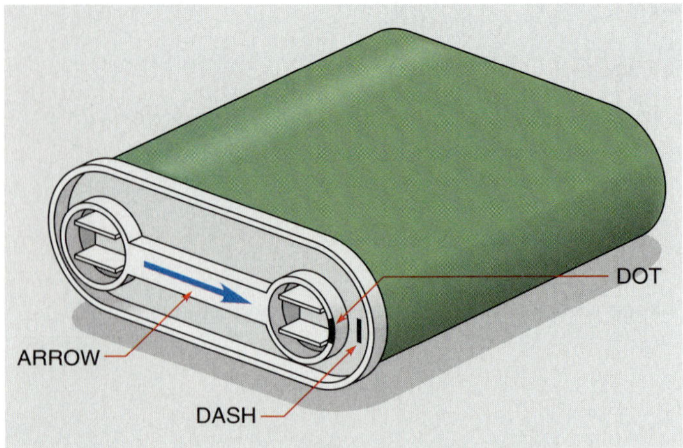

Figure 15-19 Marks indicate plate nearest capacitor case.

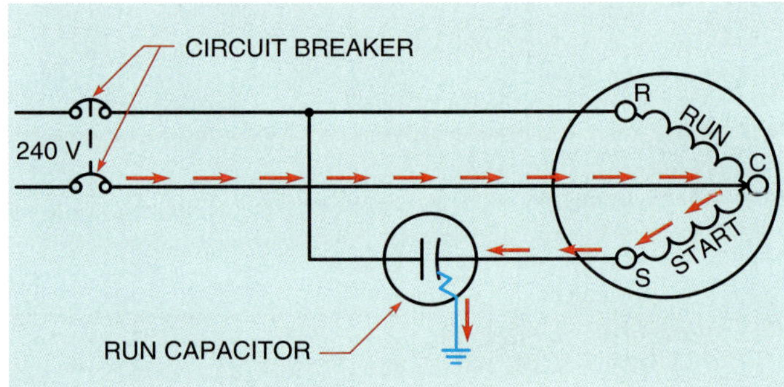

Figure 15-20 Identified capacitor terminal connected to motor start winding (incorrect connection).

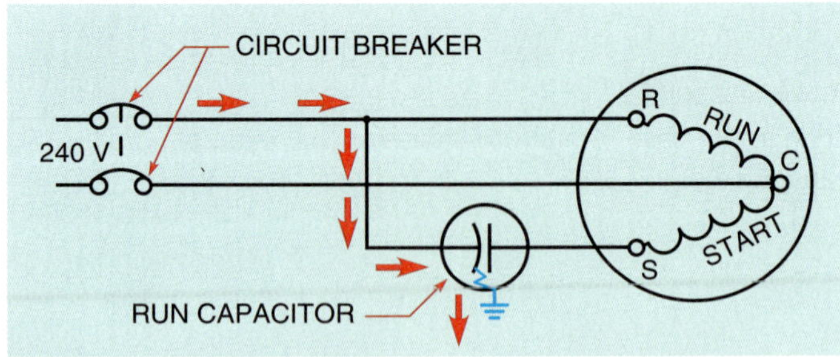

Figure 15-21 Identified terminal connected to the line (correct connections).

a dash in the capacitor can (Fig. 15-19). This identified terminal marks the connection to the plate that is located nearer to the metal container or can. It has long been known that when a capacitor's dielectric breaks down and permits a short circuit to ground, the plate nearer to the outside case most often becomes grounded. For this reason, it is generally desirable to connect the identified capacitor terminal to the line side instead of to the motor start winding.

In Figure 15-20, the run capacitor has been connected in such a manner that the identified terminal is connected to the start winding of a single-phase motor. If the capacitor should become shorted to ground, a current path exists through the motor start winding. The start winding is an inductive type load, and inductive reactance will limit the value of current flow to ground. Since the flow of current is limited, it will take the circuit breaker or fuse some time to open the circuit and disconnect the motor from the power line. This time delay can permit the start winding to overheat and become damaged.

In Figure 15-21, the run capacitor has been connected in the circuit so that the identified terminal is connected to the line side. If the capacitor should become shorted to ground, a current path exists directly to ground, bypassing the motor start winding. When the capacitor is connected in this manner, the start winding does not limit current flow and permits the fuse or circuit breaker to open the circuit almost immediately.

Polarized Capacitors

Polarized capacitors are generally referred to as **electrolytic** capacitors. These capacitors are sensitive to the polarity they are connected to and have one terminal identified as positive or negative (Fig. 15-22). Polarized capacitors can be used in DC circuits only. If their polarity connection is reversed, the capacitor can be damaged and will sometimes explode. The advantage of electrolytic capacitors is that they can have very high capacitance in a small case.

There are two basic types of electrolytic capacitors, the wet type and the dry type. The wet type electrolytic (Fig. 15-23) has a positive plate made of aluminum foil. The negative plate is actually an electrolyte made from a borax solution. A second piece of aluminum foil is placed in contact with the electrolyte and becomes the negative terminal. When a source of direct current is connected to the capacitor, the borax solution forms an insulating oxide film on the positive plate. This film is only a few molecules thick and acts as the insulator to separate the plates. The capacitance is very high because the distance between the plates is so small.

If the polarity of the wet type electrolytic capacitor becomes reversed, the oxide insulating film dissolves and the capacitor becomes shorted. If the polarity connection is corrected, the film reforms and restores the capacitor.

Figure 15-22 **Polarized capacitors.** *Courtesy of Mallory Capacitor Co.*

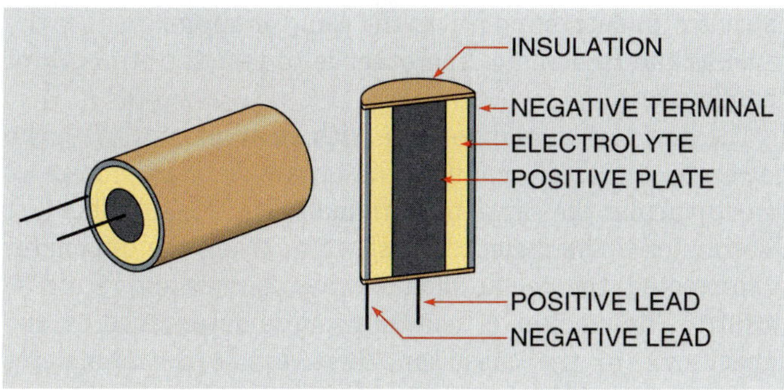

Figure 15-23 **Wet type electrolytic capacitor.**

AC Electrolytic Capacitors

The ability of the wet type electrolytic capacitor to be shorted and then reformed is the basis for a special type of nonpolarized AC electrolytic capacitor. It is used as the starting capacitor for many small single-phase motors, as the run capacitor in many ceiling fan motors, and for low-power electronic circuits when a nonpolarized capacitor with a high capacitance is required. The AC electrolytic capacitor is made by connecting two wet type electrolytic capacitors inside the same case (Fig. 15-24). In the example shown their negative terminals are connected. When alternating current is applied to the leads, one capacitor will be connected to reverse polarity and become shorted. The other capacitor will be connected to the correct polarity and will form. During the next half cycle, the polarity changes and forms the capacitor that was shorted and shorts the other capacitor. An AC electrolytic capacitor is shown in Figure 15-25.

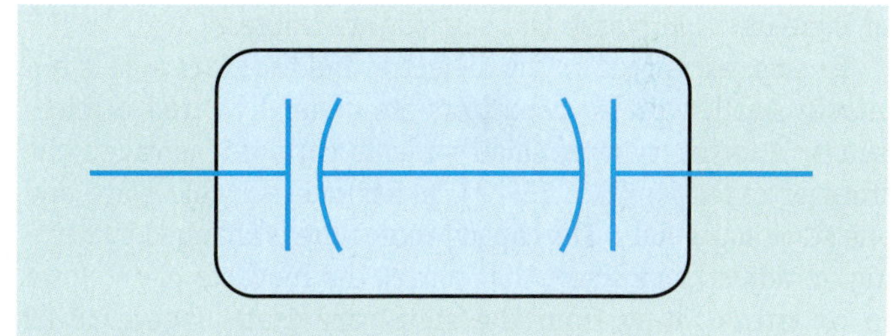

Figure 15-24 **Two wet type electrolytic capacitors connect to form an AC electrolytic capacitor.**

Figure 15-25 **An AC electrolytic capacitor.**

Dry Type Electrolytic Capacitors

The dry type electrolytic capacitor is very similar to the wet type except that gauze is used to hold the borax solution. This prevents the capacitor from leaking. Although the dry type electrolytic has the advantage of being relatively leak proof, it does have one disadvantage. If the polarity connection is reversed and the oxide film is broken down, it will not reform when connected to the proper polarity. Reversing the polarity of a dry type electrolytic capacitor will permanently damage the capacitor.

Figure 15-27 A trimmer capacitor.

Variable Capacitors

Variable capacitors are constructed so their capacitance value can be changed over a certain range. They generally contain a set of movable plates, which are connected to a shaft, and a set of stationary plates (Fig. 15-26). The movable plates can be interleaved with the stationary plates to increase or decrease the capacitance value.

Because air is used as the dielectric and the plate area is relatively small, variable capacitors are generally rated in picofarads. Another type of small variable capacitor is called the trimmer capacitor (Fig. 15-27). It has one movable plate and one stationary plate. The capacitance value is changed by turning an adjustment screw that moves the movable plate closer to or farther away from the stationary plate. Figure 15-28 shows schematic symbols used to represent variable capacitors.

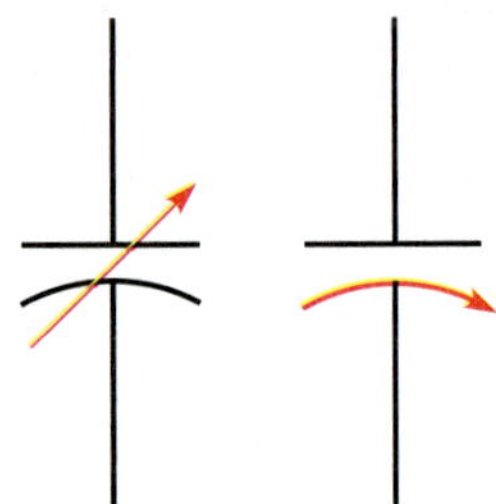

Figure 15-28 Variable capacitor symbols.

Testing Capacitors

Testing capacitors is difficult at best. Small electrolytic capacitors are generally tested for shorts with an ohmmeter. If the capacitor is not shorted, it should be tested for leakage using a variable DC power supply and a microammeter (Fig. 15-29). When rated voltage is applied to the capacitor, the microammeter should indicate zero current flow.

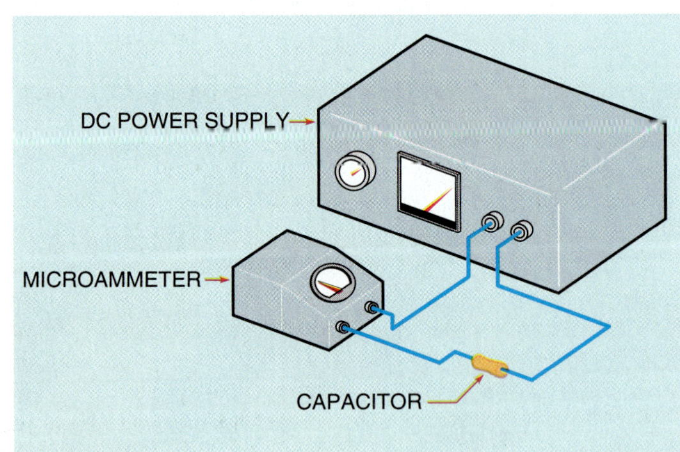

Figure 15-29 Testing a capacitor for leakage.

Large AC oil-filled capacitors can be tested in a similar manner. To test the capacitor accurately, two measurements must be made. One is to measure the capacitance value of the capacitor to determine if it is the same or approximately the same as the rate value. The other is to test the strength of the dielectric.

The first test should be made with an ohmmeter. With the power disconnected, connect the terminals of an ohmmeter directly across the capacitor terminals (Fig. 15-30). This test determines if the dielectric is shorted. When the ohmmeter is connected, the needle should swing up scale and return to infinity. The amount of needle swing is determined by the capacitance of the capacitor. Then reverse the ohmmeter connection, and the needle should move twice as far up scale and return to the infinity setting.

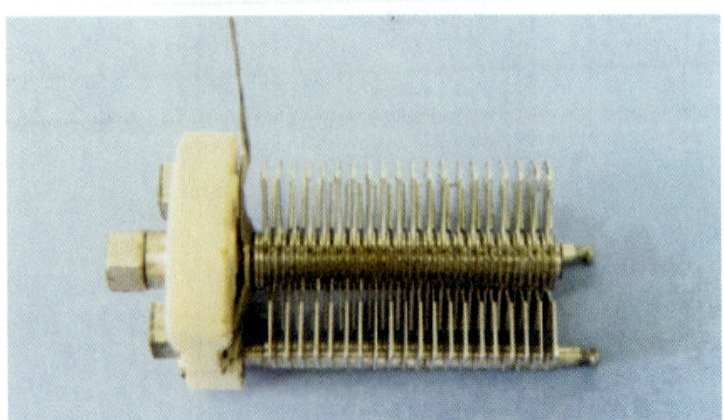

Figure 15-26 A variable capacitor.

Figure 15-30 Testing the capacitor with an ohmmeter.

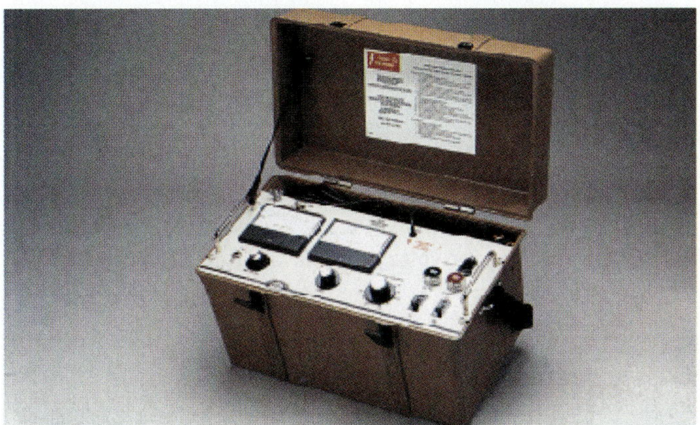

Figure 15-31 A dielectric test set. *Courtesy of Biddle Instruments.*

If the ohmmeter test is successful, the dielectric must be tested at its rated voltage with a dielectric strength test. To make this test, a dielectric test set must be used (Fig. 15-31). This device is often referred to as a **HIPOT** because of its ability to produce a high voltage or high potential. The dielectric test set contains a variable voltage control, a voltmeter, and a microammeter. To use the HIPOT, connect its terminal leads to the capacitor terminals. Increase the output voltage until rated voltage is applied to the capacitor. The microammeter indicates any current flow between the plates of the dielectric. If the capacitor is good, the microammeter should indicate zero current flow.

The capacitance value must be measured to determine if there are any open plates in the capacitor. To measure the capacitance value of the capacitor, connect some value of AC voltage across the plates of the capacitor (Fig. 15-32). This voltage must not be greater than the rated capacitor voltage. Then measure the amount of current flow in the circuit. Now that the voltage and current flow are known, the capacitive reactance of the capacitor can be computed using the formula

$$X_C = \frac{E}{I}$$

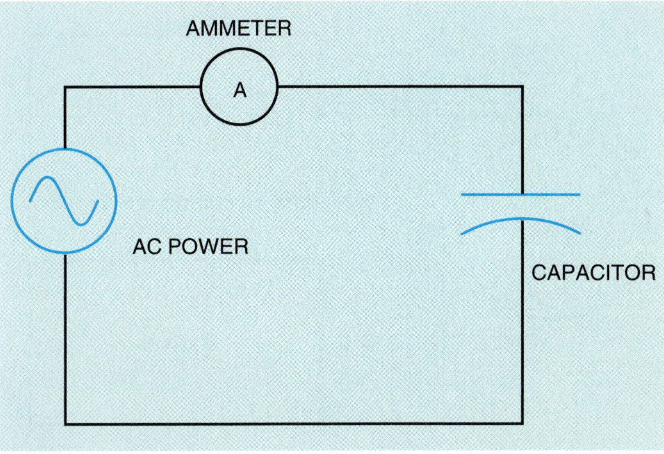

Figure 15-32 Determining the capacitance value.

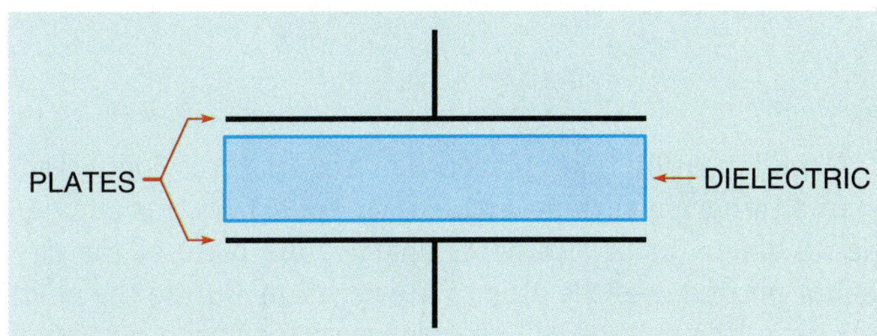

Figure 15-33 A basic capacitor.

After the capacitive reactance has been determined, the capacitance can be computed using the formula

$$C = \frac{1}{2\pi F X_C}$$

(*Note:* Capacitive reactance is measured in ohms and limits current flow in a manner similar to inductive reactance. Capacitive reactance is covered fully in this chapter.)

Connecting the Capacitor into an AC Circuit

When a capacitor (Fig. 15-33) is connected to an alternating current circuit, current will **appear to flow** through the capacitor. The reason is that in an AC circuit, the current continually changes direction and polarity. To understand this concept, consider the hydraulic circuit shown in Figure 15-34. Two tanks are connected to a common pump. Assume tank A to be full and tank B to be empty. Now assume that the pump pumps water from tank A to tank B. When tank B becomes full, the pump reverses and pumps the water from tank B back into tank A. Each time a tank is filled, the pump reverses and pumps water back into the other tank. Notice that water is continually flowing in this circuit, but there is no direct connection between the two tanks.

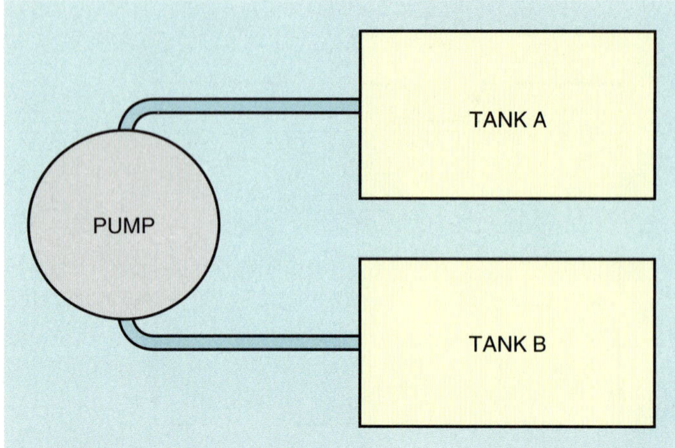

Figure 15-34 Water can flow continuously, but not between the two tanks.

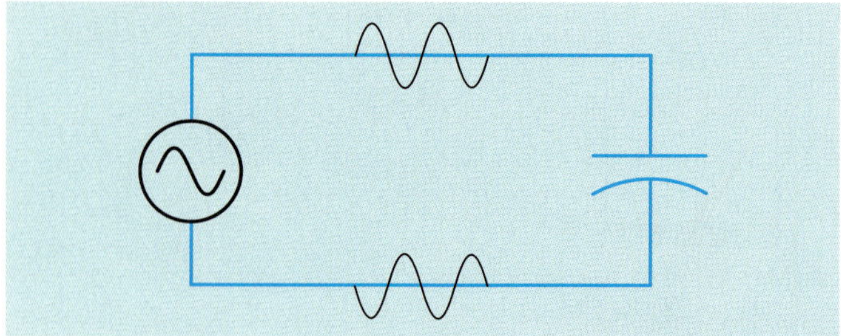

Figure 15-35 A capacitor connected to an AC circuit.

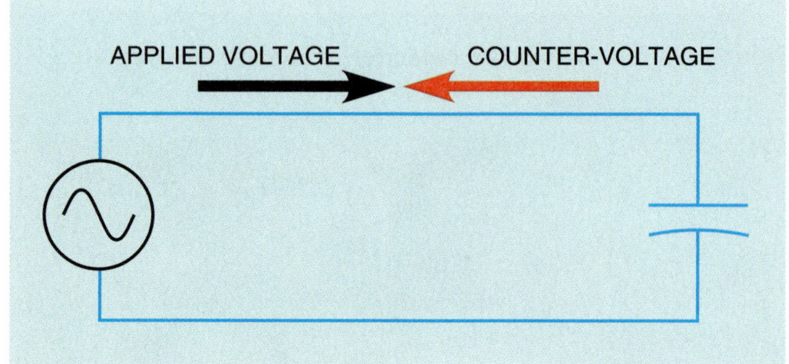

Figure 15-36 Counter-voltage limits the flow of current.

A similar action takes place when a capacitor is connected to an alternating current circuit (Fig. 15-35). In this circuit, the AC generator or alternator charges one plate of the capacitor positive and the other plate negative. During the next half cycle, the voltage will change polarity and the capacitor will discharge and recharge to the opposite polarity. As long as the voltage continues to increase, decrease, and change polarity, current will flow from one plate of the capacitor to the other. If an ammeter were placed in the circuit, it would indicate a continuous flow of current, giving the appearance that current is flowing through the capacitor.

Capacitive Reactance

As the capacitor is charged, an impressed voltage is developed across its plates as an electrostatic charge is built up (Fig. 15-36). The impressed voltage is the voltage provided by the electrostatic charge. It opposes the applied voltage and limits the flow of current in the circuit. This counter-voltage is similar to the counter-voltage produced by an inductor. The counter-voltage developed by the capacitor is called reactance also. Because it is caused by capacitance, it is called **capacitive reactance (X_C)** and is measured in ohms. The formula for finding capacitive reactance is

$$X_C = \frac{1}{2\pi FC}$$

where

X_C = capacitive reactance

π = 3.1416

F = frequency in hertz

C = capacitance in farads

Example 6

A 35-μF capacitor is connected to a 120-V 60-Hz line. How much current will flow in this circuit?

SOLUTION

The first step is to compute the capacitive reactance. Recall that the value of C in the formula is given in farads. This must be changed to the capacitive units being used—in this case, microfarads.

$$X_C = \frac{1}{2 \times 3.1416 \times 60 \times (35 \times 10^{-6})}$$

$$X_C = 75.79\ \Omega$$

Now that the value of capacitive reactance is known, it can be used like resistance in an Ohm's law formula. Since capacitive reactance is the current-limiting factor, it will replace the value of R.

$$I = \frac{E}{X_C}$$

$$I = \frac{120}{75.79}$$

$$I = 1.58\ \text{A}$$

Computing Capacitance

If the value of capacitive reactance is known, the capacitance of the capacitor can be found using the formula

$$C = \frac{1}{2\pi F X_C}$$

Example 7

A capacitor is connected into a 480-V 60-Hz circuit. An ammeter indicates a current flow of 2.6 A. What is the capacitance value of the capacitor?

SOLUTION

The first step is to compute the value of capacitive reactance. Since capacitive reactance, like resistance, limits current flow, it can be substituted for R in an Ohm's law formula.

$$X_C = \frac{E}{I}$$

$$X_C = \frac{480}{2.6}$$

$$X_C = 184.61\ \Omega$$

Now that the capacitive reactance of the circuit is known, the value of capacitance can be found.

$$C = \frac{1}{2\pi F X_C}$$

$$C = \frac{1}{2 \times 3.1416 \times 60 \times 184.61}$$

$$C = \frac{1}{69{,}596.49}$$

$$C = 0.00001437\ \text{F} = 14.37\ \mu\text{F}$$

Voltage and Current Relationships in a Pure Capacitive Circuit

Earlier in this text it was shown that the current in a pure resistive circuit is in phase with the applied voltage and that current in a pure inductive circuit lags the applied voltage by 90°. In this unit it will be shown that in a pure capacitive circuit the current will *lead* the applied voltage by 90°.

When a capacitor is connected to an alternating current, the capacitor will charge and discharge at the same rate and time as the applied voltage. The charge in coulombs is equal to the capacitance of the capacitor times the applied voltage ($Q = C \times V$). When the applied voltage is zero, the charge in coulombs and impressed voltage will be zero also. When the applied voltage reaches its maximum value—positive or negative—the charge in coulombs and impressed voltage will reach maximum also (Fig. 15-37). The impressed voltage will follow the same curve as the applied voltage.

In the wave form shown, voltage and charge are both zero at 0°. Since there is no charge on the capacitor, there is no opposition to current flow, which is shown to be maximum. As the applied voltage increases from zero toward its positive peak at 90°, the capacitor begins to charge at the same time. The charge produces an impressed voltage across the plates of the capacitor that opposes the flow of current. The impressed voltage is 180° **out of phase** with the applied voltage (Fig. 15-38). When the applied voltage reaches 90° in the positive direction, the charge reaches maximum, the impressed voltage reaches peak in the negative direction, and the current flow is zero.

As the applied voltage begins to decrease, the capacitor begins to discharge, causing the current to flow in the opposite or negative direction. When the applied voltage and charge reach zero at 180°, the impressed voltage is zero also and the current flow is maximum in the negative direction. As the applied voltage and charge increase in the negative direction, the increase of the impressed voltage across the capacitor again causes the current to decrease. The applied voltage and charge reach maximum negative after 270° of rotation. The impressed voltage reaches maximum positive

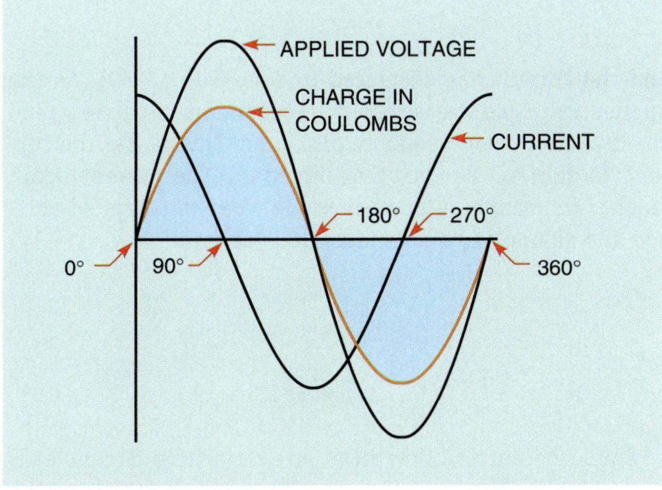

Figure 15-37 Capacitive current leads the applied voltage by 90°.

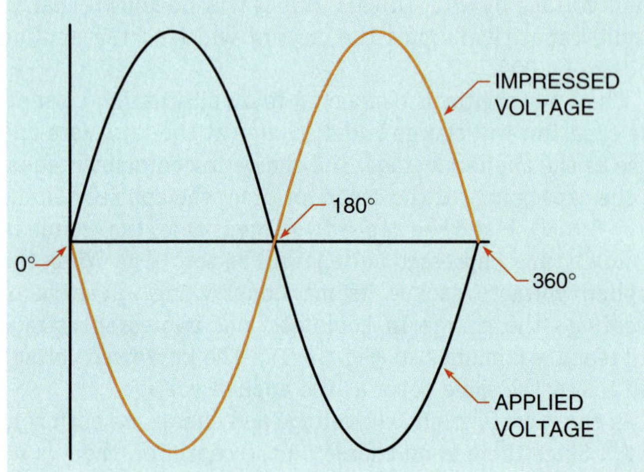

Figure 15-38 **The impressed voltage is 180° out of phase with the applied voltage.**

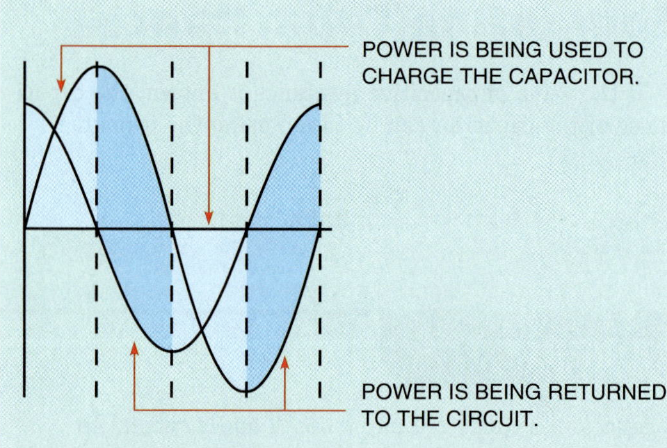

Figure 15-40 **A pure capacitive circuit has no true power (watts). The power required to charge the capacitor is returned to the circuit when the capacitor discharges.**

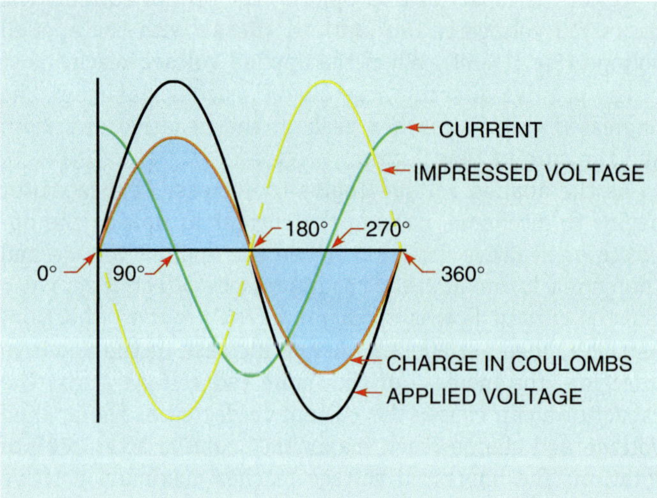

Figure 15-39 **Voltage, current, and charge relationships for a capacitive circuit.**

and the current has decreased to zero (Fig. 15-39). As the applied voltage decreases from its maximum negative value, the capacitor again begins to discharge. This causes the current to flow in the positive direction. The current again reaches its maximum positive value when the applied voltage and charge reach zero after 360° of rotation.

Power in a Pure Capacitive Circuit

Since the current flow in a pure capacitive circuit leads the applied voltage by 90°, the voltage and current have the same polarity for half of the time during one cycle and have opposite polarities the other half of the time (Fig. 15-40). During the period of time that the voltage and current have the same polarity, energy is being stored in the capacitor in the form of an electrostatic field. When the voltage and current have opposite polarities, the capacitor is discharging, and the energy is returned to the circuit. When the values of current and voltage for one full cycle are added, the sum will equal zero just as it does with pure inductive circuits. Therefore, there is no true power, or watts, produced in a pure capacitive circuit.

The power value for a capacitor is **reactive power** and is measured in **VARs,** just as it is for an inductor. Inductive VARs and capacitive VARs are 180° out of phase with each other, however (Fig. 15-41). To distinguish between inductive and capacitive VARs, inductive VARs will be shown as $VARs_L$ and capacitive VARs as $VARs_C$.

Capacitor Voltage Rating

The **voltage rating** of a capacitor is actually the voltage rating of the dielectric. **Voltage rating is extremely important concerning the life of the capacitor and should never be exceeded.** Unfortunately, there are no set standards concerning how voltage ratings are marked. It is not unusual to see capacitors marked VOLTS AC, VOLTS DC, PEAK VOLTS, and WVDC (WORKING VOLTS DC). The voltage rating of electrolytic or polarized capacitors will always be given in DC volts. The voltage rating of nonpolarized capacitors, however, can be given as AC or DC volts.

If a nonpolarized capacitor's voltage rating is given in AC volts, the voltage indicated is the RMS value. If the voltage rating is given as PEAK or as DC volts, it indicates the peak value of AC volts. If a capacitor is to be connected to an AC circuit, it is necessary to compute the peak value if the voltage rating is given as DC volts.

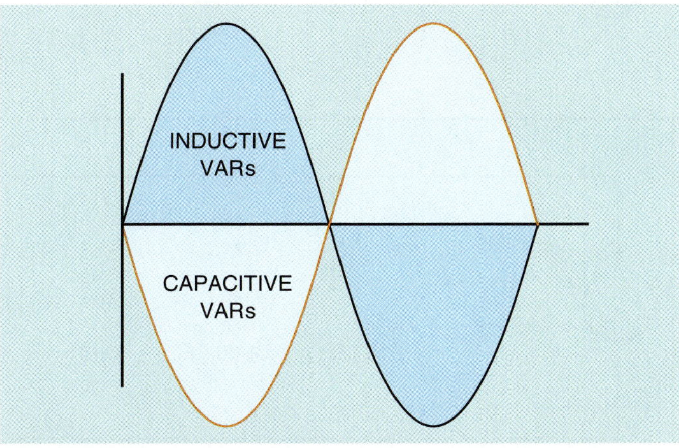

Figure 15-41 **Inductive VARs and capacitive VARs are 180° out of phase with each other.**

Example 8

An AC oil-filled capacitor has a voltage rating of 300 WVDC. Will the voltage rating of the capacitor be exceeded if the capacitor is connected to a 240-V 60-Hz line?

SOLUTION

The DC voltage rating of the capacitor indicates the peak value of voltage. To determine if the voltage rating will be exceeded, find the peak value of 240 V by multiplying by 1.414.

$$\text{Peak} = 240 \times 1.414$$

$$\text{Peak} = 339.36 \text{ V}$$

The answer is that the capacitor voltage rating will be exceeded.

CAPACITANCE	CAPACITIVE REACTANCE			
	30 Hz	60 Hz	400 Hz	1000 Hz
10 pF	530.5 M Ω	265.26 M Ω	39.79 M Ω	15.91 M Ω
350 pF	15.16 M Ω	7.58 M Ω	1.14 M Ω	454.73 k Ω
470 nF	112.88 k Ω	56.44 k Ω	8.47 k Ω	3.39 k Ω
750 nF	22.22 k Ω	11.11 k Ω	1.67 k Ω	666.67 Ω
1 μF	5.31 k Ω	2.65 k Ω	397.89 Ω	159.15 Ω
25 μF	212.21 Ω	106.1 Ω	15.915 Ω	6.37 Ω

Figure 15-42 **Capacitive reactance is inversely proportional to frequency.**

Effects of Frequency in a Capacitive Circuit

One of the factors that determine the capacitive reactance of a capacitor is the **frequency.** Capacitive reactance is inversely proportional to frequency. As the frequency increases, the capacitive reactance decreases. The chart in Figure 15-42 shows the capacitive reactance for different values of capacitance at different frequencies. Frequency has an effect on capacitive reactance because the capacitor charges and discharges faster at a higher frequency. Recall that current is a rate of electron flow. A current of 1 A is one coulomb per second.

$$I = \frac{C}{t}$$

where

I = current

C = charge in coulombs

t = time in seconds

Assume that a capacitor is connected to a 30-Hz line, and 1 coulomb of charge flows each second. If the frequency is doubled to 60 Hz, 1 coulomb of charge will flow in 0.5 s because the capacitor is being charged and discharged twice as fast (Fig. 15-43). This means that in a period of 1 s, 2 coulombs of charge will flow. Since the capacitor is being charged and discharged at a faster rate, the opposition to current flow is decreased.

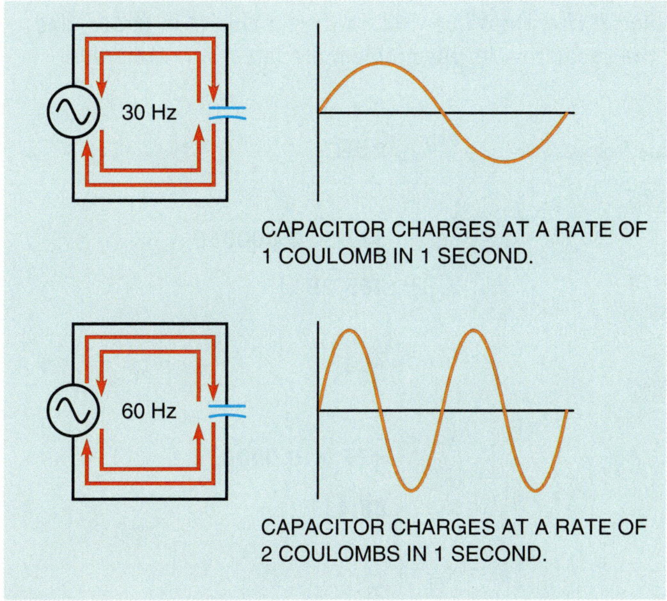

Figure 15-43 **The rate of charge increases with frequency.**

Example 9

Series Capacitors

Three capacitors with values of 10 μF, 30 μF, and 15 μF are connected in series to a 480-V 60-Hz line (Fig. 15-44). Find the following circuit values.

X_{C1} — capacitive reactance of the first capacitor
X_{C2} — capacitive reactance of the second capacitor
X_{C3} — capacitive reactance of the third capacitor
X_{CT} — total capacitive reactance for the circuit
C_T — total capacitance for the circuit
I_T — total circuit current
E_{C1} — voltage drop across the first capacitor
$VARs_{C1}$ — reactive power of the first capacitor
E_{C2} — voltage drop across the second capacitor
$VARs_{C2}$ — reactive power of the second capacitor
E_{C3} — voltage drop across the third capacitor
$VARs_{C3}$ — reactive power of the third capacitor
$VARs_{CT}$ — total reactive power for the circuit

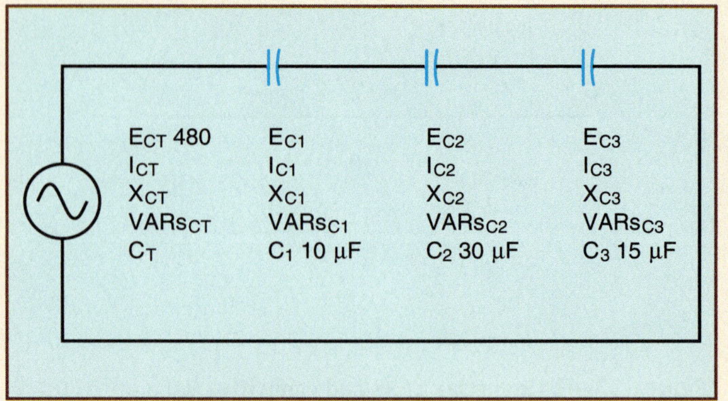

Figure 15-44 **Capacitors connected in series.**

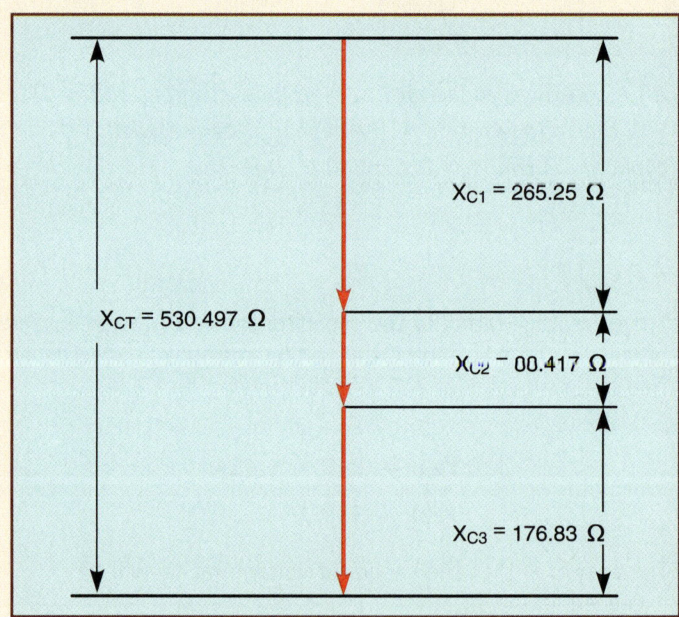

Figure 15-45 **Vector sum for capacitive reactance.**

SOLUTION

Since the frequency and the capacitance of each capacitor are known, the capacitive reactance for each capacitor can be found using the formula

$$X_C = \frac{1}{2\pi FC}$$

Recall that the value for C in the formula is in farads, and the capacitors in this problem are rated in microfarads.

$$X_{C1} = \frac{1}{2\pi FC}$$

$$X_{C1} = \frac{1}{377 \times 0.000010}$$

$$X_{C1} = 265.25\ \Omega$$

$$X_{C2} = \frac{1}{2\pi FC}$$

$$X_{C2} = \frac{1}{377 \times 0.000030}$$

$$X_{C2} = 88.417\ \Omega$$

$$X_{C3} = \frac{1}{2\pi FC}$$

$$X_{C3} = \frac{1}{377 \times 0.000015}$$

$$X_{C3} = 176.83\ \Omega$$

Since there is no phase angle shift between any of the three capacitive reactances, the total capacitive reactance will be the sum of the three reactances (Fig. 15-45).

$$X_{CT} = X_{C1} + X_{C2} + X_{C3}$$

$$X_{CT} = 265.25 + 88.417 + 176.83$$

$$X_{CT} = 530.497\ \Omega$$

Example 9 Continued

The total capacitance of a series circuit can be computed in a manner similar to that used for computing parallel resistance. Refer to the Pure Capacitive Circuits Formula section of the appendix. Total capacitance in this circuit will be computed using the formula

$$C_T = \frac{1}{\frac{1}{C_1} + \frac{1}{C_2} + \frac{1}{C_3}}$$

$$C_T = \frac{1}{\frac{1}{10} + \frac{1}{30} + \frac{1}{15}}$$

$$C_T = \frac{1}{0.2}$$

$$C_T = 5\ \mu F$$

The total current can be found by substituting the total capacitive reactance for R in an Ohm's law formula.

$$I_T = \frac{E_{CT}}{X_{CT}}$$

$$I_T = \frac{480}{530.497}$$

$$I_T = 0.905\ A$$

Because the current is the same at any point in a series circuit, the voltage drop across each capacitor can now be computed using the capacitive reactance of each capacitor and the current flowing through it.

$$E_{C1} = I_{C1} \times X_{C1}$$

$$E_{C1} = 0.905 \times 265.25$$

$$E_{C1} = 240.051\ V$$

$$E_{C2} = I_{C2} \times X_{C2}$$

$$E_{C2} = 0.905 \times 88.417$$

$$E_{C2} = 80.017\ V$$

$$E_{C3} = I_{C3} \times X_{C3}$$

$$E_{C3} = 0.905 \times 176.83$$

$$E_{C3} = 160.031\ V$$

Now that the voltage drops of the capacitors are known, the reactive power of each capacitor can be found.

$$VARs_{C1} = E_{C1} \times I_{C1}$$

$$VARs_{C1} = 240.051 \times 0.905$$

$$VARs_{C1} = 217.246$$

$$VARs_{C2} = E_{C2} \times I_{C2}$$

$$VARs_{C2} = 80.017 \times 0.905$$

$$VARs_{C2} = 72.415$$

$$VARs_{C3} = E_{C3} \times I_{C3}$$

$$VARs_{C3} = 160.031 \times 0.905$$

$$VARs_{C3} = 144.828$$

Power, whether true, apparent, or reactive, will add in any type of circuit. The total reactive power in this circuit can be found by taking the sum of all the VARs for the capacitors or by using total values of voltage and current and Ohm's law.

$$VARs_{CT} = VARs_{C1} + VARs_{C2} + VARs_{C3}$$

$$VARs_{CT} = 217.246 + 72.415 + 144.828$$

$$VARs_{CT} = 434.489$$

The circuit with all computed values is shown in Figure 15-46.

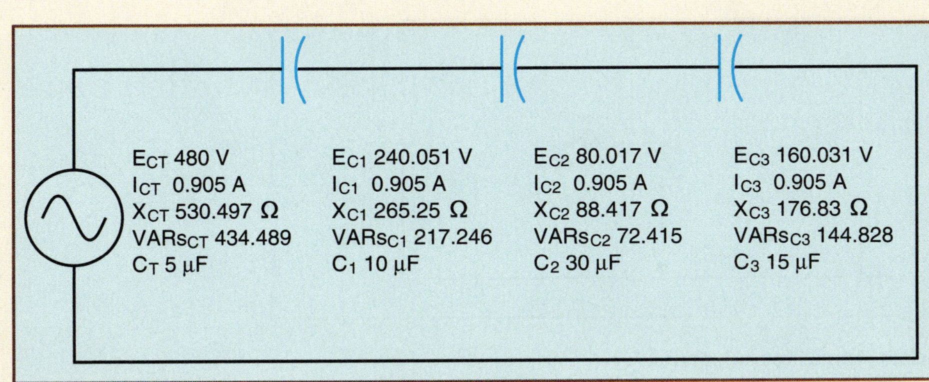

Figure 15-46 Series circuit 1 with all values.

Example 10

Parallel Capacitors

Three capacitors having values of 50 μF, 75 μF, and 20 μF are connected in parallel to a 60-Hz line. The circuit has a total reactive power of 787.08 VARs (Fig. 15-47). Find the following unknown values.

X_{C1}	—	capacitive reactance of the first capacitor
X_{C2}	—	capacitive reactance of the second capacitor
X_{C3}	—	capacitive reactance of the third capacitor
X_{CT}	—	total capacitive reactance for the circuit
E_T	—	total applied voltage
I_{C1}	—	current flow through the first capacitor
$VARs_{C1}$	—	reactive power of the first capacitor
I_{C2}	—	current flow through the second capacitor
$VARs_{C2}$	—	reactive power of the second capacitor
I_{C3}	—	current flow through the third capacitor
$VARs_{C3}$	—	reactive power of the third capacitor

Since the frequency of the circuit and the capacitance of each capacitor are known, the capacitive reactance of each capacitor can be computed using the formula

$$X_C = \frac{1}{2\pi FC}$$

(*Note:* Refer to the Pure Capacitive Circuits Formula section of the appendix.)

$$X_{C1} = \frac{1}{377 \times 0.000050}$$

$$X_{C1} = 53.05\ \Omega$$

$$X_{C2} = \frac{1}{377 \times 0.000075}$$

$$X_{C2} = 35.367\ \Omega$$

$$X_{C3} = \frac{1}{377 \times 0.000020}$$

$$X_{C3} = 132.63\ \Omega$$

The total capacitive reactance can be found in a manner similar to finding the resistance of parallel resistors.

$$X_{CT} = \frac{1}{\frac{1}{X_{C1}} + \frac{1}{X_{C2}} + \frac{1}{X_{C3}}}$$

$$X_{CT} = \frac{1}{\frac{1}{53.05} + \frac{1}{35.367} + \frac{1}{132.63}}$$

$$X_{CT} = \frac{1}{0.05466}$$

$$X_{CT} = 18.295\ \Omega$$

Now that the total capacitive reactance of the circuit is known and the total reactive power is known, the voltage applied to the circuit can be found using the formula

$$E_T = \sqrt{VARs_{CT} \times X_{CT}}$$

$$E_T = \sqrt{787.08 \times 18.295}$$

$$E_T = 120\ V$$

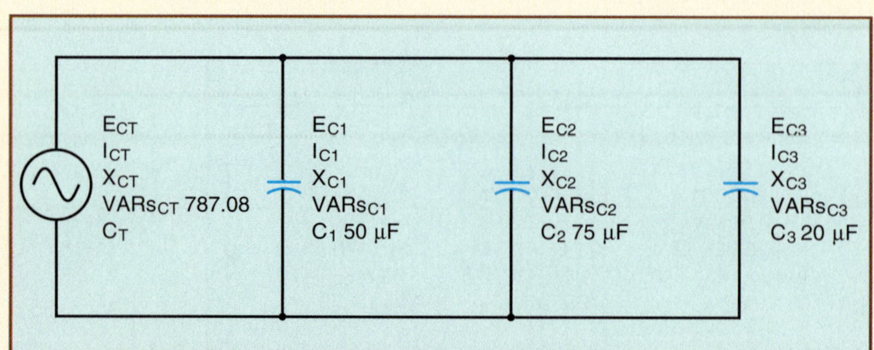

Figure 15-47 **Capacitors connected in parallel.**

Example 10 Continued

In a parallel circuit the voltage must be the same across each branch of the circuit. Therefore, 120 V is applied across each capacitor.

Now that the circuit voltage is known, the total current for the circuit and the current in each branch can be found using Ohm's law.

$$I_{CT} = \frac{E_{CT}}{X_{CT}}$$

$$I_{CT} = \frac{120}{18.295}$$

$$I_{CT} = 6.559 \text{ A}$$

$$I_{C1} = \frac{E_{C1}}{X_{C1}}$$

$$I_{C1} = \frac{120}{53.05}$$

$$I_{C1} = 2.262 \text{ A}$$

$$I_{C2} = \frac{E_{C2}}{X_{C2}}$$

$$I_{C2} = \frac{120}{35.367}$$

$$I_{C2} = 3.393 \text{ A}$$

$$I_{C3} = \frac{E_{C3}}{X_{C3}}$$

$$I_{C3} = \frac{120}{132.62}$$

$$I_{C3} = 0.905 \text{ A}$$

The amount of reactive power for each capacitor can now be computed using Ohm's law.

$$VARs_{C1} = E_{C1} \times I_{C1}$$

$$VARs_{C1} = 120 \times 2.262$$

$$VARs_{C1} = 271.442$$

$$VARs_{C2} = E_{C2} \times I_{C2}$$

$$VARs_{C2} = 120 \times 3.393$$

$$VARs_{C2} = 407.159$$

$$VARs_{C3} = E_{C3} \times I_{C3}$$

$$VARs_{C3} = 120 \times 0.905$$

$$VARs_{C3} = 108.573$$

To make a quick check of the circuit values, add the VARs for all the capacitors and see if they equal the total circuit VARs.

$$VARs_{CT} = VARs_{C1} + VARs_{C2} + VARs_{C3}$$

$$VARs_{CT} = 271.442 + 407.159 + 108.573$$

$$VARs_{CT} = 787.174$$

The slight difference in answers is caused by rounding off values. The circuit with all values is shown in Figure 15-48.

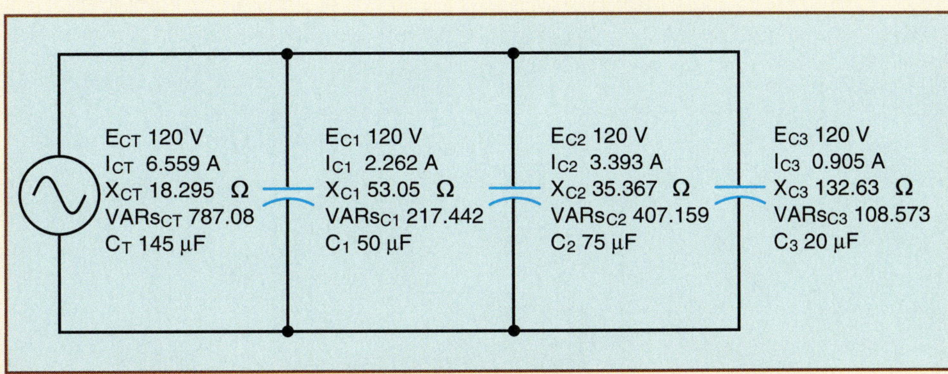

Figure 15-48 **Parallel circuit with completed values.**

Summary

- Capacitors are devices that oppose a change of voltage.
- Three factors that determine the capacitance of a capacitor are

  A. The surface area of the plates.

  B. The distance between the plates.

  C. The type of dielectric.
- A capacitor stores energy in an electrostatic field.
- Current can flow only during the time a capacitor is charging or discharging.
- Capacitors charge and discharge at an exponential rate.
- The basic unit of capacitance is the farad.
- Capacitors are generally rated in microfarads, nanofarads, or picofarads.
- When capacitors are connected in parallel, their capacitance values add.
- When capacitors are connected in series, the reciprocal of the total capacitance is equal to the sum of the reciprocals of all the capacitors.
- The charge and discharge times of a capacitor are proportional to the amount of capacitance and resistance in the circuit.
- Five time constants are required to charge or discharge a capacitor.
- Nonpolarized capacitors are often called AC capacitors.
- Nonpolarized capacitors can be connected to direct or alternating current circuits.
- Polarized capacitors are often referred to as electrolytic capacitors.
- Polarized capacitors can be connected to direct current circuits only.
- There are two basic types of electrolytic capacitors, the wet type and the dry type.
- Wet type electrolytic capacitors can be reformed by reconnecting to the correct polarity.
- Dry type electrolytic capacitors will be permanently damaged if connected to the incorrect polarity.
- To test a capacitor for leakage, a microammeter should be connected in series with the capacitor and rated voltage applied to the circuit.
- When a capacitor is connected to an alternating current circuit, current will appear to flow through the capacitor.
- Current appears to flow through a capacitor because of the continuous increase and decrease of voltage and because of the continuous change of polarity in an AC circuit.
- The current flow in a pure capacitive circuit is limited by capacitive reactance.
- Capacitive reactance is proportional to the capacitance of the capacitor and the frequency of the AC line.
- Capacitive reactance is measured in ohms.
- In a pure capacitive circuit, the current leads the applied voltage by 90°.
- There is no true power, or watts, in a pure capacitive circuit.
- Capacitive power is reactive and is measured in VARs, as is inductance.
- Capacitive and inductive VARs are 180° out of phase with each other.
- Capacitor voltage ratings are given as volts AC, peak volts, and volts DC.
- A DC voltage rating for an AC capacitor indicates the peak value of voltage.

Review Questions

1. What is the dielectric?
2. List three factors that determine the capacitance of a capacitor.
3. A capacitor uses air as a dielectric and has a capacitance of 3 μF. A dielectric material is inserted between the plates without changing the spacing, and the capacitance becomes 15 μF. What is the dielectric constant of this material?
4. In what form is the energy of a capacitor stored?
5. Four capacitors having values of 20 μF, 50 μF, 40 μF, and 60 μF are connected in parallel. What is the total capacitance of this circuit?
6. If the four capacitors in question 5 were to be connected in series, what would be the total capacitance of the circuit?
7. A 22-μF capacitor is connected in series with a 90-kΩ resistor. How long will it take this capacitor to charge?
8. A 450-pF capacitor has a total charge time of 0.5 s. How much resistance is connected in series with the capacitor?
9. Can a nonpolarized capacitor be connected to a direct current circuit?
10. Explain how an AC electrolytic capacitor is constructed.
11. What type of electrolytic capacitor will be permanently damaged if connected to the incorrect polarity?
12. A 500-nF capacitor is connected to a 300-kΩ resistor. What is the total charge time of this capacitor?
13. Can current flow through a capacitor?
14. What two factors determine the capacitive reactance of a capacitor?
15. How many degrees are the current and voltage out of phase in a pure capacitive circuit?
16. Does the current in a pure capacitive circuit lead or lag the applied voltage?
17. A 30-μF capacitor is connected into a 240-V, 60-Hz circuit. What is the current flow in this circuit?
18. A capacitor is connected into a 1250-V, 1000-Hz circuit. The current flow is 80 A. What is the capacitance of the capacitor?
19. A capacitor is to be connected into a 480-V, 60-Hz line. If the capacitor has a voltage rating of 600 VDC, will the voltage rating of the capacitor be exceeded?
20. On the average, by what factor is the life expectancy of a capacitor increased if the capacitor is operated at half its voltage rating?
21. A capacitor is connected into a 277-V, 400-Hz circuit. The circuit current is 12 A. What is the capacitance of the capacitor?
22. A capacitor has a voltage rating of 350 VAC. Can this capacitor be connected into a 450-VDC circuit without exceeding the voltage rating of the capacitor?

Practice Problems

RC Time Constants

Fill in all the missing values. Refer to the formulas given below.

Resistance	Capacitance	Time Constant	Total Time
150 kΩ	100 μF		
350 kΩ			35 s
	350 pF	0.05 s	
	0.05 μF		10 s
1.2 MΩ	0.47 μF		
	12 μF	0.05 s	
86 kΩ			1.5 s
120 kΩ	470 pF		
	250 nF		100 ms
	8 μF		150 μs
100 kΩ		150 ms	
33 kΩ	4 μF		

$$\tau = RC$$

$$R = \frac{\tau}{C}$$

$$C = \frac{\tau}{R}$$

$$\text{Total time} = \tau \times 5$$

Capacitive Circuits

Fill in all the missing values. Refer to the formulas given.

$$X_C = \frac{1}{2\pi FC}$$

$$C = \frac{1}{2\pi FX_C}$$

$$F = \frac{1}{2\pi CX_C}$$

Capacitance	X_C	Frequency
38 μF		60 Hz
	78.8 Ω	400 Hz
250 pF	4.5 kΩ	
234 μF		10 kHz
	240 Ω	50 Hz
10 μF	36.8 Ω	
560 nF		2 MHz
	15 kΩ	60 Hz
75 μF	560 Ω	
470 pF		200 kHz
	6.8 kΩ	400 Hz
34 μF	450 Ω	

Chapter 16 Single-Phase Transformers

Transformers are among the most common devices found in the electrical field. They range in size from less than one cubic inch to the size of rail cars. Their ratings can range from mVA (milli-volt-amps) to GVA (giga-volt-amps). It is imperative that anyone working in the electrical field have an understanding of transformer types and connections. This chapter will present transformers intended for use in single-phase installations. The two main types of voltage transformers, isolation transformers and autotransformers, will be discussed.

OBJECTIVES

After studying this unit, you should be able to:

- **discuss the different types of transformers.**
- **calculate values of voltage, current, and turns for single-phase transformers using formulas.**
- **calculate values of voltage, current, and turns for single-phase transformers using the turns ratio.**
- **connect a transformer and test the voltage output of different windings.**
- **discuss polarity markings on a schematic diagram.**
- **test a transformer to determine the proper polarity marks.**

Glossary of Single-Phase Transformer Terms

autotransformer a transformer that uses only one winding for both the primary and secondary

control transformer a common type of transformer used in motor control circuits to reduce the rated line voltage to the amount needed to operate the control components

distribution transformer a transformer that is generally used to lower the power company line voltage to the value needed for houses or industrial plants

excitation current the amount of current that flows in the primary winding of a transformer when no load is connected to the secondary

flux leakage the amount of magnetic flux lines that radiate into the air

inrush current the amount of current flow that occurs when power is first applied to a transformer

isolation transformers transformers that have their primary and secondary windings electrically separated from each other

laminated the process of stacking thin sheets of metal together to form the core material for a transformer

neutral conductor a conductor is generally grounded and is a common connection to other parts of a circuit

primary winding the winding of a transformer to which power is connected

secondary winding the winding of a transformer to which the load is connected

step-down transformer a transformer that produces a lower secondary voltage than primary voltage

step-up transformer a transformer that produces a higher secondary voltage than primary voltage

tape wound core a type of transformer core made by winding a long continuous metal sheet in a circular or rounded-rectangle shape

toroid core a transformer core that is in a toroid shape, which is generally round with a hole in the center similar to a donut

transformer an electrical machine used to change values of voltage, current, and impedance

turns ratio the ratio of the number of turns of wire in the primary winding compared with the number of turns in the secondary winding

volts-per-turn ratio a method of determining voltage values in a transformer by dividing the number of turns of wire in the primary winding by the voltage applied voltage

Single-Phase Transformers

A **transformer** is a magnetically operated machine that can change values of voltage, current, and impedance without a change of frequency. Transformers are the most efficient machines known. Their efficiencies commonly range from 90% to 99% at full load. Transformers can be divided into three classifications:

1. Isolation transformer.
2. Autotransformer.
3. Current transformer (discussed in Chapter 8).

All values of a transformer are proportional to its turns ratio. This does not mean that the exact number of turns of wire on each winding must be known to determine different values of voltage and current for a transformer. What must be known is the *ratio* of turns. For example, assume a transformer has two windings. One winding, the primary, has 1000 turns of wire, and the other, the secondary, has 250 turns of wire (Fig. 16-1). The **turns ratio** of this transformer is 4 to 1, or 4:1 (1000/250 = 4), because there are four turns of wire on the primary for every one turn of wire on the secondary.

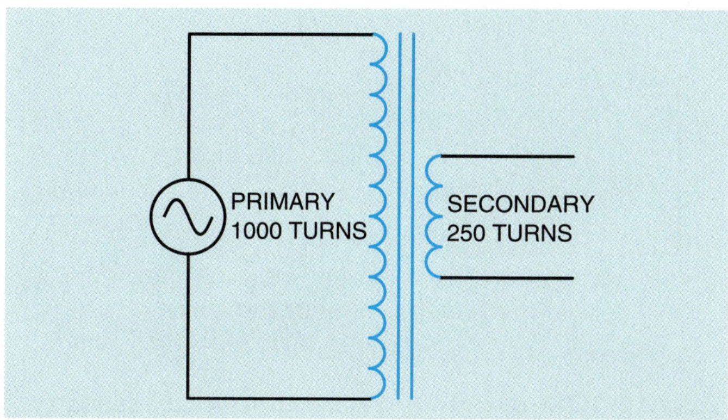

Figure 16-1 All values of a transformer are proportional to its turns ratio.

Transformer Formulas

Different formulas can be used to find the values of voltage and current for a transformer. Following is a list of standard formulas, where

N_P = **number of turns in the primary**

N_S = **number of turns in the secondary**

E_P = **voltage of the primary**

E_S = **voltage of the secondary**

I_P = **current in the primary**

I_S = **current in the secondary**

$$\frac{E_P}{E_S} = \frac{N_P}{N_S}$$

$$\frac{E_P}{E_S} = \frac{I_S}{I_P}$$

$$\frac{N_P}{N_S} = \frac{I_S}{I_P}$$

or

$$E_P \times N_S = E_S \times N_P$$

$$E_P \times I_P = E_S \times I_S$$

$$N_P \times I_P = N_S \times I_S$$

Figure 16-2 An isolation transformer has its primary and secondary winding electrically separated from each other.

The **primary winding** of a transformer is the power input winding. It is the winding that is connected to the incoming power supply. The **secondary winding** is the load winding, or output winding. It is the side of the transformer that is connected to the driven load (Fig. 16-2).

Isolation Transformers

The transformers shown in Figure 16-1 and 16-2 are **isolation transformers.** This means that the secondary winding is physically and electrically isolated from the primary winding, so there is no electrical connection between the primary and secondary winding. The transformer is magnetically coupled, not electrically coupled. This line isolation is often a very desirable characteristic. Since there is no electrical connection between the load and power supply, the transformer becomes a filter between the two. The isolation transformer will greatly reduce any voltage spikes that originate on the supply side before they are transferred to the load side. Some isolation transformers are built with a turns ratio of 1:1. A transformer of this type will have the same input and output voltage and is used for isolation only.

The isolation transformer can greatly reduce any voltage spikes before they reach the secondary because of the rise time of current through an inductor. Recall from Chapter 10 that the current in an inductor rises at an exponential rate (Fig. 16-3). As the current increases in value, the expanding magnetic field

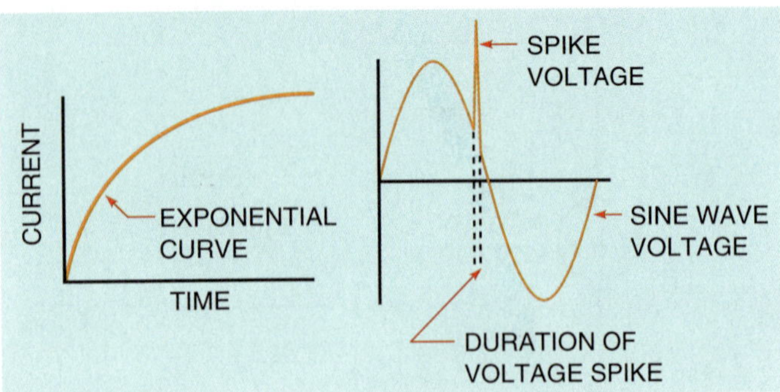

Figure 16-3 The current through an inductor rises at an exponential rate.

Figure 16-4 Voltage spikes are generally of very short duration.

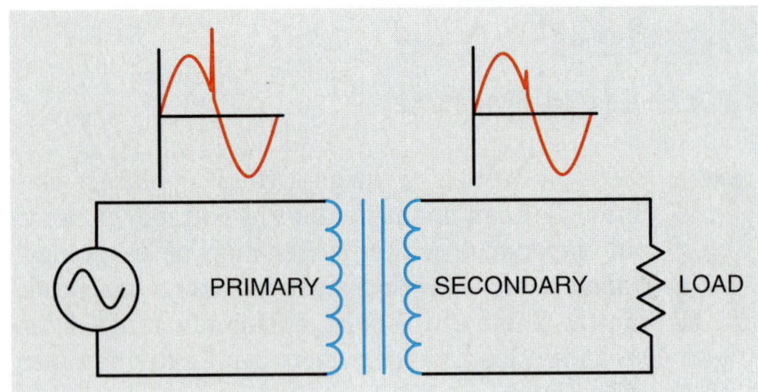

Figure 16-5 The isolation transformer greatly reduces the voltage spike.

cuts through the conductors of the coil and induces a voltage that is opposed to the applied voltage. The amount of induced voltage is proportional to the rate of change of current. This simply means that the faster the current attempts to increase, the greater the opposition to that increase will be. Spike voltages and currents are generally of very short duration, which means that they increase in value very rapidly (Fig. 16-4). This rapid change of value causes the opposition to the change to increase just as rapidly. By the time the spike has been transferred to the secondary winding of the transformer, it has been eliminated or greatly reduced (Fig. 16-5).

The basic construction of an isolation transformer is shown in Figure 16-6. A metal core is used to provide good magnetic coupling between the two windings. The core is generally made of laminations stacked together. Laminating the core helps reduce power losses caused by eddy current induction.

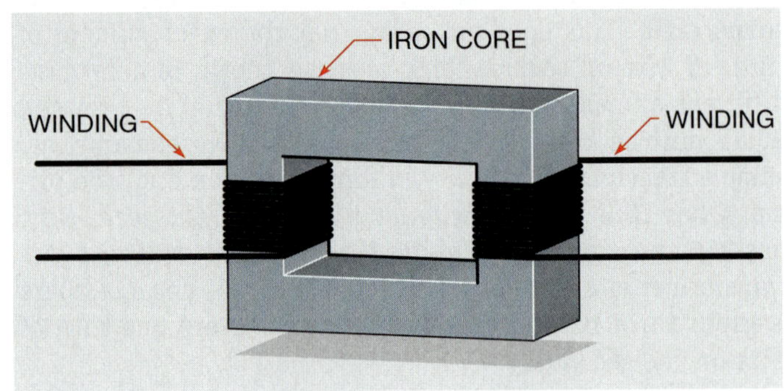

Figure 16-6 Basic construction of an isolation transformer.

Basic Operating Principles

In Figure 16-7, one winding of an isolation transformer has been connected to an alternating current supply, and the other winding has been connected to a load. As current increases from zero to its peak positive point, a magnetic field expands outward around the coil. When the current decreases from its peak positive point toward zero, the magnetic field collapses. When the current increases toward its negative peak, the magnetic field again expands, but with an opposite polarity. The field again collapses when the current decreases from its negative peak toward zero. This continually expanding and collapsing magnetic field cuts the windings of the primary and induces a voltage into it. This induced voltage opposes the applied voltage and limits the current flow of the primary. When a coil induces a voltage into itself, it is known as *self-induction*.

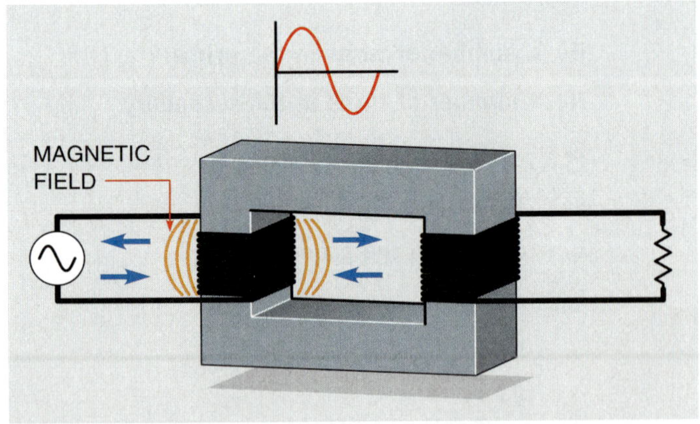

Figure 16-7 Magnetic field produced by alternating current.

Excitation Current

There will always be some amount of current flow in the primary of any voltage transformer, regardless of type or size, even if there is no load connected to the secondary. This current flow is called the **excitation current** of the transformer. The excitation current is the amount of current required to magnetize the core of the transformer. The excitation current remains constant from no load to full load. As a general rule, the excitation current is such a small part of the full load current that it is often omitted when making calculations.

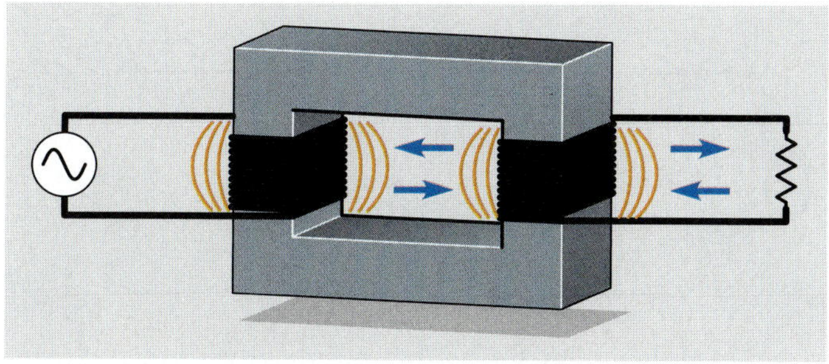

Figure 16-8 The magnetic field of the primary induces a voltage into the secondary.

Mutual Induction

Since the secondary windings of an isolation transformer are wound on the same core as the primary, the magnetic field produced by the primary winding cuts the windings of the secondary also (Fig. 16-8). This continually changing magnetic field induces a voltage into the secondary winding. The ability of one coil to induce a voltage into another coil is called *mutual induction*. The amount of voltage induced in the secondary is determined by the ratio of the number of turns of wire in the secondary to those in the primary. For example, assume the primary has 240 turns of wire and is connected to 120 VAC. This gives the transformer a **volts-per-turn ratio** of 0.5 (120 V/240 turns = 0.5 volt per turn). Now assume the secondary winding contains 100 turns of wire. Since the transformer has a volts-per-turn ratio of 0.5, the secondary voltage will be 50 V (100 × 0.5 = 50).

Transformer Calculations

In the following examples, values of voltage, current, and turns for different transformers will be computed.

Assume that the isolation transformer shown in Figure 16-2 has 240 turns of wire on the primary and 60 turns of wire on the secondary. This is a ratio of 4:1 (240/60 = 4). Now assume that 120 V is connected to the primary winding. What is the voltage of the secondary winding?

$$\frac{E_P}{E_S} = \frac{N_P}{N_S}$$

$$\frac{120}{E_S} = \frac{240}{60}$$

$$240\ E_S = 7200$$

$$E_S = 30\ \text{V}$$

The transformer in this example is known as a **step-down transformer** because it has a lower secondary voltage than primary voltage.

Now assume that the load connected to the secondary winding has an impedance of 5 Ω. The next problem is to calculate the current flow in the secondary and primary windings. The current flow of the secondary can be computed using Ohm's law since the voltage and impedance are known.

$$I = \frac{E}{Z}$$

$$I = \frac{30}{5}$$

$$I = 6\ \text{A}$$

Now that the amount of current flow in the secondary is known, the primary current can be computed using the formula

$$\frac{E_P}{E_S} = \frac{I_S}{I_P}$$

$$\frac{120}{30} = \frac{60}{I_P}$$

$$120\ I_P = 180$$

$$I_P = 1.5\ \text{A}$$

Notice that the primary voltage is higher than the secondary voltage, but the primary current is much less than the secondary current. **A good rule for any type of transformer is that power in must equal power out.** If the primary voltage and current are multiplied together, the product should equal the product of the voltage and current of the secondary.

Primary	Secondary
120 × 1.5 = 180 VA	30 × 6 = 180 VA

In this example, assume that the primary winding contains 240 turns of wire and the secondary contains 1200 turns of wire. This is a turns ratio of 1:5 (1200/240 = 5). Now assume that 120 V is connected to the primary winding. Compute the voltage output of the secondary winding.

$$\frac{E_P}{E_S} = \frac{N_P}{N_S}$$

$$\frac{120}{E_S} = \frac{240}{1200}$$

$$240\ E_S = 144{,}000$$

$$E_S = 600\ \text{V}$$

Notice that the secondary voltage of this transformer is higher than the primary voltage. This is known as a **step-up transformer.**

Now assume that the load connected to the secondary has an impedance of 2400 Ω. Find the amount of current flow in the primary and secondary windings. The current flow in the secondary winding can be computed using Ohm's law.

$$I = \frac{E}{Z}$$

$$I = \frac{600}{2400}$$

$$I = 0.25\ A$$

Now that the amount of current flow in the secondary is known, the primary current can be computed using the formula

$$\frac{E_P}{E_S} = \frac{I_S}{I_P}$$

$$\frac{120}{600} = \frac{0.25}{I_P}$$

$$120\ I_P = 150$$

$$I_P = 1.25\ A$$

Notice that the amount of power input equals the amount of power output.

Primary	Secondary
120 × 1.25 = 150 VA	600 × 0.25 = 150 VA

Calculating Isolation Transformer Values Using the Turns Ratio

As illustrated in the previous examples, transformer values of voltage, current, and turns can be computed using formulas. It is also possible to compute these values using the turns ratio. To make calculations using the turns ratio, a ratio is established that compares some number to 1, or 1 to some number. For example, assume a transformer has a primary rated at 240 V and a secondary rated at 96 V (Fig. 16-9). The turns ratio can be computed by dividing the higher voltage by the lower voltage.

$$\text{Ratio} = \frac{240}{96}$$

$$\text{Ratio} = 2.5{:}1$$

This ratio indicates that there are 2.5 turns of wire in the primary winding for every 1 turn of wire in the secondary. The side of the transformer with the lowest voltage will always have the lowest number (1) of the ratio.

Now assume that a resistance of 24 Ω is connected to the secondary winding. The amount of secondary current can be found using Ohm's law.

$$I_S = \frac{96}{24}$$

$$I_S = 4\ A$$

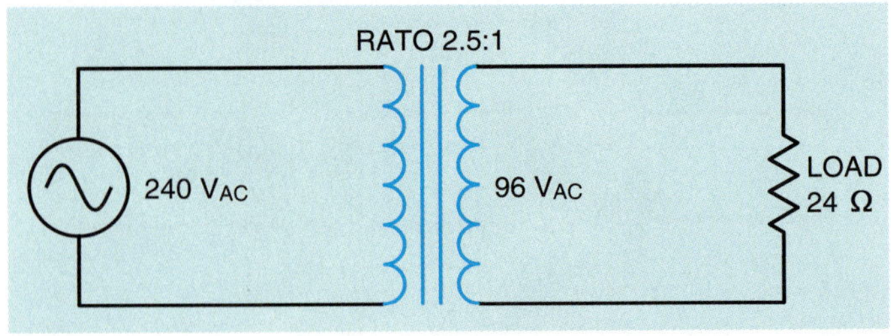

Figure 16-9 Computing transformer values using the turns ratio.

The primary current can be found using the turns ratio. Recall that the volt-amps of the primary must equal the volt-amps of the secondary. Since the primary voltage is greater, the primary current will have to be less than the secondary current.

$$I_P = \frac{I_S}{\text{turns ratio}}$$

$$I_P = \frac{4}{2.5}$$

$$I_P = 1.6\ A$$

To check the answer, find the volt-amps of the primary and secondary.

Primary	Secondary
240 × 1.6 = 384 VA	96 × 4 = 384 VA

Now assume that the secondary winding contains 150 turns of wire. The primary turns can be found by using the turns ratio also. Since the primary voltage is higher than the secondary voltage, the primary must have more turns of wire.

$$N_P = N_S \times \text{turns ratio}$$

$$N_P = 150 \times 2.5$$

$$N_P = 375\ \text{turns}$$

In the next example, assume an isolution transformer has a primary voltage of 120 V and a secondary voltage of 500 V. The secondary has a load impedance of 1200 Ω. The secondary contains 800 turns of wire (Fig. 16-10).

The turns ratio can be found by dividing the higher voltage by the lower voltage.

$$\text{Ratio} = \frac{500}{120}$$

$$\text{Ratio} = 1{:}4.17$$

The secondary current can be found using Ohm's law.

$$I_S = \frac{500}{1200}$$

$$I_S = 0.417\ A$$

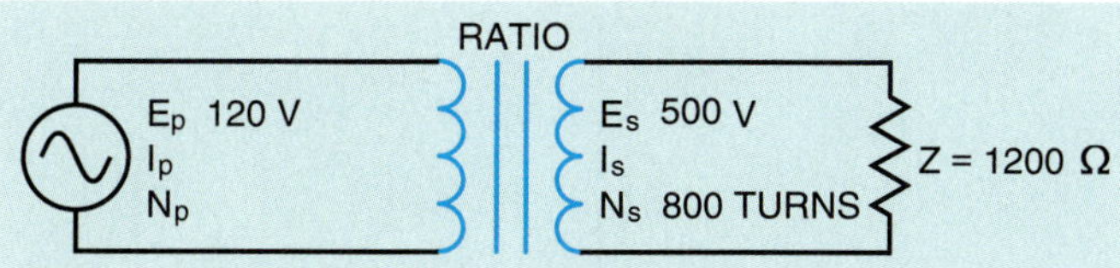

Figure 16-10 Calculating transformer values.

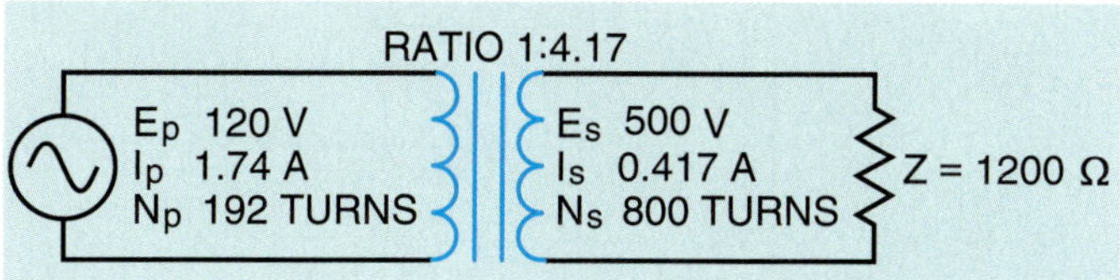

Figure 16-11 Transformer with completed values.

In this example, the primary voltage is lower than the secondary voltage. Therefore, the primary current must be higher.

$$I_P = I_S \times \text{turns ratio}$$

$$I_P = 0.417 \times 4.17$$

$$I_P = 1.74 \text{ A}$$

To check this answer, compute the volt-amps of both windings.

Primary	Secondary
$120 \times 1.74 = 208.8$ VA	$500 \times 0.417 = 208.5$ VA

The slight difference in answers is caused by rounding off values.

Since the primary voltage is less than the secondary voltage, the turns of wire in the primary will be less also.

$$N_P = \frac{N_S}{\text{turns ratio}}$$

$$N_P = \frac{800}{4.17}$$

$$N_P = 192 \text{ turns}$$

Figure 16-11 shows the transformer with all completed values.

Multiple-Tapped Windings

It is not uncommon for isolation transformers to be designed with windings that have more than one set of lead wires connected to the primary or secondary. These are called multiple-tapped windings. The transformer shown in Figure 16-12 contains a secondary winding rated at 24 V. The primary winding contains several taps, however. One of the primary lead wires is labeled C and is the common for the other leads. The other leads are labeled 120, 208, and 240. This transformer is designed so that it can be connected to different primary voltages without changing the value of the secondary voltage. In this example, it is assumed that the secondary winding has a total of 120 turns of wire. To maintain the proper turns ratio, the primary would have 600 turns of wire between C and 120, 1040 turns between C and 208, and 1200 turns between C and 240.

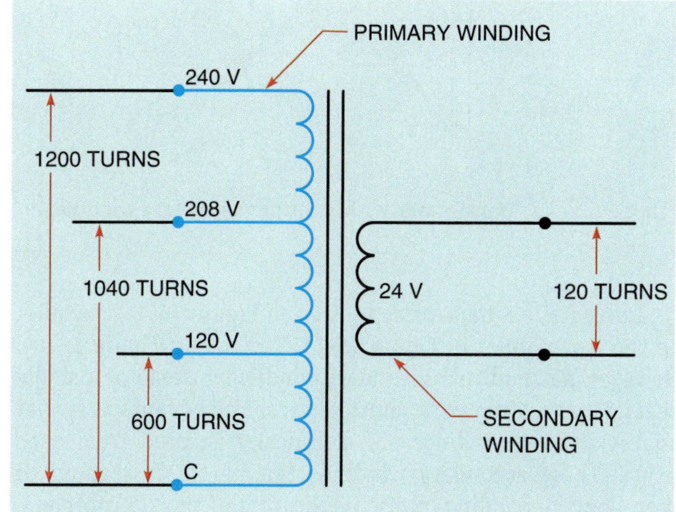

Figure 16-12 Transformer with multiple-tapped primary winding.

The isolation transformer shown in Figure 16-13 contains a single primary winding. The secondary winding, however, has been tapped at several points. One of the secondary lead wires is labeled C and is common to the other lead wires.

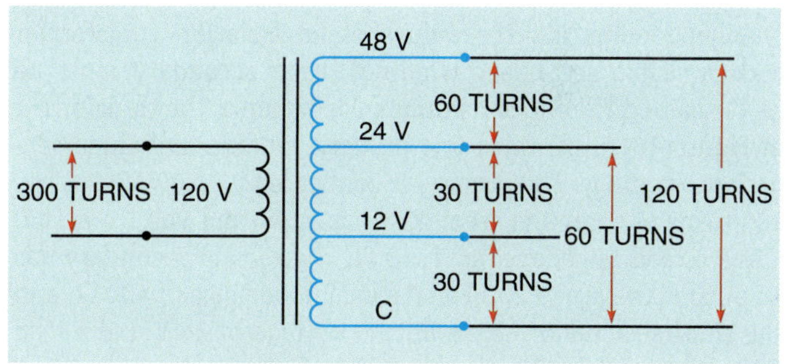

Figure 16-13 Transformer secondary with multiple taps.

When rated voltage is applied to the primary, voltages of 12, 24, and 48 V can be obtained at the secondary. It should be noted also that this arrangement of taps permits the transformer to be used as a center-tapped transformer for two of the voltages. If a load is placed across the lead wires labeled C and 24, the lead wire labeled 12 becomes a center tap. If a load is placed across the C and 48 lead wires, the 24 lead wire becomes a center tap.

In this example, it is assumed that the primary winding has 300 turns of wire. To produce the proper turns ratio would require 30 turns of wire between C and 12, 60 turns of wire between C and 24, and 120 turns of wire between C and 48.

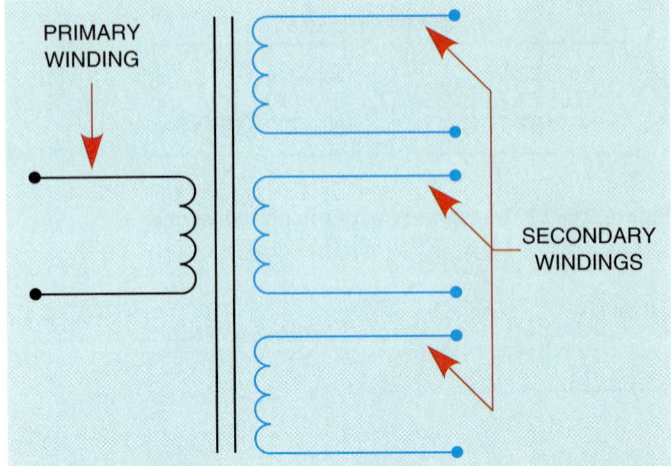

Figure 16-14 Transformer with multiple secondary windings.

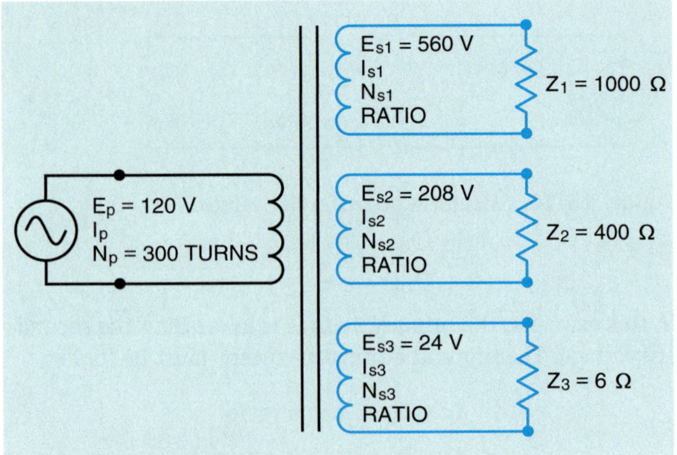

Figure 16-15 Computing values for a transformer with multiple secondary windings.

The isolation transformer shown in Figure 16-14 is similar to the transformer in Figure 16-13. The one in Figure 16-14, however, has multiple secondary windings instead of a single secondary winding with multiple taps. The advantage is that the secondary windings are electrically isolated from each other. These secondary windings can be either step-up or step-down depending on the application of the transformer.

Computing Values for Isolation Transformers with Multiple Secondaries

When computing the values of an isolation transformer with multiple secondary windings, each secondary must be treated as a different transformer. For example, the transformer in Figure 16-15 contains one primary winding and three secondary windings. The primary is connected to 120 VAC and has 300 turns of wire. One secondary has an output voltage of 560 V and a load impedance of 1000 Ω. The second secondary has an output voltage of 208 V and a load impedance of 400 Ω, and the third secondary has an output voltage of 24 V and a load impedance of 6 Ω. The current, turns of wire, and ratio for each secondary, and the current of the primary will be found.

The first step will be to compute the turns ratio of the first secondary. This can be done by dividing the smaller voltage into the larger.

$$\text{Ratio} = \frac{E_{S1}}{E_P}$$

$$\text{Ratio} = \frac{560}{120}$$

$$\text{Ratio} = 1{:}4.67$$

The current flow in the first secondary can be computed using Ohm's law.

$$I_{S1} = \frac{560}{1000}$$

$$I_{S1} = 0.56 \text{ A}$$

The number of turns of wire in the first secondary winding will be found using the turns ratio. Since this secondary has a higher voltage than the primary, it must have more turns of wire.

$$N_{S1} = N_P \times \text{turns ratio}$$

$$N_{S1} = 300 \times 4.67$$

$$N_{S1} = 1401 \text{ turns}$$

The amount of primary current needed to supply this secondary winding can be found using the turns ratio also. Since the primary has less voltage, it will require more current.

$$I_{P(\text{FIRST SECONDARY})} = I_{S1} \times \text{turns ratio}$$

$$I_{P(\text{FIRST SECONDARY})} = 0.56 \times 4.67$$

$$I_{P(\text{FIRST SECONDARY})} = 2.61 \text{ A}$$

The turns ratio of the second secondary winding will be found by dividing the higher voltage by the lower.

$$\text{Ratio} = \frac{208}{120}$$

$$\text{Ratio} = 1{:}1.73$$

The amount of current flow in this secondary can be determined using Ohm's law.

$$I_{S2} = \frac{208}{400}$$

$$I_{S2} = 0.52 \text{ A}$$

Since the voltage of this secondary is greater than the primary, it will have more turns of wire than the primary. The turns of this secondary will be found using the turns ratio.

$$N_{S2} = N_P \times \text{turns ratio}$$

$$N_{S2} = 300 \times 1.73$$

$$N_{S2} = 519 \text{ turns}$$

The voltage of the primary is less than this secondary. The primary will, therefore, require a greater amount of current. The amount of current required to operate this secondary will be computed using the turns ratio.

$$I_{P(\text{SECOND SECONDARY})} = I_{S2} \times \text{turns ratio}$$

$$I_{P(\text{SECOND SECONDARY})} = 0.52 \times 1.732$$

$$I_{P(\text{SECOND SECONDARY})} = 0.9 \text{ A}$$

The turns ratio of the third secondary winding will be computed in the same way as the other two. The larger voltage will be divided by the smaller.

$$\text{Ratio} = \frac{120}{24}$$

$$\text{Ratio} = 5{:}1$$

The primary current will be found using Ohm's law.

$$I_{S3} = \frac{24}{6}$$

$$I_{S3} = 4 \text{ A}$$

The output voltage of the third secondary is less than the primary. The number of turns of wire will, therefore, be less than the primary turns.

$$N_{S3} = \frac{N_P}{\text{turns ratio}}$$

$$N_{S3} = \frac{300}{5}$$

$$N_{S3} = 60 \text{ turns}$$

The primary has a higher voltage than this secondary. The primary current will, therefore, be less by the amount of the turns ratio.

$$I_{P(\text{THIRD SECONDARY})} = \frac{I_{S3}}{\text{turns ratio}}$$

$$I_{P(\text{THIRD SECONDARY})} = \frac{4}{5}$$

$$I_{P(\text{THIRD SECONDARY})} = 0.8 \text{ A}$$

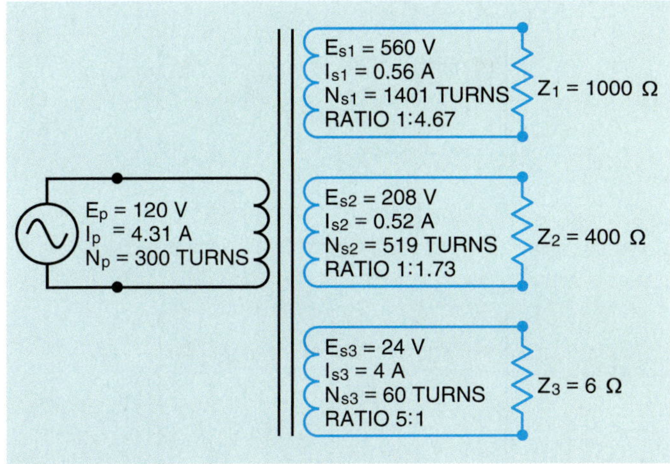

Figure 16-16 The transformer with all computed values.

The primary must supply current to each of the three secondary windings. Therefore, the total amount of primary current will be the sum of the currents required to supply each secondary.

$$I_{P(\text{TOTAL})} = I_{P1} + I_{P2} + I_{P3}$$

$$I_{P(\text{TOTAL})} = 2.61 + 0.9 + 0.8$$

$$I_{P(\text{TOTAL})} = 4.31 \text{ A}$$

The transformer with all computed values is shown in Figure 16-16.

Distribution Transformers

A common type of isolation transformer is the **distribution transformer,** Figure 16-17. This transformer changes the high voltage of power company distribution lines to the common 240/120 V that supplies power to most homes and many businesses. In this example, it is assumed that the primary is connected to a 7200-V line. The secondary is 240 V with a center tap. The center tap is grounded and becomes the **neutral conductor** or common conductor. If voltage is measured across the entire secondary, a voltage of 240 V will be seen. If voltage is measured from either line to the center tap, half of the secondary voltage, or 120 V, will be seen (Fig. 16-18). This occurs because the grounded neutral conductor becomes the center point of two in-phase voltages. A vector diagram drawn to illustrate this condition, shows that the grounded neutral conductor is connected to the center point of the two in-phase voltages (Fig. 16-19). Loads that are intended to operate on 240 V, such as water heaters, electric-resistance heating units, and central air conditioners are connected directly across the lines of the secondary (Fig. 16-20).

Loads that are intended to operate on 120 V connect from the center tap, or neutral, to one of the secondary

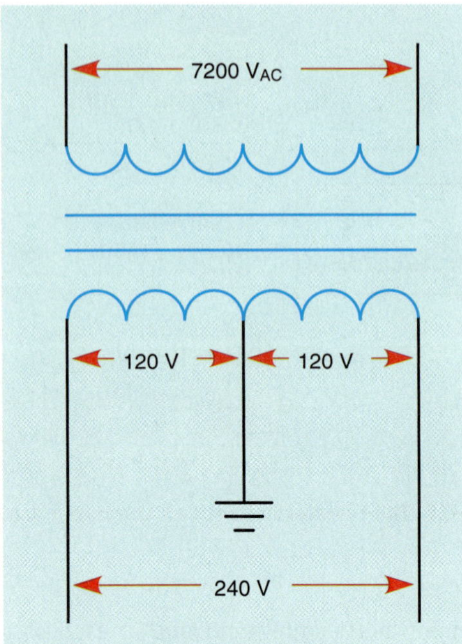

Figure 16-17 Distribution transformer.

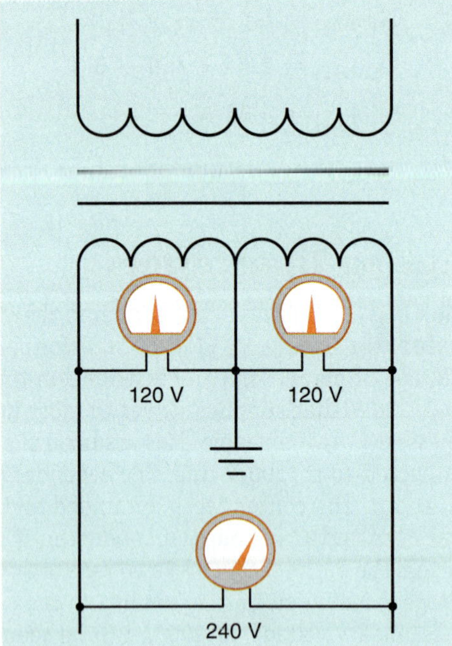

Figure 16-18 The voltage from either line to neutral is 120 V. The voltage across the entire secondary winding is 240 V.

lines. The function of the neutral is to carry the difference in current between the two secondary lines and maintain a balanced voltage. In Figure 16-21, one of the secondary lines has a current flow of 30 A and the other has a current flow of 24 A. The neutral conducts the sum of the unbalanced load. In this example, the neutral current will be 6 A (30 − 24 = 6).

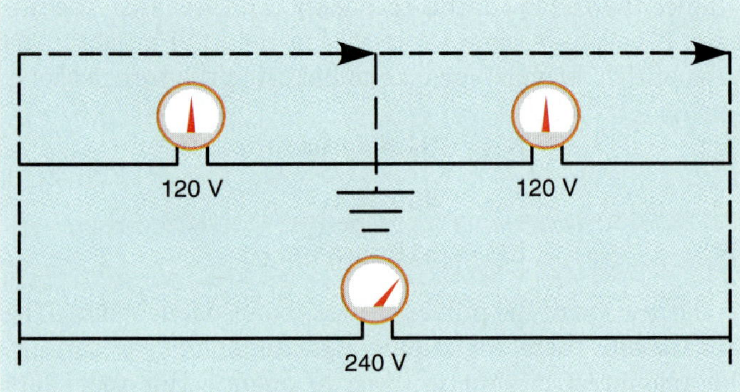

Figure 16-19 The voltages across the secondary are in phase with each other.

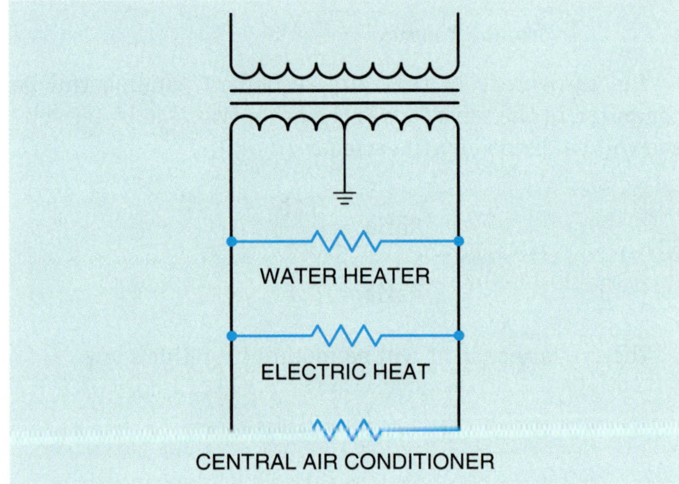

Figure 16-20 240-V loads connect directly across the secondary winding.

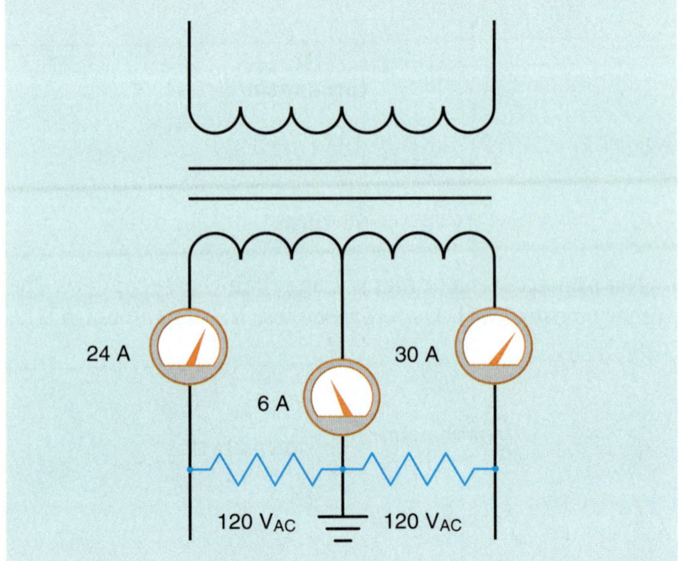

Figure 16-21 The neutral carries the sum of the unbalanced load.

Control Transformers

Another common type of isolation transformer found throughout industry is the **control transformer** (Fig. 16-22). The control transformer reduces line voltage to the value needed to operate control circuits. The most common type of control transformer contains two primary windings and one secondary. The primary windings are generally rated at 240 V each, and the secondary at 120 V. This arrangement provides a 2:1 turns ratio between each of the primary windings and the secondary. For example, assume that each of the primary windings contains 200 turns of wire. The secondary will contain 100 turns of wire.

One of the primary windings in Figure 16-23 is labeled H_1 and H_2. The other is labeled H_3 and H_4. The secondary winding is labeled X_1 and X_2. If the primary of the transformer is to be connected to 240 V, the two primary windings will be connected in parallel by connecting H_1 and H_3 together, and H_2 and H_4 together. When the primary windings are connected in parallel, the same voltage is applied across both windings. This has the same effect as using one primary winding with a total of 200 turns of wire. A turns ratio of 2:1 is maintained, and the secondary voltage is 120 V.

If the transformer is to be connected to 480 V, the two primary windings will be connected in series by connecting H_2 and H_3 together (Fig. 16-24). The incoming power is connected to H_1 and H_4. Series-connecting the primary windings increases the number of turns in the primary to 400. This produces a turns ratio of 4:1. When 480 V is connected to the primary, the secondary voltage remains at 120.

The primary leads of a control transformer are generally cross-connected, as shown in Figure 16-25, so metal links can be used to connect the primary for 240- or 480-V operation. If the primary is to be connected for 240-V operation, the metal links are connected under screws as shown in Figure 16-26.

Figure 16-22 Control transformer with fuse protection added to the secondary winding. *Courtesy of Hevi-Duty Electric.*

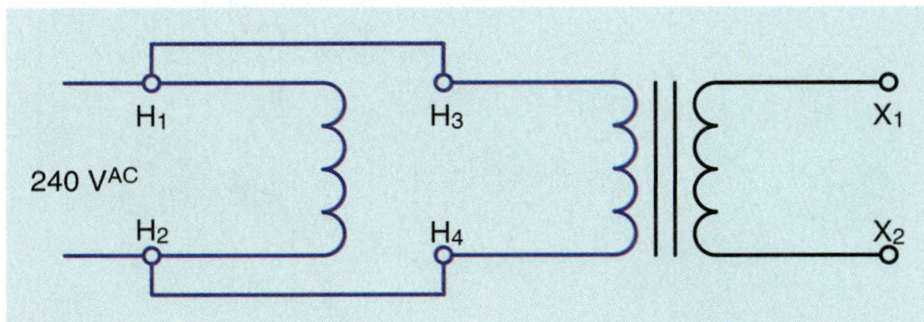

Figure 16-23 Control transformer connected for 240-V operation.

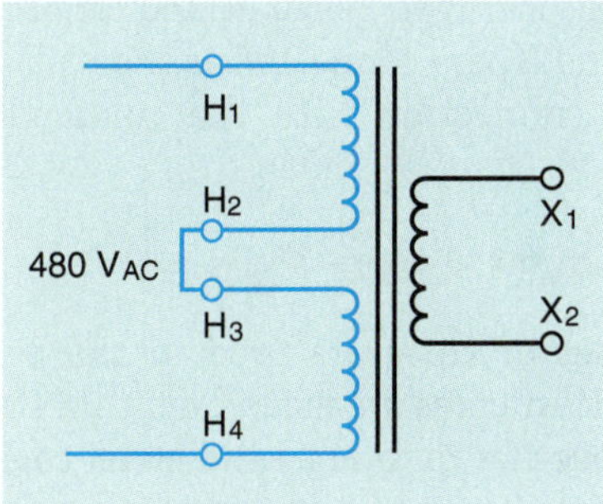

Figure 16-24 Control transformer connected for 480-V operation.

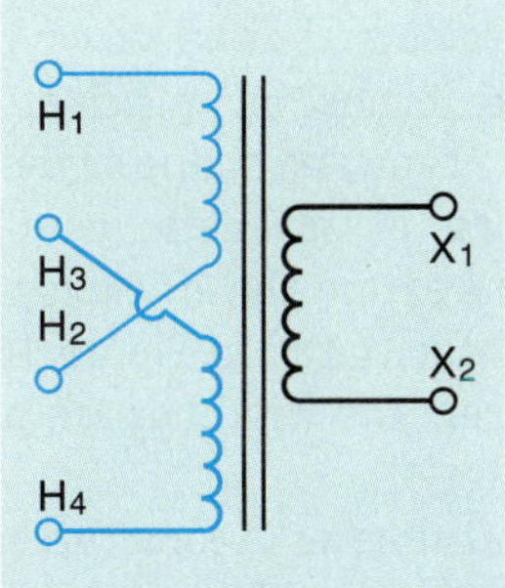

Figure 16-25 The primary windings of a control transformer are crossed.

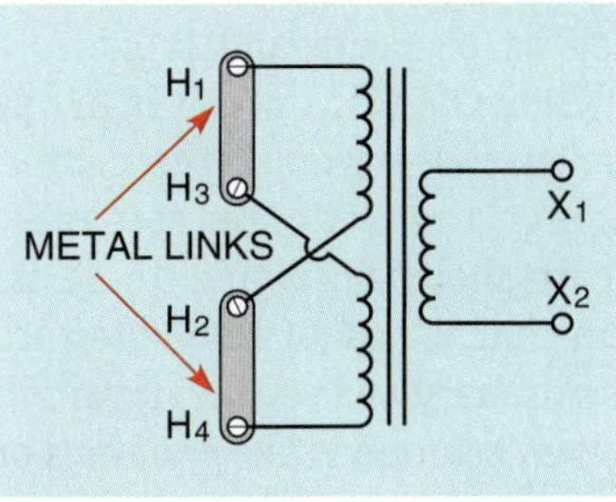

Figure 16-26 Metal links connect transformer for 240-V operation.

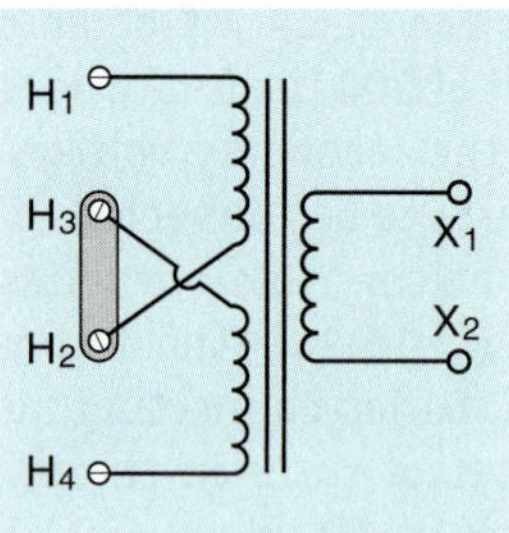

Figure 16-27 Control transformer connected for 480-V operation.

Figure 16-28 Core of a 600-MVA three-phase transformer.
Courtesy of Houston Lighting and Power.

Notice that leads H_1 and H_3 are connected together and leads H_2 and H_4 are connected together. Compare this connection with the connection shown in Figure 16-23.

If the transformer is to be connected for 480-V operation, terminals H_2 and H_3 are connected as shown in Figure 16-27. Compare this connection with the connection shown in Figure 16-24.

Transformer Core Types

Several types of cores are used in the construction of transformers. Most cores are made from thin steel punchings **laminated** together to form a solid metal core. The core for a 600-MVA (mega-volt-amp) three-phase transformer is shown in Figure 16-28. Laminated cores are preferred because a thin layer of oxide forms on the surface of each lamination and acts as an insulator to reduce the formation of eddy currents inside the core material. The amount of core material needed for a particular transformer is determined by the power rating of the transformer, but it must be sufficient to prevent saturation at full load. The type and shape of the core generally determine the amount of magnetic coupling between the windings and to some extent the efficiency of the transformer.

The transformer illustrated in Figure 16-29 is known as a core type transformer. The windings are placed around each end of the core material.

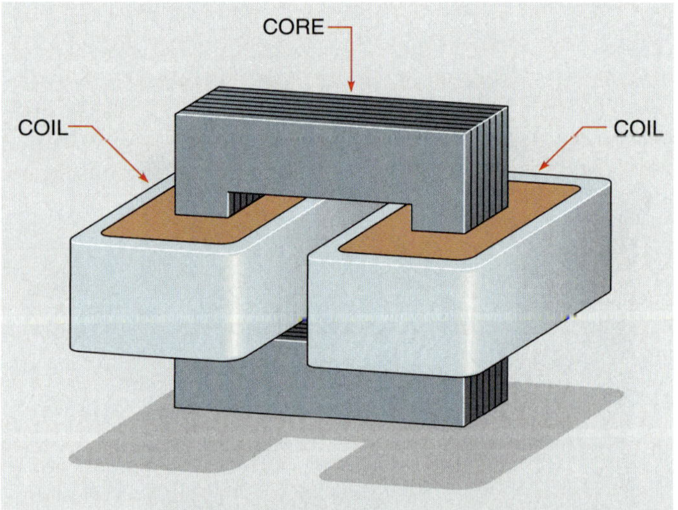

Figure 16-29 A core type transformer.

The shell type transformer is constructed in a similar manner to the core type, except that the shell type has a metal core piece through the middle of the window (Fig. 16-30). The primary and secondary windings are wound around the center core piece with the low-voltage winding being closest to the metal core. This arrangement permits the transformer to be surrounded by the core and provides excellent magnetic coupling. When the transformer is in operation, all the magnetic flux must pass through the center core piece. It then divides through the two outer core pieces.

The H type core shown in Figure 16-31 is similar to the shell type core in that it has an iron core through its center around which the primary and secondary windings are wound. The H core, however, surrounds the windings on four sides instead of two. This extra metal helps reduce stray leakage flux and improve the efficiency of the transformer. The H type core is often found on high-voltage distribution transformers.

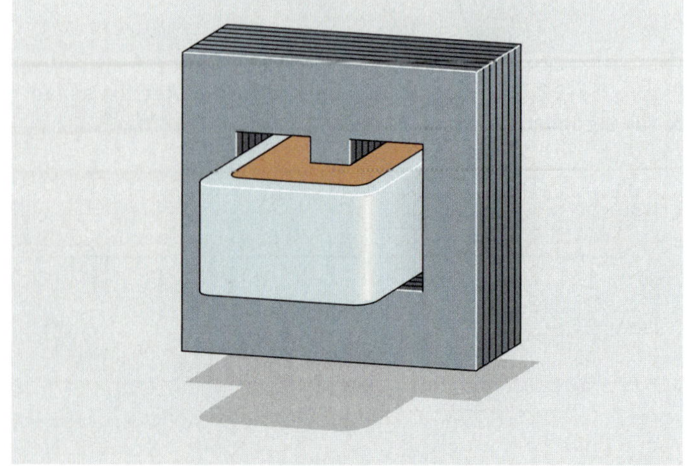

Figure 16-30 A shell type transformer.

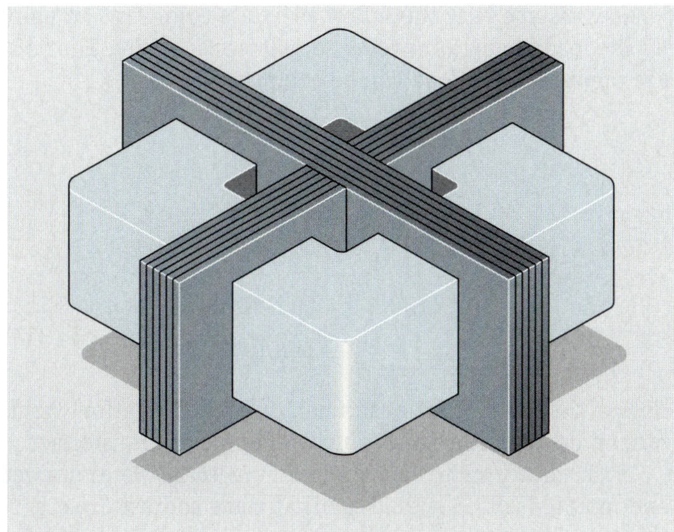

Figure 16-31 A transformer with an H type core.

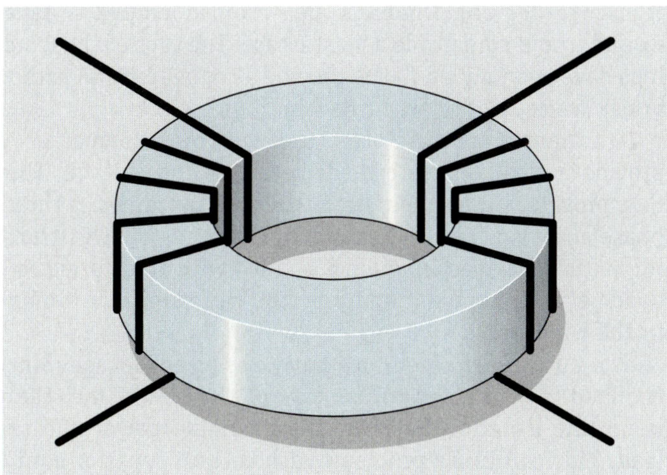

Figure 16-32 A toroid transformer.

The **tape wound core** or **toroid core** (Fig. 16-32) is constructed by tightly winding one long continuous silicon steel tape into a spiral. The tape may or may not be housed in a plastic container, depending on the application. This type of core does not require steel punchings laminated together. Since the core is one continuous length of metal, **flux leakage** is kept to a minimum. Flux leakage is the magnetic flux lines that do not follow the metal core and are lost to the surrounding air. The tape wound core is one of the most efficient core designs available.

Transformer Inrush Current

A reactor is an inductor used to add inductance to the circuit. Although transformers and reactors are both inductive devices, there is a great difference in their operating characteristics. Reactors are often connected in series with a low-impedance load to prevent **inrush current** (the amount of current that flows when power is initially applied to the circuit) from becoming excessive (Fig. 16-33). Transformers, however, can produce extremely high inrush currents when power is first applied to the primary winding. The type of core used when constructing inductors and transformers is primarily responsible for this difference in characteristics.

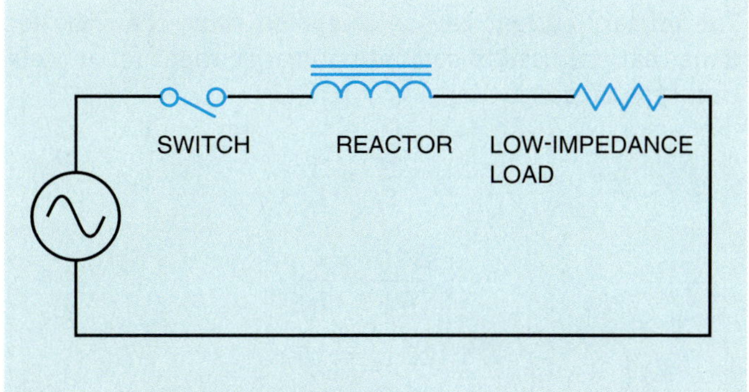

Figure 16-33 Reactors help prevent inrush current from becoming excessive when power is first turned on.

Autotransformers

Autotransformers are one-winding transformers. They use the same winding for both the primary and secondary. The primary winding in Figure 16-34 is between points B and N and has a voltage of 120 V applied to it. Between points B and N, there are 120 turns of wire. Now assume that the selector switch is set to point D. The load is now connected between points D and N. The secondary of this transformer contains 40 turns of wire. If the amount of voltage applied to the load is to be computed, the following formula can be used.

$$\frac{E_P}{E_S} = \frac{N_P}{N_S}$$

$$\frac{120}{E_S} = \frac{120}{40}$$

$$120\ E_S = 4800$$

$$E_S = 40\text{ V}$$

Assume that the load connected to the secondary has an impedance of 10 Ω. The amount of current flow in the secondary circuit can be computed using the formula

$$I = \frac{E}{Z}$$

$$I = \frac{40}{10}$$

$$I = 4\text{ A}$$

The primary current can be computed using the same formula that was used to compute primary current for an isolation type of transformer.

$$\frac{E_P}{E_S} = \frac{I_S}{I_P}$$

$$\frac{120}{40} = \frac{4}{I_P}$$

$$120\ I_P = 160$$

$$I_P = 1.333\ A$$

The amount of power input and output for the autotransformer must be the same, just as they are in an isolation transformer.

Primary	Secondary
$120 \times 1.333 = 160$ VA	$40 \times 4 = 160$ VA

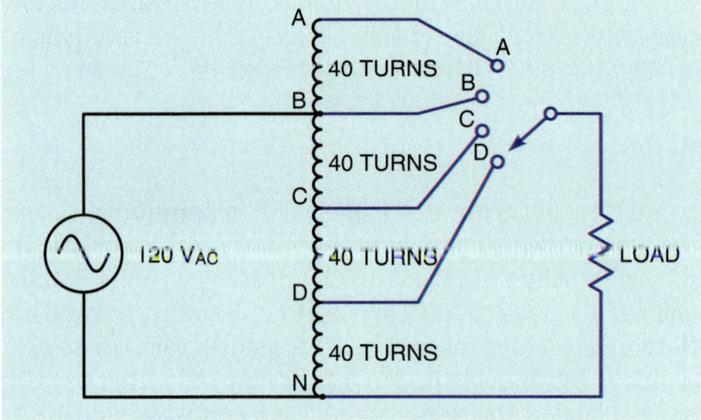

Figure 16-34 **An autotransformer has only one winding, which is used for both the primary and secondary.**

Now assume that the rotary switch is connected to point A. The load is now connected to 160 turns of wire. The voltage applied to the load can be computed by

$$\frac{E_P}{E_S} = \frac{N_P}{N_S}$$

$$\frac{120}{E_S} = \frac{120}{160}$$

$$120\ E_S = 19{,}200$$

$$E_S = 160\ V$$

Notice that the autotransformer, like the isolation transformer, can be either a step-up or step-down transformer.

If the rotary switch shown in Figure 16-34 were removed and replaced with a sliding tap that made contact directly to the transformer winding, the turns ratio could be adjusted continuously. This type of transformer is commonly referred to as a Variac or Powerstat, depending on the manufacturer. A cutaway view of a variable autotransformer is shown in Figure 16-35. The windings are wrapped around a tape-wound toroid core inside a plastic case. The tops of the windings have been milled flat to provide a commutator. A carbon brush makes contact with the windings.

Autotransformers are often used by power companies to provide a small increase or decrease to line voltage. They help provide voltage regulation to large power lines. A three-phase autotransformer is shown in Figure 16-36. This transformer is contained in a housing filled with transformer oil, which acts as a coolant and prevents moisture from forming in the windings.

The autotransformer does have one disadvantage. Since the load is connected to one side of the power line, there is no line isolation between the incoming power and the load. This can cause problems with certain types of equipment and must be a consideration when designing a power system.

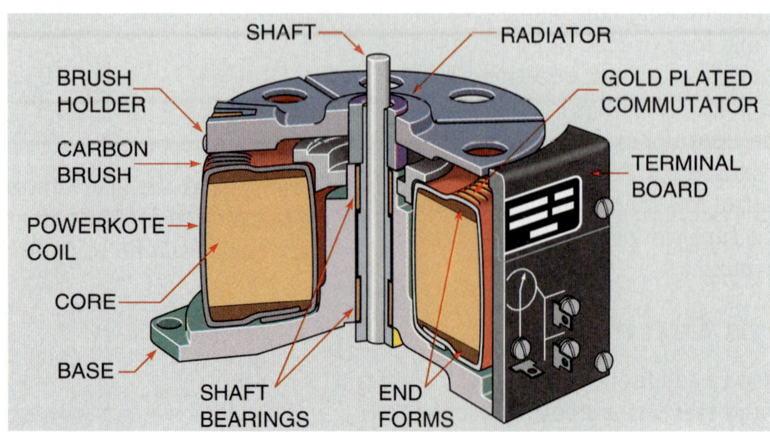

Figure 16-35 **Cutaway view of a powerstat.**

Figure 16-36 Three-phase autotransformer. *Courtesy of Magnatek.*

Transformer Polarities

To understand transformer polarity, the voltage produced across a winding must be considered during some point in time. In a 60-Hz AC circuit, the voltage changes polarity 60 times per second. When discussing transformer polarity, it is necessary to consider the relationship between the different windings at the same point in time. It will, therefore, be assumed that this point in time is when the peak positive voltage is being produced across the winding.

Polarity Markings on Schematics

When a transformer is shown on a schematic diagram it is common practice to indicate the polarity of the transformer windings by placing a dot beside one end of each winding, as shown in Figure 16-37. These dots signify that the polarity is the same at that point in time for each winding. For example, assume the voltage applied to the primary winding is at its peak positive value at the terminal indicated by the dot. The voltage at the dotted lead of the secondary will be at its peak positive value at the same time.

This same type of polarity notation is used for transformers that have more than one primary or secondary winding. An example of a transformer with a multisecondary is shown in Figure 16-38.

Additive and Subtractive Polarities

The polarity of transformer windings can be determined by connecting them as an autotransformer and testing for additive or subtractive polarity, often referred to as a boost or buck connection. This is done by connecting one lead of the secondary to one lead of the primary and measuring the voltage across both windings (Fig. 16-39). The transformer shown in the example has a primary voltage rating of 120 V and a secondary voltage rating of 24 V. This same circuit has been redrawn in Figure 16-40 to show the connection more clearly.

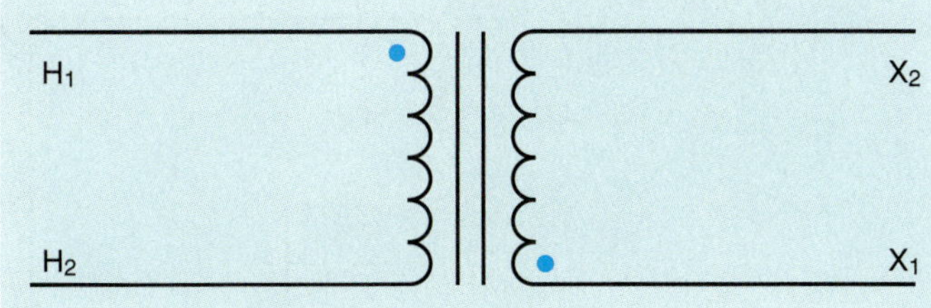

Figure 16-37 Transformer polarity dots.

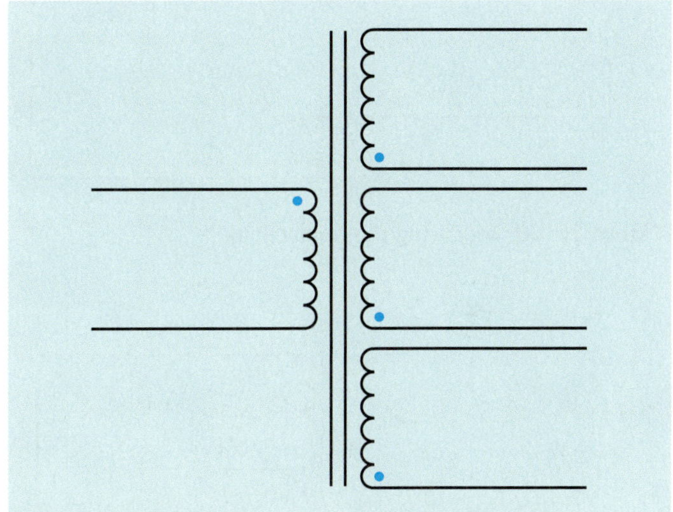

Figure 16-38 Polarity marks for multiple secondaries.

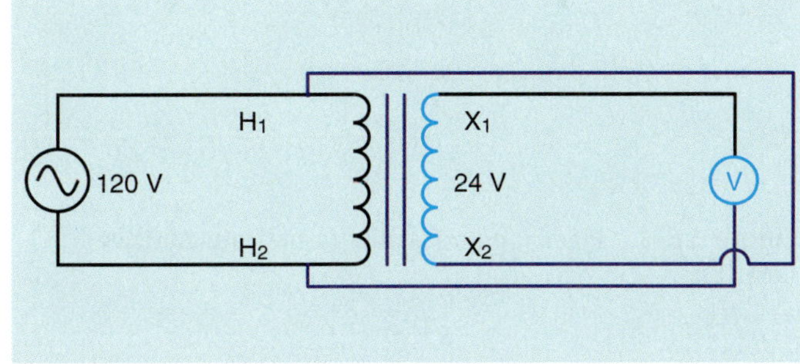

Figure 16-39 Connecting the secondary and primary windings forms an autotransformer.

Notice that the secondary winding has been connected in series with the primary winding. The transformer now contains only one winding and is, therefore, an autotransformer. When 120 V is applied to the primary winding, the voltmeter connected across the secondary will indicate either the *sum* of the two voltages or the *difference* between the two voltages. If this voltmeter indicates 144 V (120 + 24 = 144) the windings are connected additive (boost), and polarity dots can be placed as shown in Figure 16-41. Notice in this connection that the secondary voltage is added to the primary voltage.

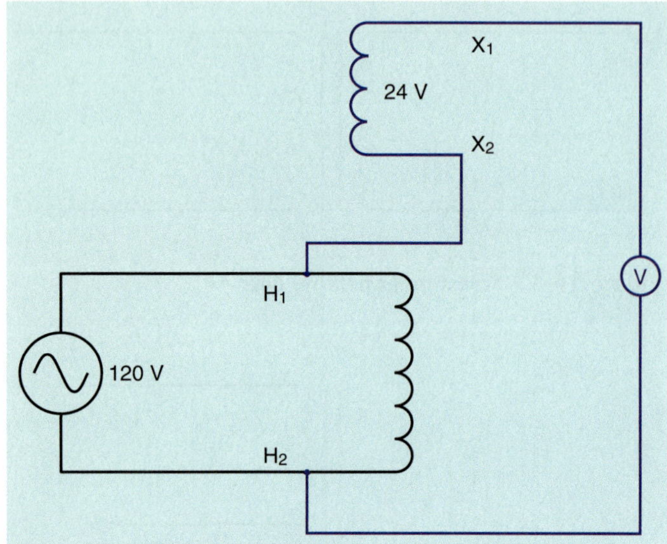

Figure 16-40 Redrawing the connection.

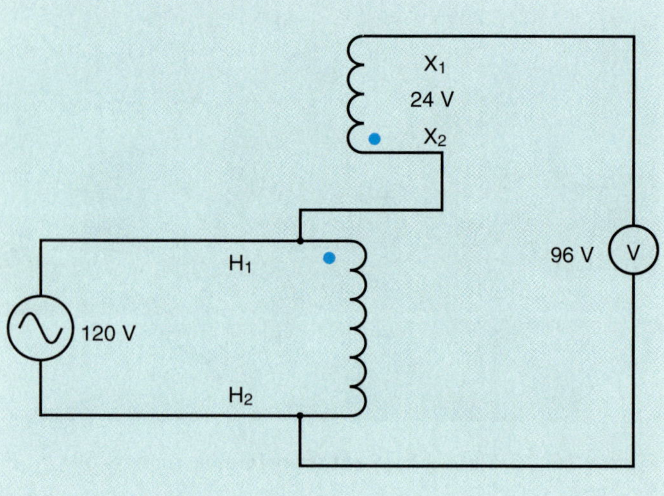

Figure 16-42 Polarity dots indicate subtractive polarity.

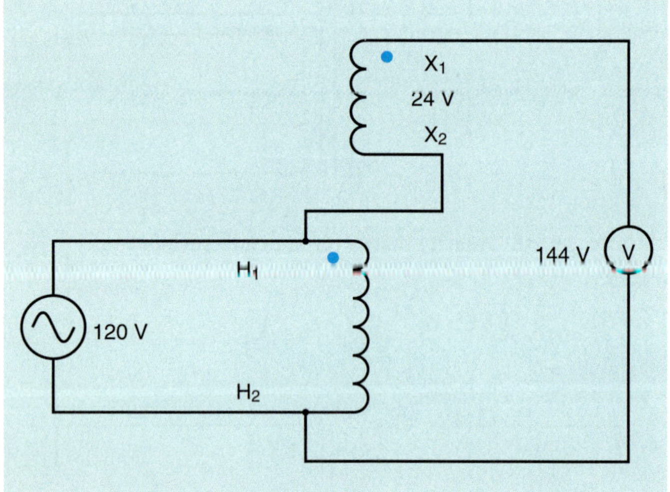

Figure 16-41 Placing polarity dots to indicate additive polarity.

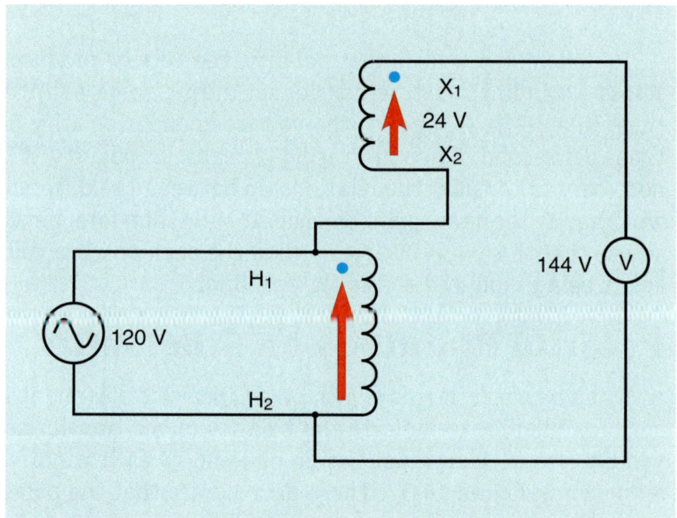

Figure 16-43 Arrows help indicate the placement of the polarity dots.

If the voltmeter connected to the secondary winding indicates a voltage of 96 V (120 − 24 = 96) the windings are connected subtractive (buck), and polarity dots are placed as shown in Figure 16-42.

Using Arrows to Place Dots

To help in the understanding of additive and subtractive polarity, arrows can be used to indicate a direction of greater than or less than values. In Figure 16-43, arrows have been added to indicate the direction in which the dot is to be placed. In this example, the transformer is connected additive, or boost, and both of the arrows point in the same direction. Notice that the arrow points to the dot. In Figure 16-44 it is seen that values of the two arrows add to produce 144 V.

In Figure 16-45, arrows have been added to a subtractive, or buck, connection. In this instance, the arrows point in opposite directions, and the voltage of one tries to cancel the voltage of the other. The result is that the smaller value is eliminated, and the larger value is reduced as shown in Figure 16-46.

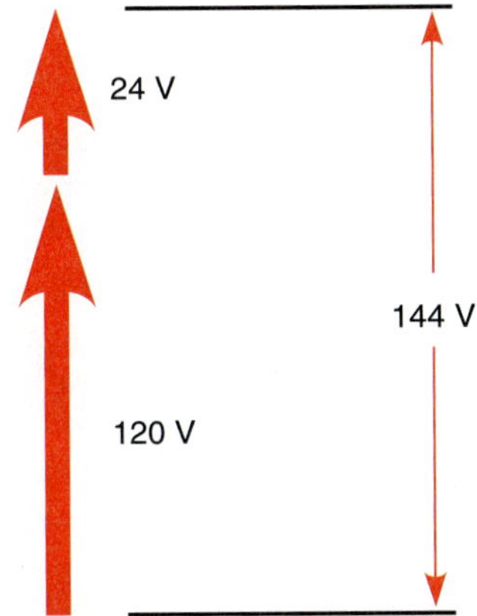

Figure 16-44 **The values of the arrows add to indicate additive polarity (boost connection).**

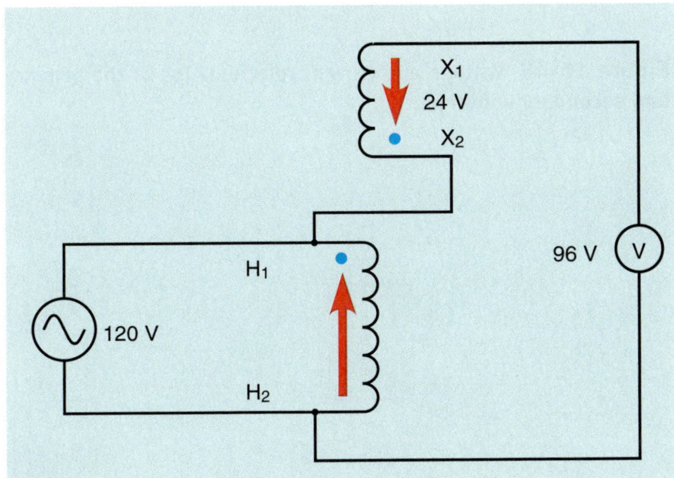

Figure 16-45 **The values of the arrows subtract (buck connection).**

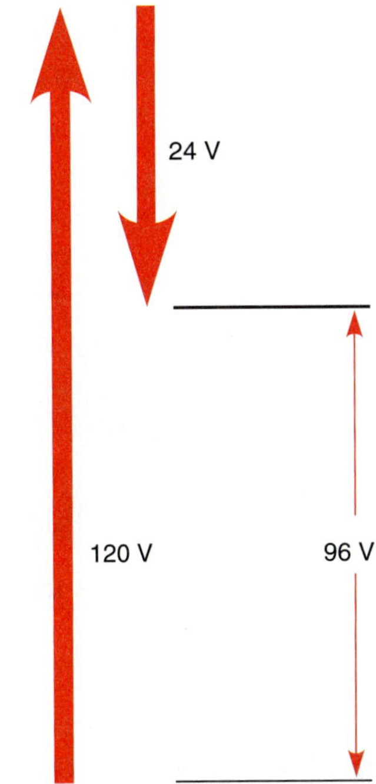

Figure 16-46 **The arrows help indicate subtractive polarity.**

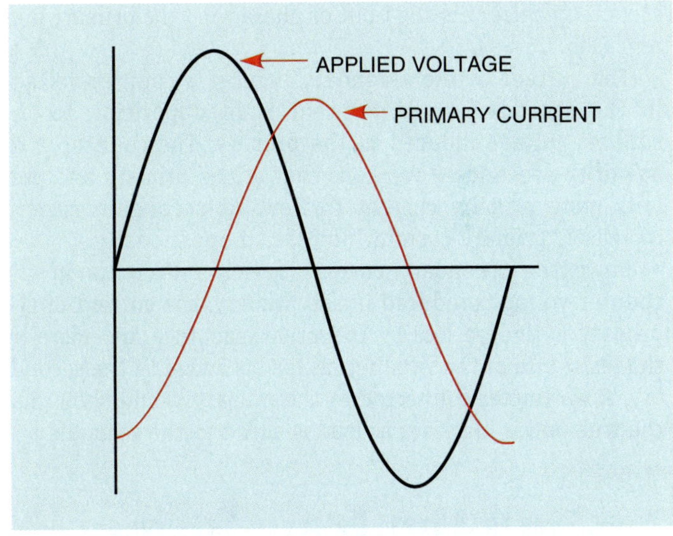

Figure 16-47 **At no load, the primary current lags the voltage by 90°.**

Voltage and Current Relationships in a Transformer

When the primary of a transformer is connected to power but there is no load connected to the secondary, current is limited by the inductive reactance of the primary. At this time, the transformer is essentially an inductor, and the excitation current is lagging the applied voltage by 90° (Fig. 16-47). The primary current induces a voltage in the secondary. This induced voltage is proportional to the rate of change of current. The secondary voltage will be maximum during the periods when the primary current is changing the most (0°, 180°, and 360°), and it will be zero when the primary current is not changing (90° and 270°). A plot of the primary current and secondary voltage shows that the secondary voltage lags the primary current by 90° (Fig. 16-48). Because the secondary voltage lags the primary current by 90° and the applied voltage leads the primary current by 90°, the secondary voltage is 180° out of phase with the applied voltage and in phase with the induced voltage in the primary.

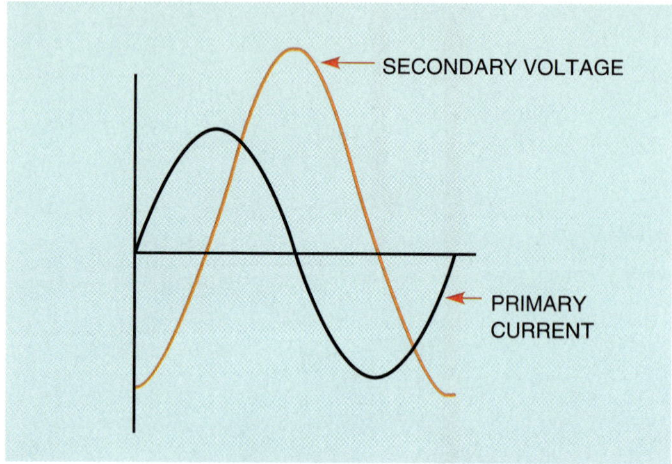

Figure 16-48 **The secondary voltage lags the primary current by 90°.**

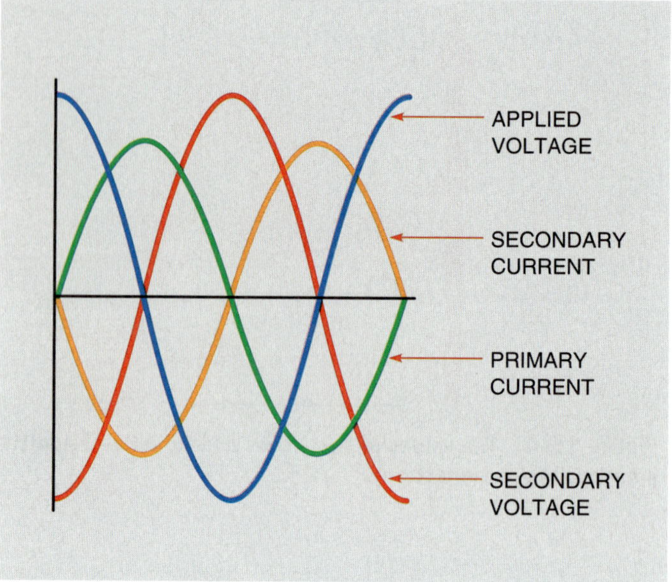

Figure 16-49 **Voltage and current relationships of the primary and secondary windings.**

Adding Load to the Secondary

When a load is connected to the secondary, current begins to flow. Because the transformer is an inductive device, the secondary current lags the secondary voltage by 90°. Since the secondary voltage lags the primary current by 90°, the secondary current is 180° out of phase with the primary current (Fig. 16-49).

The current of the secondary induces a counter-voltage in the secondary windings that is in opposition to the counter-voltage induced in the primary. The counter-voltage of the secondary weakens that of the primary and permits more primary current to flow. As secondary current increases, primary current increases proportionally.

Since the secondary current causes a decrease in the counter-voltage produced in the primary, the current of the primary is limited less by inductive reactance and more by the resistance of the windings as load is added to the secondary. A wattmeter connected to the primary would show that the true power increases as load is added to the secondary.

Testing the Transformer

Several tests can be made to determine the condition of the transformer. A simple test for grounds, shorts, or opens can be made with an ohmmeter (Fig. 16-50). Ohmmeter A is connected to one lead of the primary and one lead of the secondary. This test checks for shorted windings between the primary and secondary. The ohmmeter should indicate infinity. If there is more than one primary or secondary winding, all isolated windings should be tested for shorts. Ohmmeter B illustrates testing the windings for grounds. One lead of the ohmmeter is connected to the case of the transformer, and the other is connected to the winding. All windings should be tested for grounds, and the ohmmeter should indicate infinity for each winding. Ohmmeter C illustrates testing the windings for continuity. The wire resistance of the winding should be indicated by the ohmmeter.

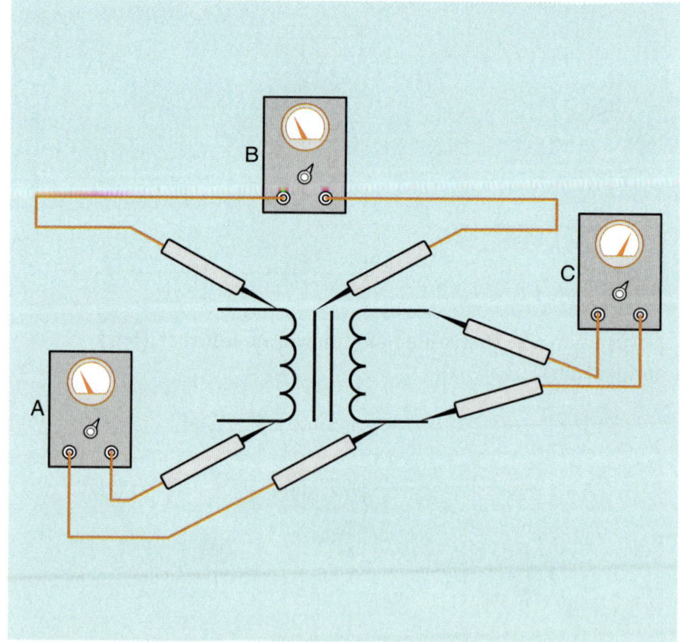

Figure 16-50 **Testing a transformer with an ohmmeter.**

If the transformer appears to be in good condition after the ohmmeter test, it should then be tested for shorts and grounds with a megohm-meter. A MEGGER® will reveal problems of insulation breakdown that an ohmmeter will not. The condition of the dielectric oil in large oil-filled transformers should be tested at periodic intervals. This involves taking a sample of the oil and performing tests for dielectric strength and contamination.

Transformer Ratings

Most transformers have a nameplate with information concerning the transformer. The information listed is generally determined by the size, type, and manufacturer. Almost all nameplates list the primary voltage, secondary voltage, and KVA (kilo-volt-amps) rating. Transformers are rated in kilo-volt-amps and not kilowatts because the true power is determined by the power factor of the load. Other information that may or may not be listed is frequency, temperature rise in C°, impedance, type of insulating oil, gallons of insulating oil, serial number, type number, model number, and whether the transformer is single-phase or three-phase.

Determining Maximum Current

The nameplate does not list the current rating of the windings. Since power input must equal power output, the current rating for a winding can be determined by dividing the KVA rating by the winding voltage. For example, assume a transformer has a KVA rating of 0.5 KVA, a primary voltage of 480 V, and a secondary voltage of 120 V. To determine the maximum current that can be supplied by the secondary, divide the KVA rating by the secondary voltage.

$$I_S = \frac{KVA}{E_S}$$

$$I_S = \frac{500}{120}$$

$$I_S = 4.16 \text{ A}$$

The primary current can be computed in the same way.

$$I_P = \frac{KVA}{E_P}$$

$$I_P = \frac{500}{480}$$

$$I_P = 1.04 \text{ A}$$

Transformers with multiple secondary windings generally have the current rating listed with the voltage rating.

Transformer Impedance

Transformer impedance is determined by the physical construction of the transformer. Factors such as the amount and type of core material, wire size used to construct the windings, number of turns, and degree of magnetic coupling between the windings greatly affect the transformer's impedance. Impedance is expressed as a percent (%Z or %IZ) and is measured by connecting a short circuit across the low-voltage winding of the transformer and then connecting a variable voltage source to the high-voltage winding, Figure 16-51. The variable voltage is then increased until rated current flows in the low-voltage winding. The transformer impedance is determined by calculating the percentage of variable voltage compared with the rated voltage of the high-voltage winding.

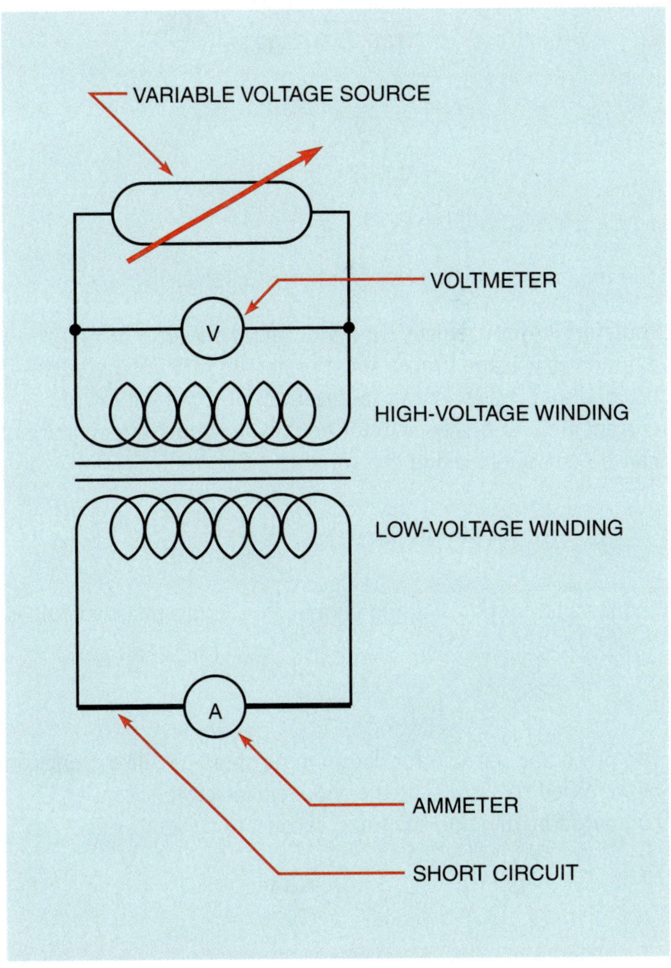

Figure 16-51 **Determining transformer impedance.**

Example 1

Assume that the transformer shown in Figure 16-51 is a 2400/480 volt 15 kVA transformer. To determine the impedance of the transformer, first compute the full-load current rating of the secondary winding.

$$I = \frac{VA}{E}$$

$$I = \frac{15{,}000}{480}$$

$$I = 31.25\ A$$

Next, increase the source voltage connected to the high-voltage winding until a current of 31.25 amperes flows in the low-voltage winding. Assume that the voltage value is 138 volts. Finally, determine the percentage of applied voltage compared with the rated voltage.

$$\%Z = \frac{\text{source voltage}}{\text{rated voltage}} \times 100$$

$$\%Z = \frac{138}{2400} \times 100$$

$$\%Z = 0.0575 \times 100$$

$$\%Z = 5.75$$

The impedance of this transformer is 5.75%.

Transformer impedance is a major factor in determining the amount of voltage drop a transformer will exhibit between no load and full load and in determining the amount of current flow in a short-circuit condition. Short-circuit current can be computed using the formula

$$(\text{Single phase})\ I_{SC} = \frac{VA}{E \times \%Z}$$

The formula for determining current in a single-phase circuit is

$$I = \frac{VA}{E}$$

The preceding formula for determining short-circuit current can be modified to show that the short-circuit current can be computed by dividing the rated secondary current by the %Z.

$$I_{SC} = \frac{I_{Rated}}{\%Z}$$

Example 2

A single-phase transformer is rated at 50 kVA and has a secondary voltage of 240 volts. The nameplate reveals that the transformer has an internal impedance (%IZ) of 2.5%. What is the short-circuit current for this transformer?

$$I_{Secondary} = \frac{50{,}000}{240}$$

$$I_{Secondary} = 208.3\ \text{amperes}$$

$$I_{Shortcircuit} = \frac{208.3}{\%Z}$$

$$I_{Shortcircuit} = \frac{208.3}{0.025}$$

$$I_{Shortcircuit} = 8{,}333.3\ \text{amperes}$$

It is sometimes necessary to compute the amount of short-circuit current when determining the correct fuse rating for a circuit. The fuse must have a high enough "interrupt" rating to clear the fault in the event of a short circuit.

Summary

- All values of voltage, current, and impedance in a transformer are proportional to the turns ratio.
- Transformers can change values of voltage, current, and impedance, but cannot change the frequency.
- The primary winding of a transformer is connected to the power line.
- The secondary winding is connected to the load.
- A transformer that has a lower secondary voltage than primary voltage is a step-down transformer.
- A transformer that has a higher secondary voltage than primary voltage is a step-up transformer.
- An isolation transformer has its primary and secondary windings electrically and mechanically separated from each other.
- When a coil induces a voltage into itself, it is known as self-induction.
- When a coil induces a voltage into another coil, it is known as mutual induction.
- Transformers can have very high inrush current when first connected to the power line because of the magnetic domains in the core material.

- Inductors provide an air gap in their core material that causes the magnetic domains to reset to a neutral position.
- Autotransformers have only one winding, which is used as both the primary and secondary.
- Autotransformers have a disadvantage in that they have no line isolation between the primary and secondary winding.
- Isolation transformers help filter voltage and current spikes between the primary and secondary side.
- Polarity dots are often added to schematic diagrams to indicate transformer polarity.
- Transformers can be connected as additive or subtractive polarity.

Review Questions

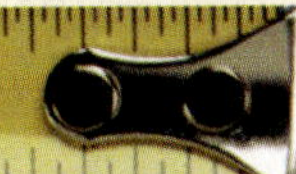

1. What is a transformer?
2. What are common efficiencies for transformers?
3. What is an isolation transformer?
4. All values of a transformer are proportional to its ____________ ____________.
5. What is an autotransformer?
6. What is a disadvantage of an autotransformer?
7. Explain the difference between a step-up and a step-down transformer.
8. A transformer has a primary voltage of 240 V and a secondary voltage of 48 V. What is the turns ratio of this transformer?
9. A transformer has an output of 750 VA. The primary voltage is 120 V. What is the primary current?
10. A transformer has a turns ratio of 1:6. The primary current is 18 A. What is the secondary current?
11. What do the dots shown beside the terminal leads of a transformer represent on a schematic?
12. A transformer has a primary voltage rating of 240 V and a secondary voltage rating of 80 V. If the windings were connected subtractive, what voltage would appear across the entire connection?
13. If the windings of the transformer in question 12 were to be connected additive, what voltage would appear across the entire winding?
14. The primary leads of a transformer are labeled 1 and 2. The secondary leads are labeled 3 and 4. If polarity dots are placed beside leads 1 and 4, which secondary lead would be connected to terminal 2 to make the connection additive?

Practice Problems

Refer to Figure 16-52 to answer the following questions. Find all the missing values.

1

E_P 120	E_S 24
I_P ______	I_S ______
N_P 300	N_S ______
Ratio ______	$Z = 3\ \Omega$

2

E_P 240	E_S 320
I_P ______	I_S ______
N_P ______	N_S 280
Ratio ______	$Z = 500\ \Omega$

3

E_P ______	E_S 160
I_P ______	I_S ______
N_P ______	N_S 80
Ratio 1:2.5	$Z = 12\ \Omega$

4

E_P 48	E_S 240
I_P ______	I_S ______
N_P 220	N_S ______
Ratio ______	$Z = 360\ \Omega$

5

E_P ______	E_S ______
I_P 16.5 ______	I_S 3.25 ______
N_P ______	N_S 450
Ratio ______	$Z = 56\ \Omega$

6

E_P 480	E_S ______
I_P ______	I_S ______
N_P 275	N_S 525
Ratio ______	$Z = 1.2\ k\Omega$

Refer to Figure 16-53 to answer the following questions. Find all the missing values.

7

E_P 208	E_{S1} 320	E_{S2} 120	E_{S3} 24
I_P ______	I_{S1} ______	I_{S2} ______	I_{S3} ______
N_P 800	N_{S1} ______	N_{S2} ______	N_{S3} ______
	Ratio 1:	Ratio 2:	Ratio 3:
	R_1 12 kΩ	R_2 6 Ω	R_3 8 Ω

8

E_P 277	E_{S1} 480	E_{S2} 208	E_{S3} 120
I_P ______	I_{S1} ______	I_{S2} ______	I_{S3} ______
N_P 350	N_{S1} ______	N_{S2} ______	N_{S3} ______
	Ratio 1:	Ratio 2:	Ratio 3:
	R_1 200 Ω	R_2 60 Ω	R_3 24 Ω

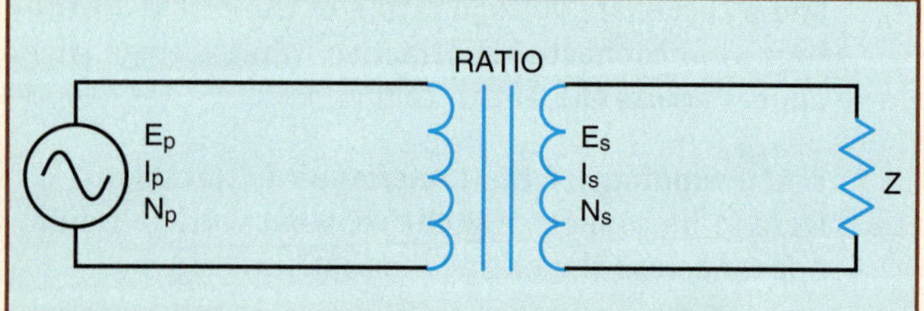

Figure 16-52 Isolation transformer practice problems.

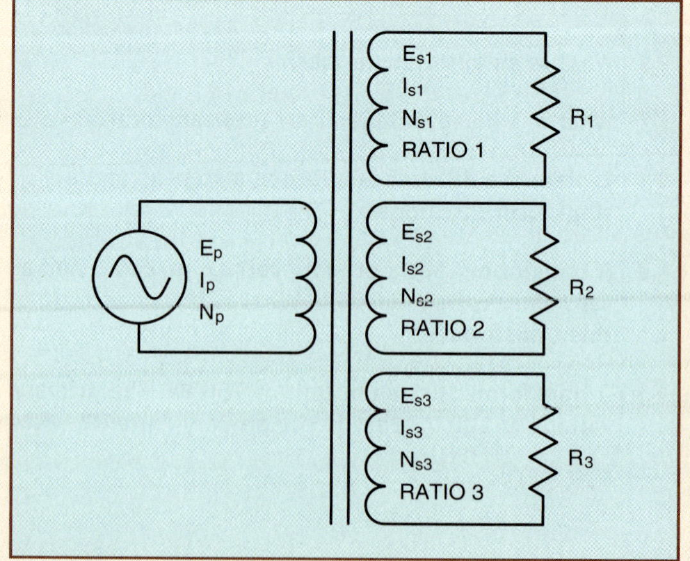

Figure 16-53 Single-phase transformer with multi-secondaries.

Chapter 17 Single-Phase Motors

Although most of the large motors used in industry are three-phase, at times single-phase motors must be used. Single-phase motors are used almost exclusively to operate home appliances such as air conditioners, refrigerators, well pumps, and fans. They are generally designed to operate on 120 V or 240 V. They range in size from fractional horsepower to several horsepower, depending on the application.

OBJECTIVES

After studying this unit, you should be able to:

- list the different types of split-phase motors.
- discuss the operation of split-phase motors.
- reverse the direction of rotation of a split-phase motor.
- discuss the operation of multispeed split-phase motors.
- discuss the operation of shaded-pole type motors.
- discuss the operation of repulsion type motors.
- discuss the operation of stepping motors.
- discuss the operation of universal motors.

Glossary of Single-Phase Motor Terms

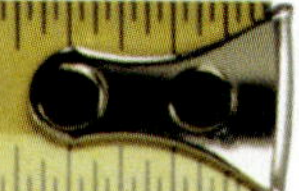

centrifugal switch a switch used to disconnect the start windings in a split-phase motor after the motor has accelerated to approximately 75% of rated speed

compensating winding a winding used in universal motors to counteract the inductive reactance in the armature windings

conductive compensation accomplished by connecting the compensating winding of a universal motor in series with the field winding

Holtz motor a type of single-phase synchronous motor that operates at a speed of 1200 RPM

inductive compensation accomplished by shorting the compensating winding leads together and permitting induced voltage to supply current to the winding

multispeed motors motors designed to operate at more than one full load speed

neutral plane the point at which no voltage is induced in the armature winding

run winding one of the windings in a split-phase motor

shaded-pole induction motor a single-phase motor that produces a rotating magnetic field by shading one side of each pole piece; shading is accomplished by placing a loop of large copper wire around one side of the pole piece

shading coil the loop of large wire used to form a shaded pole

split-phase motor a type of single-phase motor that splits the current flow through two separate windings to produce a rotating magnetic field

start winding one of the windings used in a split-phase motor

synchronous motors motors that operate at a constant speed from no load to full load

synchronous speed the speed of the rotating magnetic field of an AC induction motor

two phase a power system that produces two separate phase voltages 90° apart

universal motor a type of single-phase motor that can operate on either direct or alternating current

Warren motor a type of single-phase synchronous motor that operates at a speed of 3600 RPM

Split-Phase Motors

Split-phase motors fall into three general classifications:

1. The resistance-start induction-run motor.
2. The capacitor-start induction-run motor.
3. The capacitor-start capacitor-run motor.

Although these motors have different operating characteristics, they are similar in construction and use the same operating principle. **Split-phase motors** receive their name from the way they operate on the principle of a rotating magnetic field. A rotating magnetic field, however, cannot be produced with only one phase. Split-phase motors, therefore, split the current flow through two separate windings to simulate a two-phase power system. A rotating magnetic field can be produced with a two-phase system.

The Two-Phase System

In some parts of the world, two-phase power is produced. A **two-phase** system is produced by having an alternator with two sets of coils wound 90° apart (Fig. 17-1). The voltages of a two-phase system are, therefore, 90° out of phase with each other. These two out-of-phase voltages can produce a rotating magnetic field. Because there have to be two voltages or currents out of phase with each other to produce a rotating magnetic field, split-phase motors use two separate windings to create a phase difference between the currents in the two windings. These motors literally split one phase and produce a second phase, hence the name split-phase motor.

Stator Windings

The stator of a split-phase motor contains two separate windings, the **start winding** and the **run winding.** The start winding is made of small wire and is placed near the top of the stator core. The run winding is made of relatively large wire and is placed in the bottom of the stator core. Figures 17-2A and 2B show photographs of two split-phase stators. The stator in A is used for a resistance-start induction-run motor, or a capacitor-start, induction-run motor. The stator in B is used for a capacitor-start, capacitor-run motor. Both stators contain four poles, and the start winding is placed at a 90° angle from the run winding.

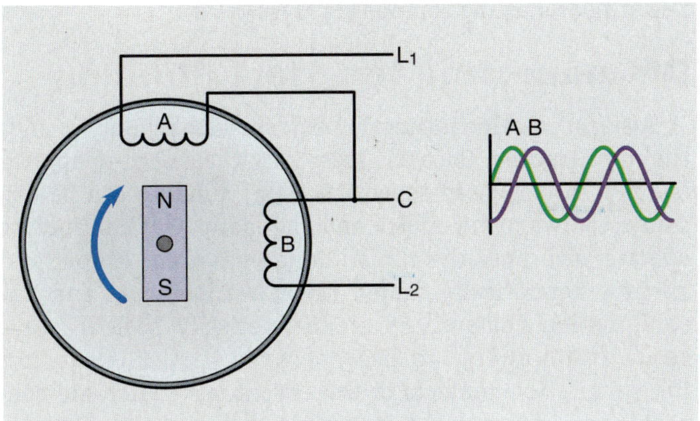

Figure 17-1 A two-phase alternator produces voltages that are 90° out of phase with each other.

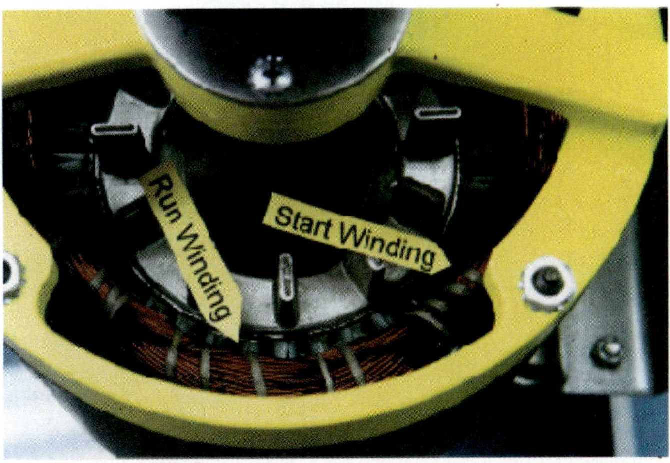

Figure 17-2A Stator windings used in single-phase motors.

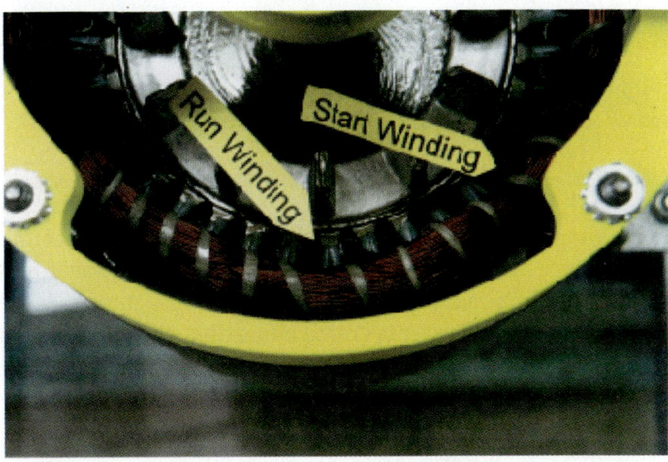

Figure 17-2B Stator windings used in single-phase motors.

Notice the difference in size and position of the two windings of the stator shown in Figure 17-2A. The start winding is made from small wire and placed near the top of the stator core. This causes it to have a higher resistance than the run winding. The start winding is located between the poles of the run winding. The run winding is made with larger wire and placed near the bottom of the core. This gives it higher inductive reactance and less resistance than the start winding. These two windings are connected in parallel with each other (Fig. 17-3).

When power is applied to the stator, current flows through both windings. Because the start winding is more resistive, the current flow through it will be more in phase

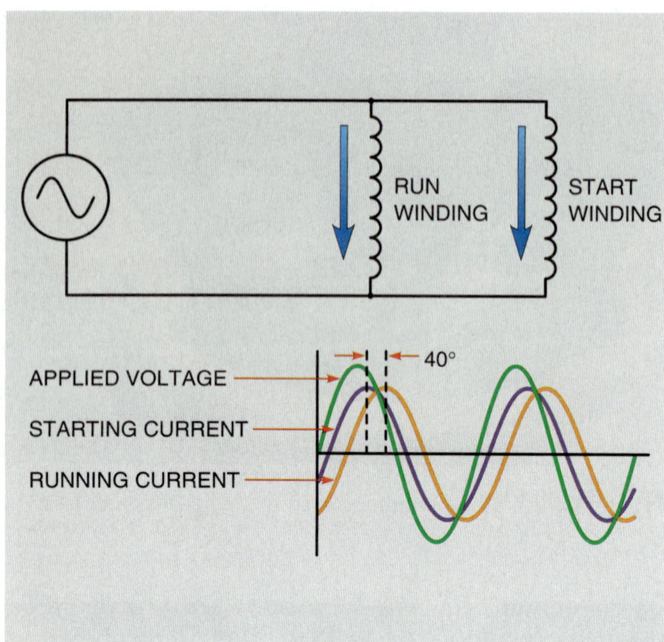

Figure 17-3 The start and run windings are connected in parallel with each other.

with the applied voltage than will the current flow through the run winding. The current flow through the run winding will lag the applied voltage due to inductive reactance. These two out-of-phase currents create a rotating magnetic field in the stator. The speed of this rotating magnetic field is called **synchronous speed** and is determined by two factors:

1. number of stator poles
2. frequency of the applied voltage.

The speed of the rotating magnetic field can be determined using the formula

$$S = \frac{120\ F}{P}$$

where

S = RPM

F = Frequency in hertz

P = Number of stator poles

Example

A single-phase motor contains six stator poles and is connected to a 60 Hz line. What is the speed of the rotating magnetic field?

$$S = \frac{120 \times 60}{6}$$

$$S = 1200 \text{ RPM}$$

Table 17-1 RPM at 60 Hz

Stator Poles	RPM
2	3600
4	1800
6	1200
8	900

The power line frequency throught the United States is 60 Hertz. Table 17-1 lists the revolutions per minute (RPM) for motors with different numbers of stator poles.

Resistance-Start Induction-Run Motors

The resistance-start induction-run motor is so named because the out-of-phase condition between start and run winding current is caused by the start winding being more resistive than the run winding. The amount of starting torque produced by a split-phase motor is determined by three factors:

1. The strength of the magnetic field of the stator.
2. The strength of the magnetic field of the rotor.
3. The phase angle difference between current in the start winding and current in the run winding. (Maximum torque is produced when these two currents are 90° out of phase with each other.)

Although these two currents are out of phase with each other, they are not 90° out of phase. The run winding is more inductive than the start winding, but it does have some resistance, which prevents the current from being 90° out of phase with the voltage. The start winding is more resistive than the run winding, but it does have some inductive reactance, preventing the current from being in phase with the applied voltage. Therefore, a phase angle difference of 35° to 40° is produced between these two currents, resulting in a rather poor starting torque (Fig. 17-4).

Disconnecting the Start Winding

A stator rotating magnetic field is necessary only to start the rotor turning. Once the rotor has accelerated to approximately 75% of rated speed, the start winding can be disconnected from the circuit and the motor will continue to operate with only the run winding energized. Motors that are not hermetically sealed (most refrigeration and air-conditioning compressors are hermetically sealed) use a **centrifugal switch** to disconnect the start windings from the circuit. The contacts of the centrifugal switch are connected in series with the start winding (Fig. 17-5). The centrifugal switch contains a set of spring-loaded weights. When the shaft is not turning, the springs hold a fiber washer in

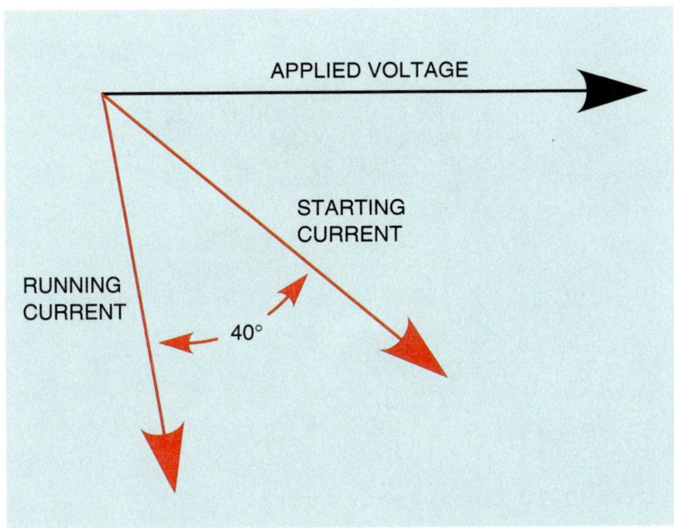

Figure 17-4 Running current and starting current are 35° to 40° out of phase with each other.

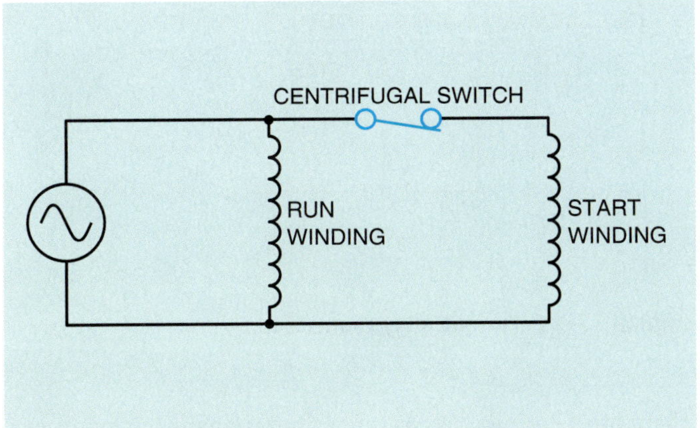

Figure 17-5 A centrifugal switch is used to disconnect the start winding from the circuit.

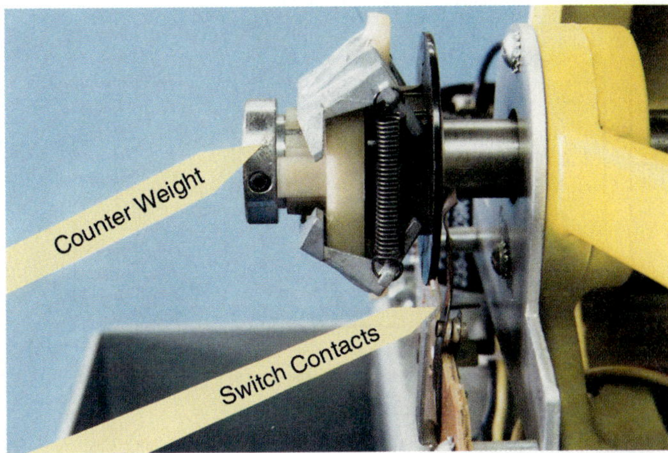

Figure 17-6 The centrifugal switch is closed when the rotor is not turning.

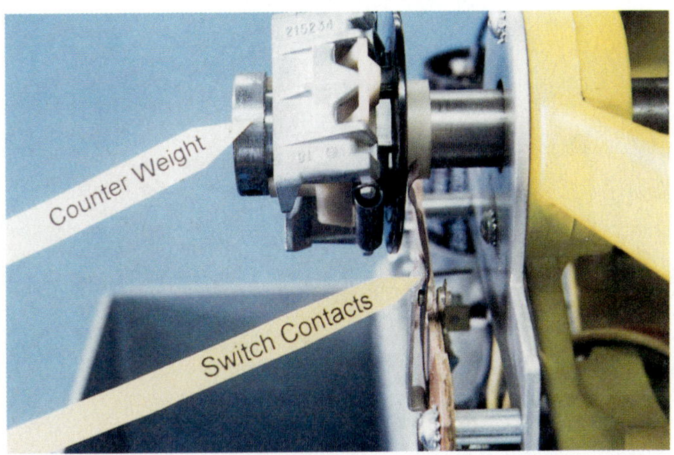

Figure 17-7 The contact opens when the rotor reaches about 75% of rated speed.

contact with the movable contact of the switch (Fig. 17-6). The fiber washer causes the movable contact to complete a circuit with a stationary contact.

When the rotor accelerates to about 75% of rated speed, centrifugal force causes the weights to overcome the force of the springs. The fiber washer retracts and permits the contacts to open and disconnect the start winding from the circuit (Fig. 17-7). The start winding of this type motor is intended to be energized only during the period of time that the motor is actually starting. If the start winding is not disconnected, it will be damaged by excessive current flow.

Starting Relays

Resistance-start induction-run and capacitor-start induction-run motors are sometimes hermetically sealed, such as with air conditioning and refrigeration compressors. When they are hermetically sealed, a centrifugal switch cannot be used to disconnect the start winding. A device that can be mounted externally is needed to disconnect the start windings from the circuit. Starting relays perform this function.

There are three basic types of starting relays used with the resistance-start and capacitor-start motors:

1. Hot-wire relay.
2. Current relay.
3. Solid state starting relay.

The *hot-wire relay* functions as both a starting relay and an overload relay. In the circuit shown in Figure 17-8, it is assumed that a thermostat controls the operation of the motor. When the thermostat closes, current flows through a resistive wire and through two normally closed contacts connected to the start and run windings of the motor. The high starting current of the motor rapidly heats the resistive wire, causing it to expand. The expansion of the wire causes the spring-loaded start winding contact to open and disconnect the

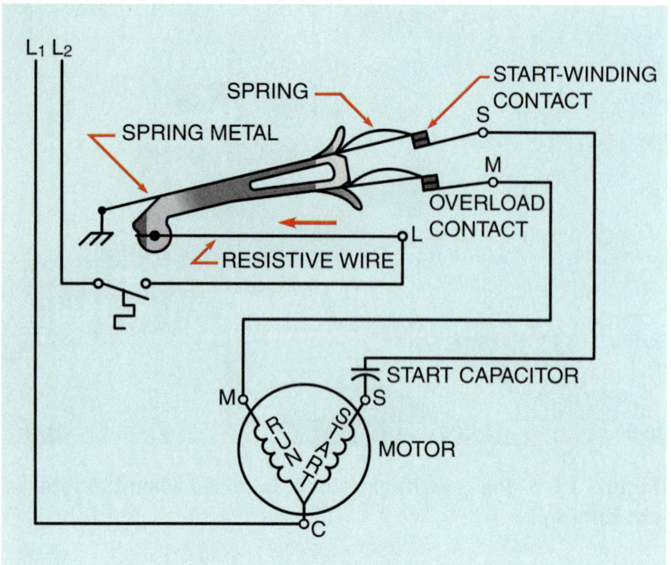

Figure 17-8 **Hot-wire relay connection.**

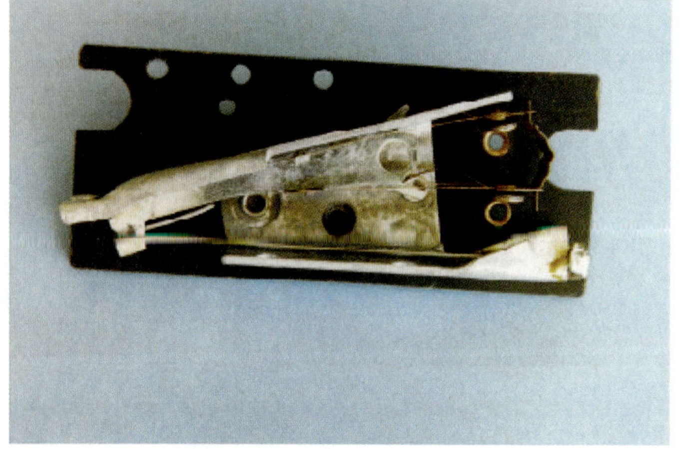

Figure 17-9 **Hot-wire type of starting relay.**

start winding from the circuit, reducing motor current. If the motor is not overloaded, the resistive wire never becomes hot enough to cause the overload contact to open, and the motor continues to run. If the motor becomes overloaded, however, the resistive wire expands enough to open the overload contact and disconnect the motor from the line. A photograph of a hot-wire starting relay is shown in Figure 17-9.

The *current relay* also operates by sensing the amount of current flow in the circuit. This type of relay operates on the principle of a magnetic field instead of expanding metal. The current relay contains a coil with a few turns of large wire and a set of normally open contacts, Figure 17-10. The coil of the relay is connected in series with the run winding of the motor, and the contacts are connected in series with the start winding, as shown in Figure 17-11. When the thermostat contact closes, power is applied to the run winding of

Figure 17-10 **Current type of starting relay.**

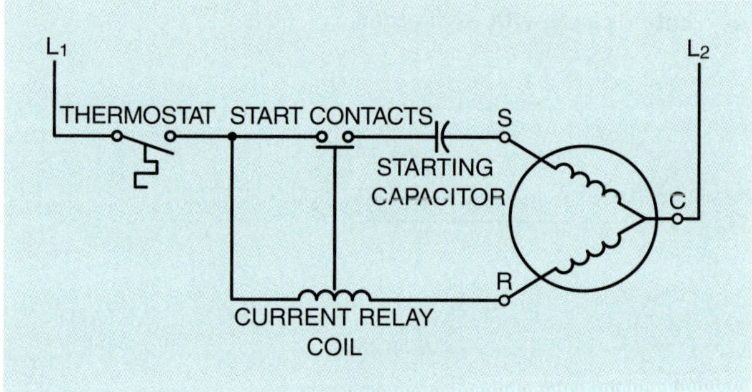

Figure 17-11 **Current relay connection.**

the motor. Because the start winding is open, the motor cannot start, causing a high current to flow in the run winding circuit. This high current flow produces a strong magnetic field in the coil of the relay, causing the normally open contacts to close and connect the start winding to the circuit. When the motor starts, the run-winding current is greatly reduced, permitting the start contacts to reopen and disconnect the start winding from the circuit.

The *solid state starting relay,* Figure 17-12, performs the same basic function as the current relay and in many cases is replacing both the current relay and the centrifugal switch. The solid state starting relay is generally more reliable and less expensive than the current relay or the centrifugal switch. The solid state starting relay is actually an electronic component known as a *thermistor*. A thermistor is a device that exhibits a change of resistance with a change of temperature. This particular thermistor has a positive coefficient of temperature, which means that when its temperature increases, its resistance increases also. The schematic diagram in Figure 17-13 illustrates the connection of the solid state starting relay. The thermistor is connected in series with the start winding of the motor. When the motor is not in operation, the thermistor is at a low temperature

Figure 17-12 Solid state starting relay.

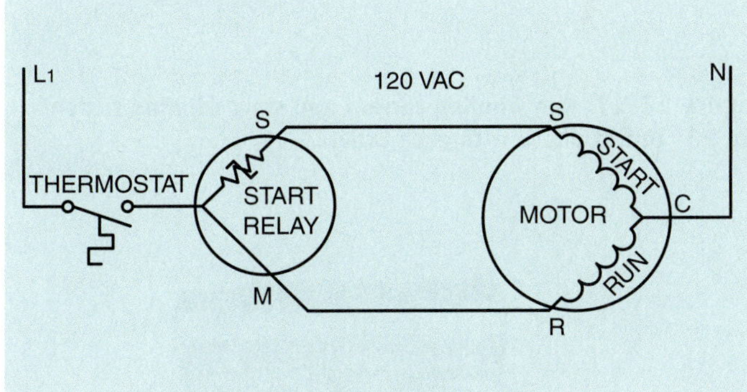

Figure 17-13 Solid state starting relay connection.

Figure 17-14 Squirrel-cage rotor used in a split-phase motor.
Courtesy of Bodine Electric Co.

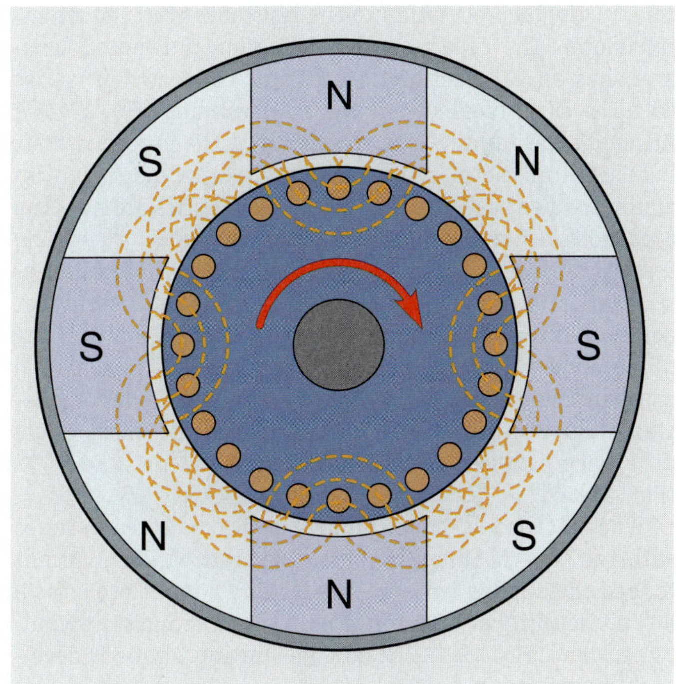

Figure 17-15 A rotating magnetic field is produced by the stator and rotor flux.

and its resistance is low, typically 3 or 4 ohms. When the thermostat contact closes, current flows to both the run and start windings of the motor. The current flowing through the thermistor causes an increase in temperature. This increased temperature causes the resistance of the thermistor to suddenly change to a high value of several thousand ohms. The change of temperature is so sudden that it has the effect of opening a set of contacts. Although the start winding is never completely disconnected from the power line, the amount of current flow through it is very small, typically 0.03 to 0.05 amperes, and does not affect the operation of the motor. This small amount of *leakage current* maintains the temperature of the thermistor and prevents it from returning to a low value of resistance. After the motor is disconnected from the power line, a cooldown time of 2 to 3 minutes should be allowed to permit the thermistor to return to a low resistance before the motor is restarted.

Relationship of Stator and Rotor Fields

The split-phase motor contains a squirrel-cage rotor, (Fig. 17-14). When power is connected to the stator windings, the rotating magnetic field induces a voltage into the bars of the squirrel-cage rotor. The induced voltage causes current to flow in the rotor, and a magnetic field is produced around the rotor bars. The magnetic field of the rotor is attracted to the stator field, and the rotor begins to turn in the direction of the rotating magnetic field. After the centrifugal switch opens, only the run winding induces voltage into the rotor. This induced voltage is in phase with the stator current. The inductive reactance of the rotor is high, causing the rotor current to be almost 90° out of phase with the induced voltage. This causes the pulsating magnetic field of the rotor to lag the pulsating magnetic field of the stator by 90°. Magnetic poles, located midway between the stator poles, are created in the rotor (Fig. 17-15). These two pulsating magnetic fields produce a rotating magnetic field of their own, and the rotor continues to rotate.

Direction of Rotation

The direction of rotation for the motor is determined by the direction of rotation of the rotating magnetic field created by the run and start windings when the motor is first started. The direction of motor rotation can be changed by reversing the connection of either the start winding or the run winding, but not both. If the start winding is disconnected, the motor can be operated in either direction by manually turning the rotor shaft in the desired direction of rotation.

Capacitor-Start Induction-Run Motors

The capacitor-start induction-run motor is very similar in construction and operation to the resistance-start induction-run motor. The capacitor-start induction-run motor, however, has an AC electrolytic capacitor connected in series with the centrifugal switch and start winding (Fig. 17-16). Although the running characteristics of the capacitor-start induction-run motor and the resistance-start induction-run motor are identical, the starting characteristics are not. The capacitor-start induction-run motor produces a starting torque that is substantially higher than that of the resistance-start induction-run motor. Recall that one of the factors that determines the starting torque for a split-phase motor is the phase angle difference between start winding current and run winding current. The starting torque of a resistance-start induction-run motor is low because the phase angle difference between these two currents is only about 40° (Fig. 17-16).

When a capacitor of the proper size is connected in series with the start winding, it causes the start winding current to lead the applied voltage. This leading current produces a 90° phase shift between run winding current and start winding current (Fig. 17-17). Maximum starting torque is developed at this point.

Although the capacitor-start induction-run motor has a high starting torque, the motor should not be started more than about eight times per hour. Frequent starting can damage the start capacitor due to overheating. If the capacitor must be replaced, care should be taken to use a capacitor of the correct microfarad rating. If a capacitor with too little capacitance is used, the starting current will be less than 90° out of phase with the running current, and the starting torque will be reduced. If the capacitance value is too great, the starting current will be more than 90° out of phase with the running current, and the starting torque will again be reduced. A capacitor-start induction-run motor is shown in Figure 17-18.

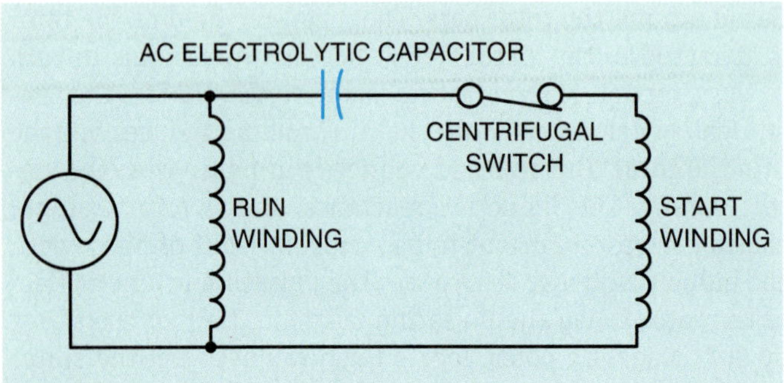

Figure 17-16 **An AC electrolytic capacitor is connected in series with the start winding.**

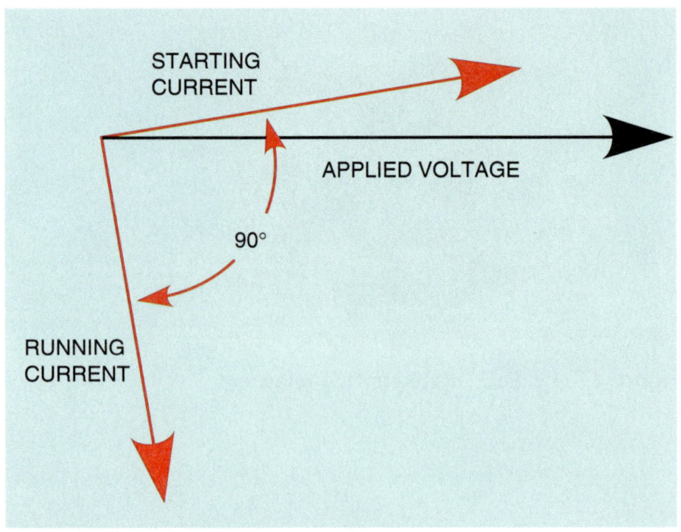

Figure 17-17 **Run winding current and start winding current are 90° out of phase with each other.**

Figure 17-18 **Capacitor-start induction-run motor.** *From Duff and Herman,* Alternating Current Fundamentals, *4th ed., copyright 1991 by Delmar Publishers Inc.*

Dual-Voltage Split-Phase Motors

Many split-phase motors are designed for operation on 120 or 240 V. Figure 17-19 shows the schematic diagram of a split-phase motor designed for dual-voltage operation. This particular motor contains two run windings and two start windings. The lead numbers for single-phase motors are numbered in a standard manner. One of the run windings has lead numbers of T_1 and T_2. The other run winding has its leads numbered T_3 and T_4. This motor uses two different sets of start winding leads. One set is labeled T_5 and T_6, and the other set is labeled T_7 and T_8.

If the motor is to be connected for high-voltage operation, the run windings and start windings will be connected in series as shown in Figure 17-20. The start windings are then connected in parallel with the run windings. If the opposite direction of rotation is desired, T_5 and T_8 will be changed.

For low-voltage operation, the windings must be connected in parallel as shown in Figure 17-21. This connection is made by first connecting the run windings in parallel by hooking T_1 and T_3 together and T_2 and T_4 together. The start windings are paralleled by connecting T_5 and T_7 together and T_6 and T_8 together. The start windings are then connected in parallel with the run windings. If the opposite direction of rotation is desired, T_5 and T_6 should be reversed along with T_7 and T_8.

Not all dual-voltage single-phase motors contain two sets of start windings. Figure 17-22 shows the schematic diagram of a motor that contains two sets of run windings and only

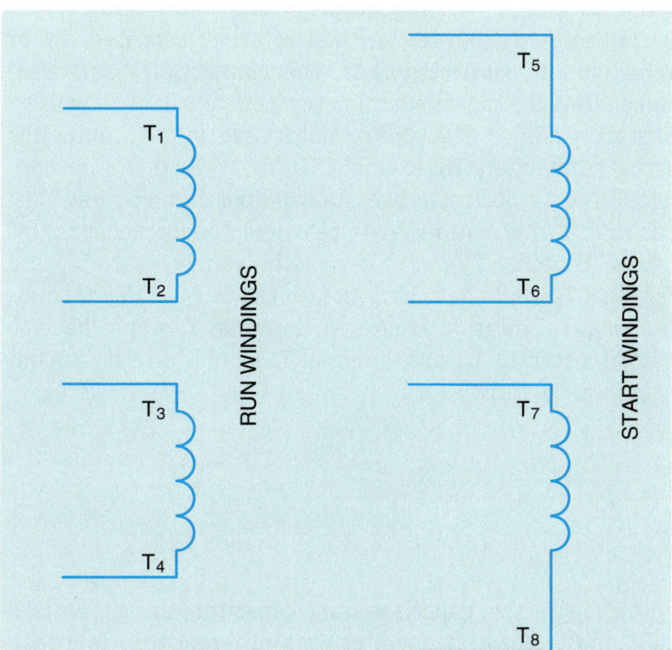

Figure 17-19 Dual-voltage windings for a split-phase motor.

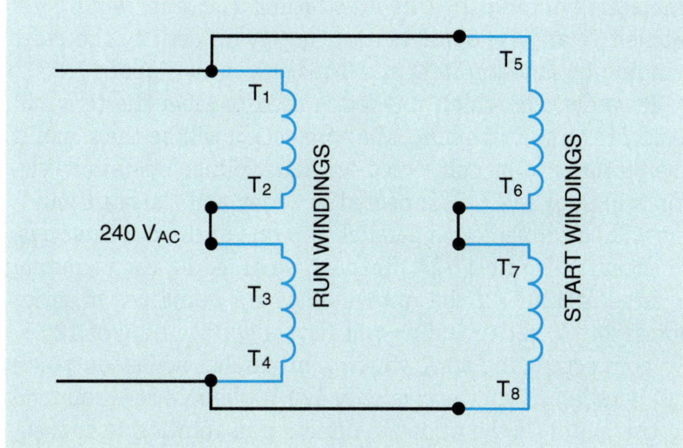

Figure 17-20 High-voltage connection for a split-phase motor with two run and two start windings.

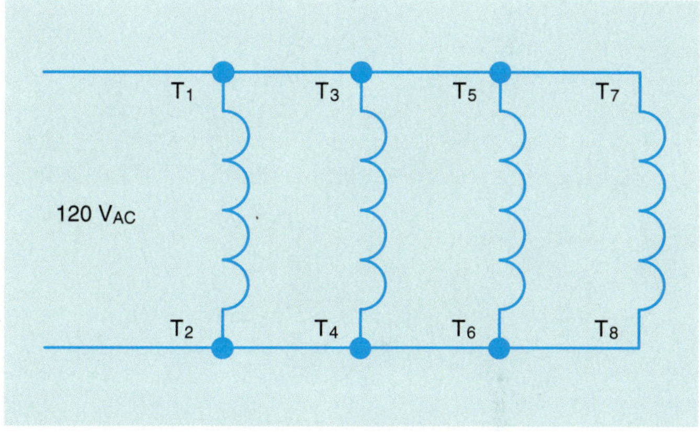

Figure 17-21 Low-voltage connection for a split-phase motor with two run and two start windings.

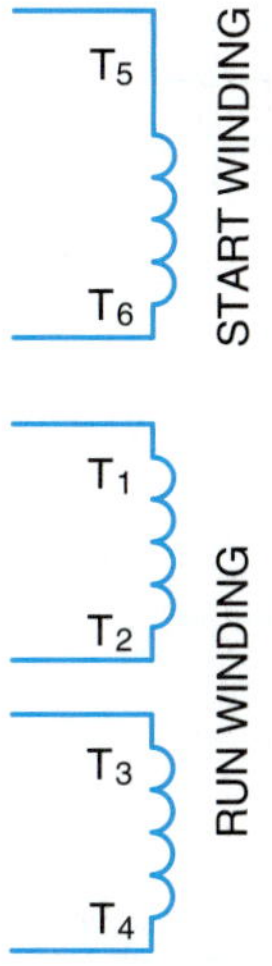

Figure 17-22 Dual-voltage motor with one start winding labeled T_5 and T_6.

one start winding. In this illustration, the start winding is labeled T_5 and T_6. Some motors, however, identify the start winding by labeling it T_5 and T_8 as shown in Figure 17-23.

Regardless of which method is used to label the terminal leads of the start winding, the connection will be the same. If the motor is to be connected for high-voltage operation, the run windings will be connected in series and the start winding will be connected in parallel with one of the run windings, as shown in Figure 17-24. In this type of motor, each winding is rated at 120 V. If the run windings are connected in series across 240 V, each winding will have a voltage drop of 120 V. By connecting the start winding in parallel across only one run winding, it will receive only 120 V when power is applied to the motor. If the opposite direction of rotation is desired, T_5 and T_8 should be changed.

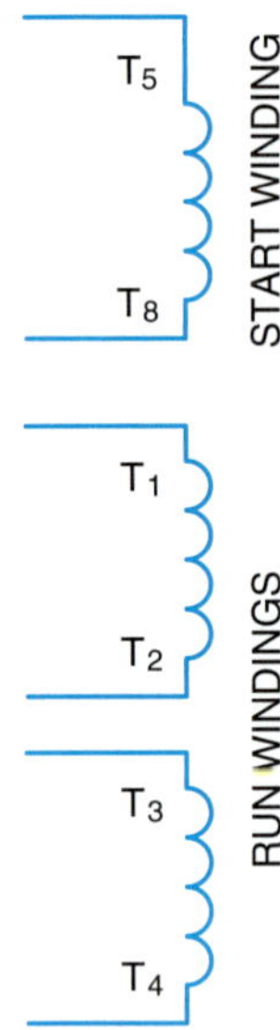

Figure 17-23 Dual-voltage motor with one start winding labeled T_5 and T_8.

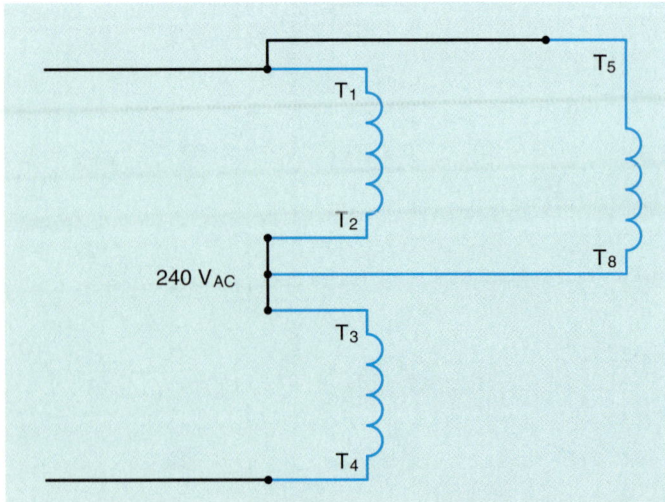

Figure 17-24 High-voltage connection with one start winding.

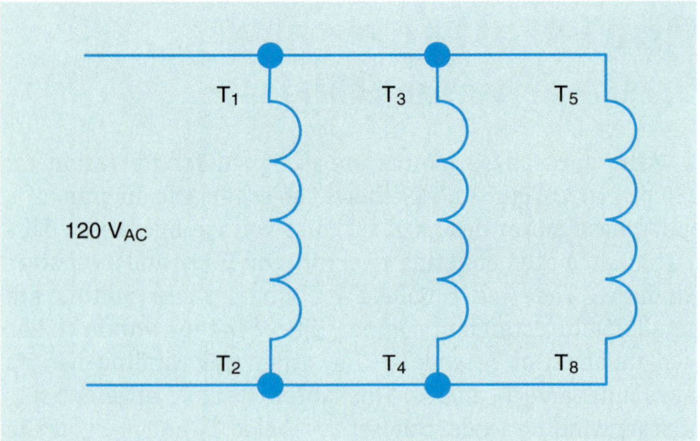

Figure 17-25 Low-voltage connection for a split-phase motor with one start winding.

If the motor is to be operated on low voltage, the windings are connected in parallel as shown in Figure 17-25. Since all windings are connected in parallel, each will receive 120 V when power is applied to the motor.

Determining the Direction of Rotation for Split-Phase Motors

The direction of rotation of a single-phase motor can generally be determined when the motor is connected. The direction of rotation is determined by facing the back or rear of the motor. Figure 17-26 shows a connection diagram for rotation. If clockwise rotation is desired, T_5 should be connected to T_1. If counterclockwise rotation is desired, T_8 (or T_6) should be connected to T_1. This connection diagram assumes that the motor contains two sets of run and two sets of start windings. The type of motor used will determine the actual connection. For example, Figure 17-24 shows the connection of a motor with two run windings and only one start winding. If this motor were to be connected for clockwise rotation, terminal T_5 would have to be connected to T_1, and terminal T_8 would have to be connected to T_2 and T_3. If counterclockwise rotation is desired, terminal T_8 would have to be connected to T_1, and terminal T_5 would have to be connected to T_2 and T_3.

Capacitor-Start Capacitor-Run Motors

Although the capacitor-start capacitor-run motor is a split-phase motor, it operates on a different principle than the resistance-start induction-run motor or the capacitor-start induction-run motor. The capacitor-start capacitor-run

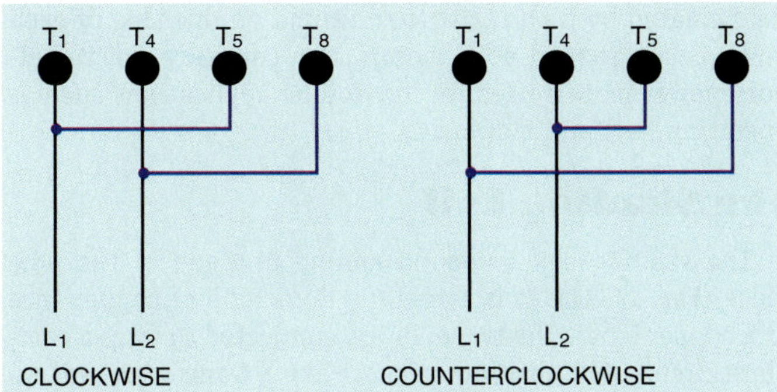

Figure 17-26 **Determining direction of rotation for a split-phase motor.**

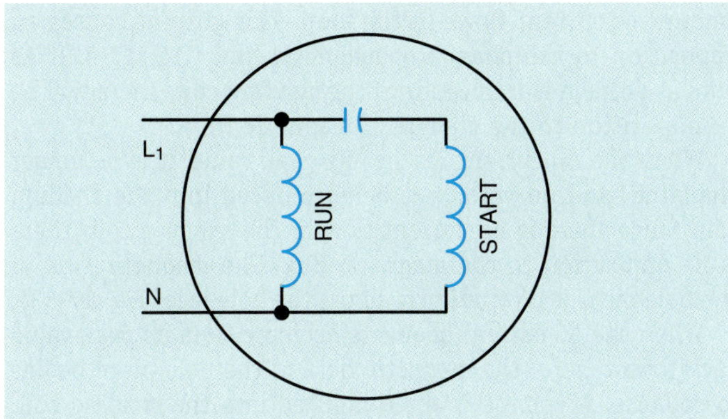

Figure 17-27 **A capacitor-start capacitor-run motor.**

motor is designed in such a manner that its start winding remains energized at all times. A capacitor is connected in series with the winding to provide a continuous leading current in the start winding (Fig. 17-27). Since the start winding remains energized at all times, no centrifugal switch is needed to disconnect the start winding as the motor approaches full speed. The capacitor used in this type of motor will generally be of the oil-filled type since it is intended for continuous use. An exception to this general rule are small fractional-horsepower motors used in reversible ceiling fans. These fans have a low current draw and use an AC electrolytic capacitor to help save space.

The capacitor-start capacitor-run motor actually operates on the principle of a rotating magnetic field in the stator. Since both run and start windings remain energized at all times, the stator magnetic field continues to rotate and the motor operates as a two-phase motor. This motor has excellent starting and running torque. It is quiet in operation and has a high efficiency. Since the capacitor remains connected in the circuit at all times, the motor power factor is close to unity.

Although the capacitor-start capacitor-run motor does not require a centrifugal switch to disconnect the capacitor from the start winding, some motors use a second capacitor

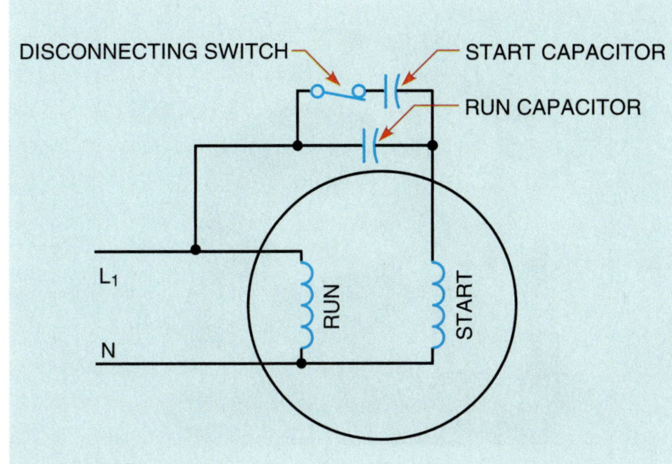

Figure 17-28 **Capacitor-start capacitor-run motor with additional starting capacitor.**

during the starting period to help improve starting torque (Fig. 17-28). A good example of this can be found on the compressor of a central air conditioning unit designed for operation on single-phase power. If the motor is not hermetically sealed, a centrifugal switch is used to disconnect the start capacitor from the circuit when the motor reaches approximately 75% of rated speed. Hermetically sealed motors, however, must use some type of external switch to disconnect the start capacitor from the circuit.

The capacitor-start capacitor-run motor, or permanent split capacitor motor as it is generally referred to in the air conditioning and refrigeration industry, generally employs a potential starting relay to disconnect the starting capacitor when a centrifugal switch cannot be used. The potential starting relay, Figure 17-29A and B, operates by sensing an increase in the voltage developed in the start winding when the motor is operating. A schematic diagram of a potential starting relay circuit is shown in Figure 17-30. In this circuit, the potential relay is used to disconnect the starting capacitor from the circuit when the motor reaches about 75% of its full speed. The starting relay coil, SR, is connected in parallel with the start winding of the motor. A normally closed SR contact is connected in series with the starting capacitor. When the thermostat contact closes, power is applied to both the run and start windings. At this point, both the start and run capacitors are connected in the circuit.

As the rotor begins to turn, its magnetic field induces a voltage into the start winding, producing a higher voltage across the start winding than the applied voltage. When the motor has accelerated to about 75% of its full speed, the voltage across the start winding is high enough to energize the coil of the potential relay. This causes the normally closed SR contact to open and disconnect the start capacitor from the circuit. Because the start winding of this motor is never disconnected from the power line, the coil of the potential starting relay remains energized as long as the motor is in operation.

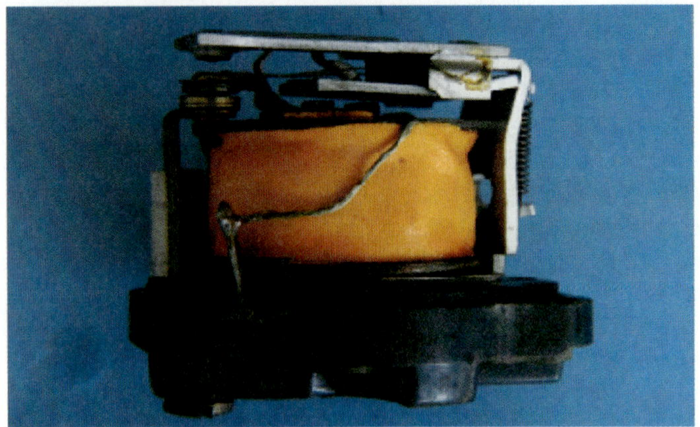

Figure 17-29 Potential starting relays.

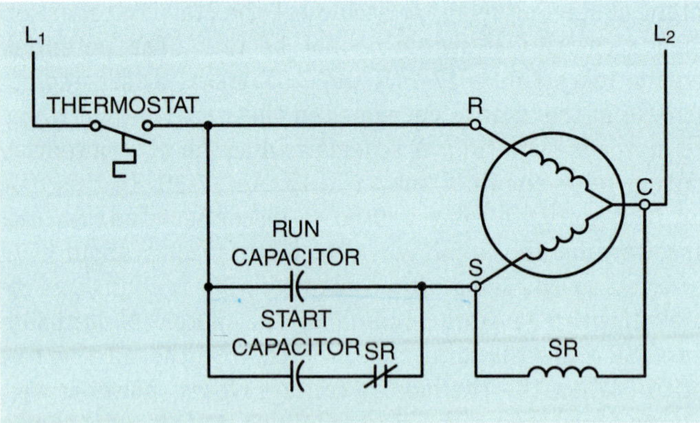

Figure 17-30 Potential relay connection.

Shaded-Pole Induction Motors

The **shaded-pole induction motor** is popular because of its simplicity and long life. This motor contains no start windings or centrifugal switch. It contains a squirrel-cage rotor and operates on the principle of a rotating magnetic field created by a **shading coil** wound on one side of each pole piece. Shaded-pole motors are generally fractional-horsepower motors used for low-torque applications such as operating fans and blowers.

The Shading Coil

The shading coil is wound around one end of the pole piece (Fig. 17-31). It is actually a large loop of copper wire or a copper band. The two ends are connected to form a complete circuit. The shading coil acts like a transformer with a shorted secondary winding. When the current of the AC wave form increases from zero toward its positive peak, a magnetic field is created in the pole piece. As magnetic lines of flux cut through the shading coil, a voltage is induced in the coil. Since the coil is a low-resistance short circuit, a large amount of current flows in the loop. This current causes an opposition to the change of magnetic flux (Fig. 17-32). As long as voltage is induced into the shading coil, there will be an opposition to the change of magnetic flux.

When the AC current reaches its peak value, it is no longer changing, and no voltage is being induced into the shading coil. Since there is no current flow in the shading coil, there is no opposition to the magnetic flux. The magnetic flux of the pole piece is now uniform across the pole face (Fig. 17-33).

When the AC current begins to decrease from its peak value back toward zero, the magnetic field of the pole piece begins to collapse. A voltage is again induced into the shading coil.

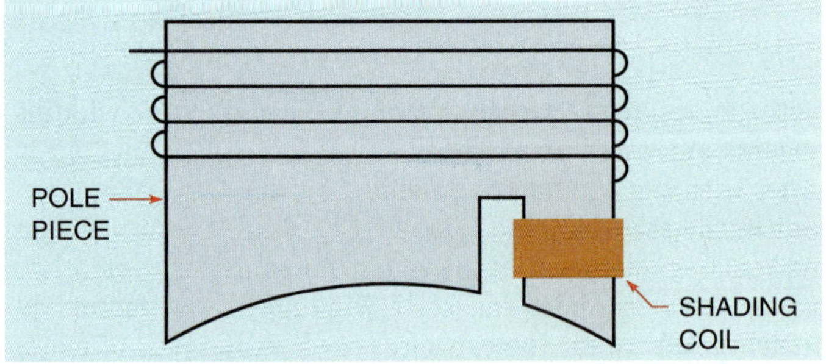

Figure 17-31 A shaded pole.

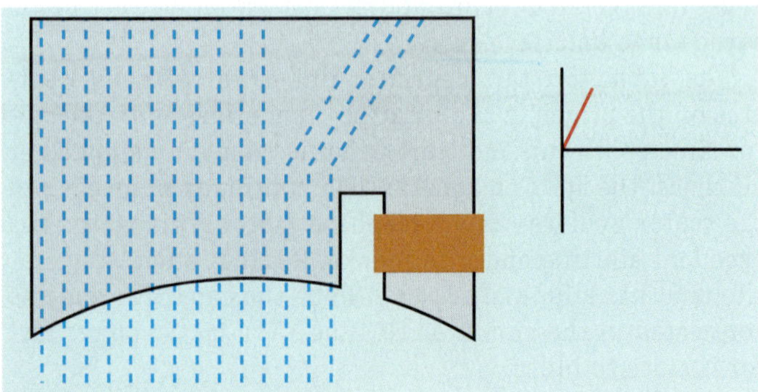

Figure 17-32 The shading coil opposes a change of flux as current increases.

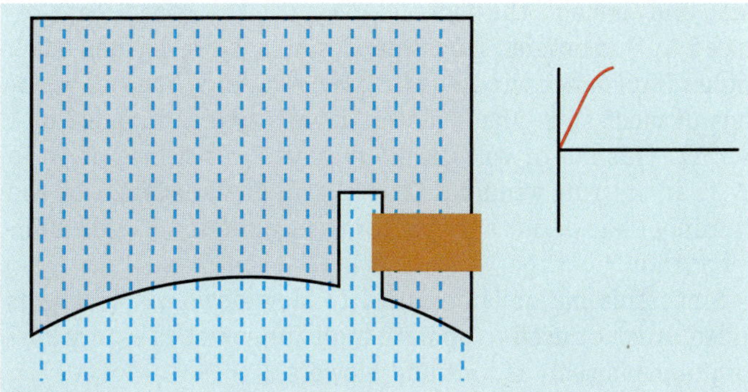

Figure 17-33 **There is opposition to magnetic flux when the current is not changing.**

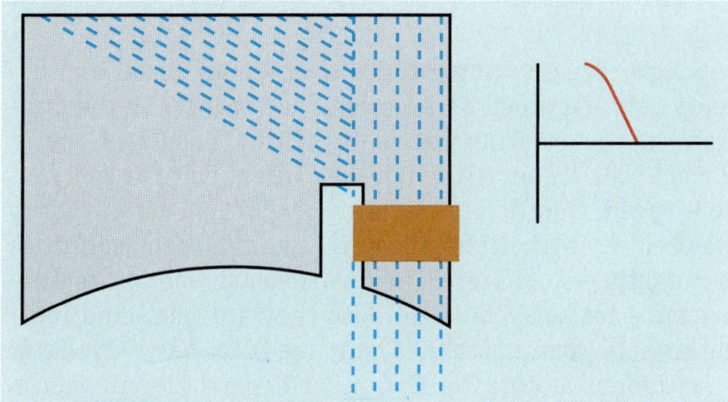

Figure 17-34 **The shading coil opposes a change of flux when the current decreases.**

This induced voltage creates a current that opposes the change of magnetic flux (Fig. 17-34). This causes the magnetic flux to be concentrated in the shaded section of the pole piece.

When the AC current passes through zero and begins to increase in the negative direction, the same set of events happens, except that the polarity of the magnetic field is reversed. If these events were to be viewed in rapid order, the magnetic field would be seen to rotate across the face of the pole piece.

Speed

The speed of the shaded-pole induction motor is determined by the same factors that determine the synchronous speed of other induction motors: frequency and number of stator poles. Shaded-pole motors are commonly wound as four- or six-pole motors. Figure 17-35 shows a drawing of a four-pole shaded-pole induction motor.

General Operating Characteristics

The shaded-pole motor contains a standard squirrel-cage rotor. The amount of torque produced is determined by the strength of the magnetic field of the stator, the strength of the magnetic field of the rotor, and the phase angle difference between rotor and stator flux. The shaded-pole induction motor has low starting and running torque.

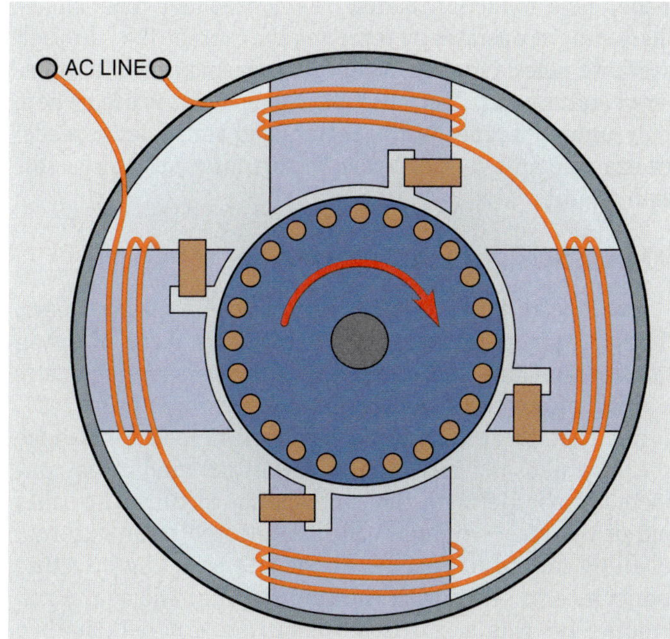

Figure 17-35 **Four-pole shaded-pole induction motor.**

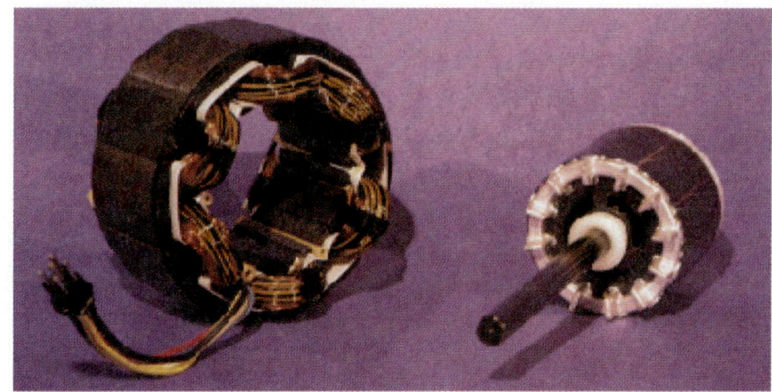

Figure 17-36 **Stator winding and rotor of a shaded-pole induction motor.** *Courtesy of A. O. Smith.*

The direction of rotation is determined by the direction in which the rotating magnetic field moves across the pole face. The rotor turns in the direction shown by the arrow in Figure 17-35. The direction can be changed by removing the stator winding and turning it around. This is not a common practice, however. As a general rule the shaded-pole induction motor is considered to be nonreversible. Figure 17-36 shows a stator winding and rotor of a shaded-pole induction motor.

Multispeed Motors

There are two basic types of **multispeed** single-phase **motors.** One is the consequent pole type and the other is a specially wound capacitor-start capacitor-run motor or shaded-pole induction motor. The consequent pole single-phase **motor** operates by reversing the current flow through alternate poles and increasing or decreasing the total number of stator poles. The consequent pole motor is used where high running torque must be maintained at different speeds; for example, in two-speed compressors for central air conditioning units.

Multispeed Fan Motors

Multispeed fan motors have been used for many years. They are generally wound for two to five steps of speed and operate fans and squirrel-cage blowers. A schematic drawing of a three-speed motor is shown in Figure 17-37. Notice that the run winding has been tapped to produce low, medium, and high speed. The start winding is connected in parallel with the run winding section. The other end of the start winding lead is connected to an external oil-filled capacitor. This motor changes speed by inserting inductance in series with the run winding. The actual run winding for this motor is between the terminals marked *high* and *common*. The winding shown between *high* and *medium* is connected in series with the main run winding. When the rotary switch is connected to the medium speed position, the inductive reactance of this coil limits the amount of current flow through the run winding. When the current of the run winding is reduced, the strength of its magnetic field is reduced and the motor produces less torque. This causes a greater amount of slip, and the motor speed decreases.

If the rotary switch is changed to the *low* position, more inductance is inserted in series with the run winding. This causes less current to flow through the run winding and another reduction in torque. When the torque is reduced, the motor speed decreases again.

Common speeds for a four-pole motor of this type are 1625, 1500, and 1350 RPM. Notice that this motor does not have wide ranges between speeds as would be the case with a consequent pole motor. Most induction motors would overheat and damage the motor winding if the speed were reduced to this extent. This type of motor, however, has much higher impedance windings than most motors. The run windings of most split-phase motors have a wire resistance of 1 to 4 Ω. This motor will generally have a resistance of 10 to 15 Ω in its run winding. It is the high impedance of the windings that permits the motor to be operated in this manner without damage.

Since this motor is designed to slow down when load is added, it is not used to operate high-torque loads—only low-torque loads such as fans and blowers.

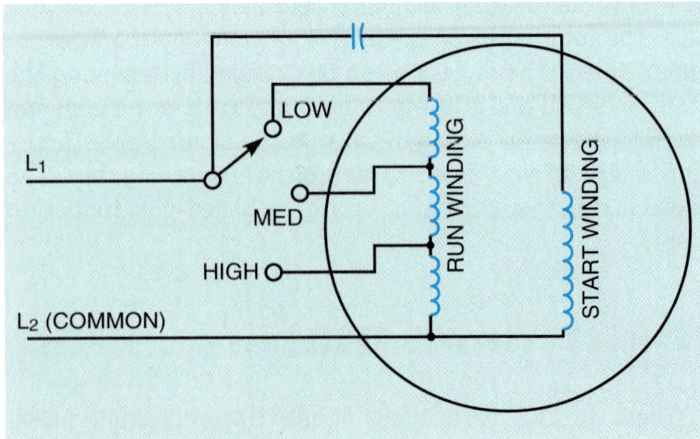

Figure 17-37 A three-speed motor.

Single-Phase Synchronous Motors

Single-phase **synchronous motors** are small and develop only fractional horsepower. They operate on the principle of a rotating magnetic field developed by a shaded-pole stator. Although they will operate at synchronous speed, they do not require DC excitation current. They are used where constant speed is required, such as in clock motors, timers, and recording instruments, and as the driving force for small fans, because they are small and inexpensive to manufacture. There are two basic types of synchronous motor: the Warren, or General Electric motor, and the Holtz motor. These motors are also referred to as hysteresis motors.

Warren Motors

The **Warren motor** is constructed with a laminated stator core and a single coil. The coil is generally wound for 120-V AC operation. The core contains two poles, which are divided into two sections each. One-half of each pole piece contains a shading coil to produce a rotating magnetic field (Fig. 17-38). Since the stator is divided into two poles, the synchronous field speed is 3600 RPM when connected to 60 Hz.

The difference between the Warren and Holtz motor is the type of rotor used. The rotor of the Warren motor is constructed by stacking hardened steel laminations onto the rotor shaft. These disks have high hysteresis loss. The laminations form two crossbars for the rotor. When power is connected to the motor, the rotating magnetic field induces a voltage into the rotor, and a strong starting torque is developed, causing the rotor to accelerate to near-synchronous speed. Once the motor has accelerated to near-synchronous speed, the flux of the rotating magnetic field follows the path of minimum reluctance (magnetic resistance) through the two crossbars. This causes the rotor to lock in step with the rotating magnetic field, and the motor operates at 3600 RPM. These motors are often used with small geartrains to reduce the speed to the desired level.

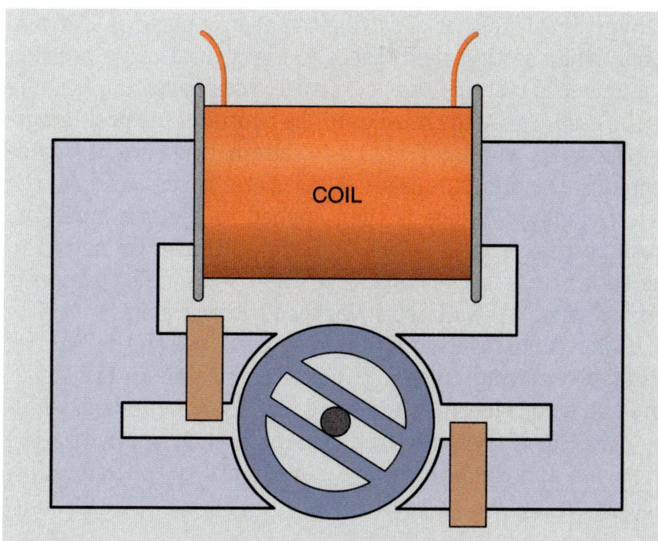

Figure 17-38 A Warren motor.

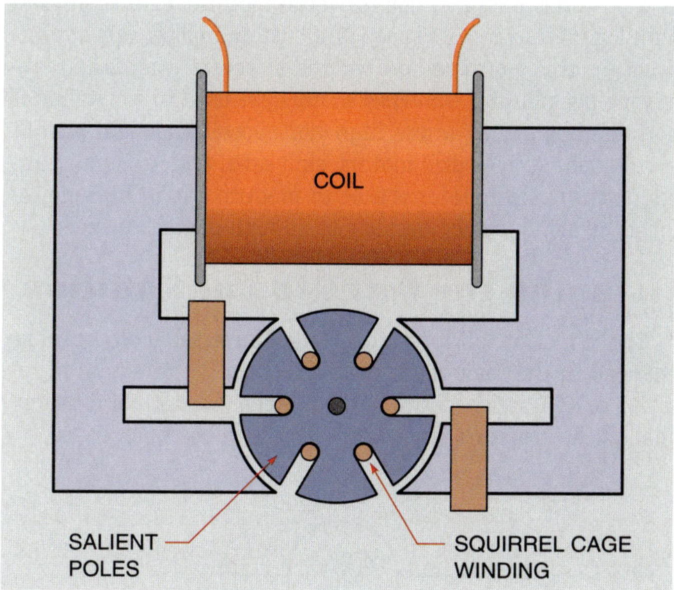

Figure 17-39 A Holtz motor.

Holtz Motors

The **Holtz motor** uses a different type of rotor (Fig. 17-39). This rotor is cut in such a manner that six slots are formed. These slots form six salient (projecting or jutting) poles for the rotor. A squirrel-cage winding is constructed by inserting a metal bar at the bottom of each slot. When power is connected to the motor, the squirrel-cage winding provides the torque necessary to start the rotor turning. When the rotor approaches synchronous speed, the salient poles will lock in step with the field poles each half cycle. This produces a rotor speed of 1200 RPM (one-third of synchronous speed) for the motor.

Universal Motors

The **universal motor** is often referred to as an AC series motor. It is very similar to a DC series motor in its construction in that it contains a wound armature and brushes (Fig. 17-40). The universal motor, however, has the addition of a **compensating winding.** If a DC series motor were connected to alternating current, the motor would operate poorly for several reasons. The armature windings would have a large amount of inductive reactance when connected to alternating current. In addition, the field poles of most DC machines contain solid metal pole pieces. If the field were connected to AC, a large amount of power would be lost to eddy current induction in the pole pieces. Universal motors contain a laminated core to help prevent this problem. The compensating winding is wound around the stator and functions to counteract the inductive reactance in the armature winding.

The universal motor is so named because it can be operated on AC or DC voltage. When if is operated on direct current, the compensating winding is connected in series with the series field winding (Fig. 17-41).

Figure 17-40 Armature and brushes of a universal motor.

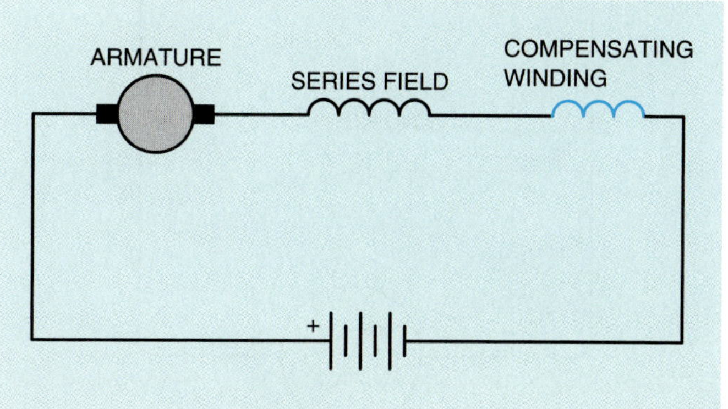

Figure 17-41 The compensating winding is connected in series with the series field winding.

Connecting the Compensating Winding for AC Current

When the universal motor is operated with AC power, the compensating winding can be connected in two ways. If it is connected in series with the armature, as shown in Figure 17-42, it is known as **conductive compensation.**

The compensating winding can also be connected by shorting its leads together as shown in Figure 17-43. When connected in this manner, the winding acts like a shorted secondary winding of a transformer. Induced current permits the winding to operate when connected in this manner. This connection is known as **inductive compensation.** Inductive compensation cannot be used when the motor is connected to direct current.

The Neutral Plane

Since the universal motor contains a wound armature, commutator, and brushes, the brushes should be set at the **neutral plane** position. This can be done in the universal motor in a manner similar to that of setting the neutral plane of a DC machine. When setting the brushes to the neutral plane position in a universal motor, either the series or compensating winding can be used. To set the brushes to the neutral plane position using the series winding (Fig. 17-44) alternating current is connected to the armature leads. A voltmeter is connected to the series winding. Voltage is then applied to the armature. The brush position is then moved until the voltmeter connected to the series field reaches a null position. (The null position is reached when the voltmeter reaches its lowest point.)

If the compensating winding is used to set the neutral plane, alternating current is again connected to the armature and a voltmeter is connected to the compensating winding (Fig. 17-45). Alternating current is then applied to the armature, and the brushes are moved until the voltmeter indicates its highest or peak voltage.

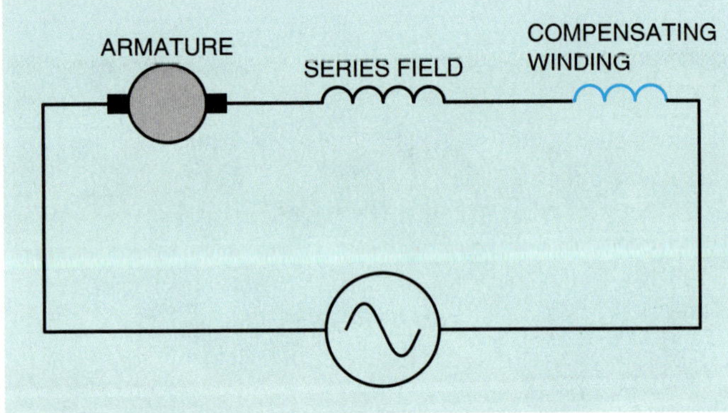

Figure 17-42 Conductive compensation.

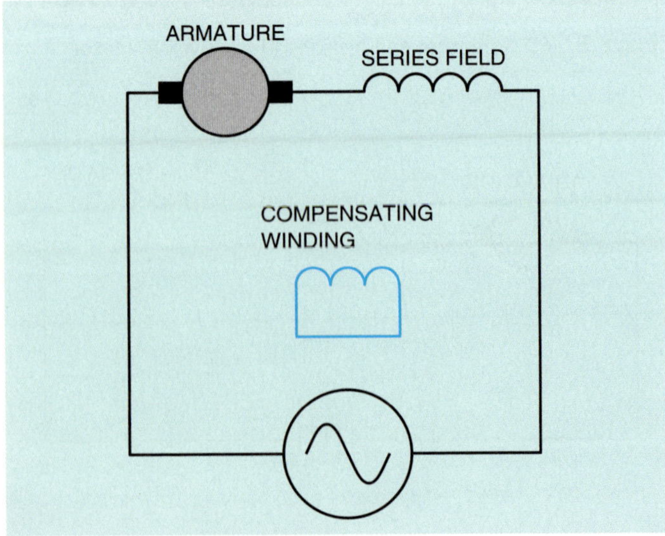

Figure 17-43 Inductive compensation.

Speed Regulation

The speed regulation of the universal motor is very poor. Since it is a series motor, it has the same poor speed regulation as a DC series motor. If the universal motor is connected to a light load or no load, its speed is almost unlimited. It is not unusual for this motor to be operated at several thousand revolutions per minute. Universal motors are used in a number of portable appliances where high horsepower and light weight are needed, such as drill motors, skill saws, and vacuum cleaners. The universal motor is able to produce a high horsepower for its size and weight because of its high operating speed.

Changing the Direction of Rotation

The direction of rotation of the universal motor can be changed in the same manner as changing the direction of rotation of a DC series motor. To change the direction of rotation, change the armature leads with respect to the field leads.

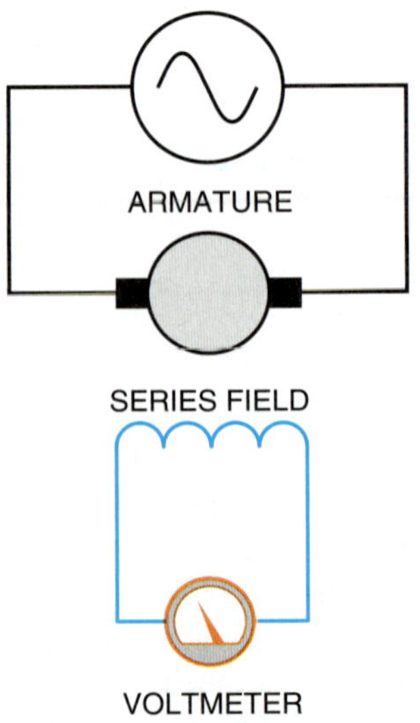

Figure 17-44 Using the series field to set the brushes at the neutral plane position.

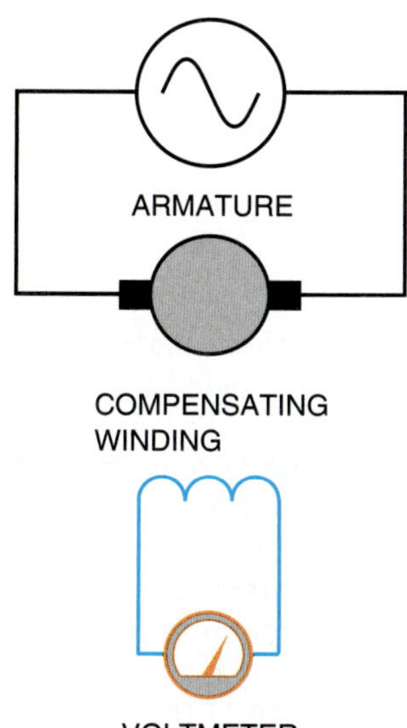

Figure 17-45 Using the compensating winding to set the brushes to the neutral plane position.

Summary

- Not all single-phase motors operate on the principle of a rotating magnetic field.
- Split-phase motors start as two-phase motors by producing an out-of-phase condition for the current in the run winding and the current in the start winding.
- The resistance of the wire in the start winding of a resistance-start induction-run motor is used to produce a phase angle difference between the current in the start winding and the current in the run winding.
- The capacitor-start induction-run motor uses an AC electrolytic capacitor to increase the phase angle difference between starting and running current. This causes an increase in starting torque.
- Maximum starting torque for a split-phase motor is developed when the start winding current and run winding current are 90° out of phase with each other.
- Most resistance-start induction run motors and capacitor-start induction-run motors use a centrifugal switch to disconnect the start windings when the motor reaches approximately 75% of full-load speed.
- The capacitor-start capacitor-run motor operates like a two-phase motor because both the start and run windings remain energized during motor operation.
- Most capacitor-start capacitor-run motors use an AC oil-filled capacitor connected in series with the start winding.
- The capacitor of the capacitor-start capacitor-run motor helps correct the power factor.
- Shaded-pole induction motors operate on the principle of a rotating magnetic field.
- The rotating magnetic field of a shaded-pole induction motor is produced by placing shading loops or coils on one side of the pole piece.
- The synchronous field speed of a single-phase motor is determined by the number of stator poles and the frequency of the applied voltage.
- Consequent pole motors are used when a change of motor speed is desired and high torque must be maintained.
- Multispeed fan motors are constructed by connecting windings in series with the main run winding.
- Multispeed fan motors have high-impedance stator windings to prevent them from overheating when their speed is reduced.
- The direction of rotation for split-phase motors is changed by reversing the start winding in relation to the run winding.
- Shaded-pole motors are generally considered to be nonreversible.
- There are two types of single-phase synchronous motor: the Warren and the Holtz.
- Single-phase synchronous motors are sometimes called hysteresis motors.
- The Warren motor operates at a speed of 3600 RPM.
- The Holtz motor operates at a speed of 1200 RPM.
- Universal motors operate on direct or alternating current.
- Universal motors contain a wound armature and brushes.
- Universal motors are also called AC series motors.
- Universal motors have a compensating winding that helps overcome inductive reactance.
- The direction of rotation for a universal motor can be changed by reversing the armature leads with respect to the field leads.

Review Questions

1. What are the three basic types of split-phase motors?
2. The voltages of a two-phase system are how many degrees out of phase with each other?
3. How are the start and run windings of a split-phase motor connected in relation to each other?
4. In order to produce maximum starting torque in a split-phase motor, how many degrees out of phase should the start and run winding currents be with each other?
5. What is the advantage of the capacitor-start induction-run motor over the resistance-start induction-run motor?
6. On the average, how many degrees out of phase with each other are the start and run winding currents in a resistance-start induction-run motor?
7. What device is used to disconnect the start windings for the circuit in most nonhermetically sealed capacitor-start induction-run motors?
8. Why does a split-phase motor continue to operate after the start windings have been disconnected from the circuit?
9. How can the direction of rotation of a split-phase motor be reversed?
10. If a dual-voltage split-phase motor is to be operated on high voltage, how are the run windings connected in relation to each other?
11. When determining the direction of rotation for a split-phase motor, should you face the motor from the front or from the rear?
12. What type of split-phase motor does not generally contain a centrifugal switch?
13. What is the principle of operation of a capacitor-start capacitor-run motor?
14. What causes the magnetic field to rotate in a shaded-pole induction motor?
15. How can the direction of rotation of a shaded-pole induction motor be changed?
16. How is the speed of a consequent pole motor changed?
17. Why can a multispeed fan motor be operated at lower speed than most induction motors without harm to the motor windings?
18. What is the speed of operation of the Warren motor?
19. What is the speed of operation of the Holtz motor?
20. Why is the AC series motor often referred to as a universal motor?
21. What is the function of the compensating winding?
22. How is the direction of rotation of the universal motor reversed?
23. When the motor is connected to DC voltage, how must the compensating winding be connected?
24. Explain how to set the neutral plane position of the brushes using the series field.
25. Explain how to set the neutral plane position using the compensating winding.

Appendix A Trigonometric Functions—Natural Sines and Cosines

NOTE: For cosines, use right-hand column of degrees and lower line of tenths.

Deg.	°0.0	°0.1	°0.2	°0.3	°0.4	°0.5	°0.6	°0.7	°0.8	°0.9	°1.0	
0°	0.0000	0.0017	0.0035	0.0052	0.0070	0.0087	0.0105	0.0122	0.0140	0.0157	0.0175	89
1	0.0175	0.0192	0.0209	0.0227	0.0244	0.0262	0.0279	0.0297	0.0314	0.0332	0.0349	88
2	0.0349	0.0366	0.0384	0.0401	0.0419	0.0436	0.0454	0.0471	0.0488	0.0506	0.0523	87
3	0.0523	0.0541	0.0558	0.0576	0.0593	0.0610	0.0628	0.0645	0.0663	0.0680	0.0698	86
4	0.0698	0.0715	0.0732	0.0750	0.0767	0.0785	0.0802	0.0819	0.0837	0.0854	0.0872	85
5	0.0872	0.0889	0.0906	0.0924	0.0941	0.0958	0.0976	0.0993	0.1011	0.1028	0.1045	84
6	0.1045	0.1063	0.1080	0.1097	0.1115	0.1132	0.1149	0.1167	0.1184	0.1201	0.1219	83
7	0.1219	0.1236	0.1253	0.1271	0.1288	0.1305	0.1323	0.1340	0.1357	0.1374	0.1392	82
8	0.1392	0.1409	0.1426	0.1444	0.1461	0.1478	0.1495	0.1513	0.1530	0.1547	0.1564	81
9	0.1564	0.1582	0.1599	0.1616	0.1633	0.1650	0.1668	0.1685	0.1702	0.1719	0.1736	80°
10°	0.1736	0.1754	0.1771	0.1788	0.1805	0.1822	0.1840	0.1857	0.1874	0.1891	0.1908	79
11	0.1908	0.1925	0.1942	0.1959	0.1977	0.1994	0.2011	0.2028	0.2045	0.2062	0.2079	78
12	0.2079	0.2096	0.2113	0.2130	0.2147	0.2164	0.2181	0.2198	0.2215	0.2232	0.2250	77
13	0.2250	0.2267	0.2284	0.2300	0.2317	0.2334	0.2351	0.2368	0.2385	0.2402	0.2419	76
14	0.2419	0.2436	0.2453	0.2470	0.2487	0.2504	0.2521	0.2538	0.2554	0.2571	0.2588	75
15	0.2588	0.2605	0.2622	0.2639	0.2656	0.2672	0.2689	0.2706	0.2723	0.2740	0.2756	74
16	0.2756	0.2773	0.2790	0.2807	0.2823	0.2840	0.2857	0.2874	0.2890	0.2907	0.2924	73
17	0.2924	0.2940	0.2957	0.2974	0.2990	0.3007	0.3024	0.3040	0.3057	0.3074	0.3090	72
18	0.3090	0.3107	0.3123	0.3140	0.3156	0.3173	0.3190	0.3206	0.3223	0.3239	0.3256	71
19	0.3256	0.3272	0.3289	0.3305	0.3322	0.3338	0.3355	0.3371	0.3387	0.3404	0.3420	70°
20°	0.3420	0.3437	0.3453	0.3469	0.3486	0.3502	0.3518	0.3535	0.3551	0.3567	0.3584	69
21	0.3584	0.3600	0.3616	0.3633	0.3649	0.3665	0.3681	0.3697	0.3714	0.3730	0.3746	68
22	0.3746	0.3762	0.3778	0.3795	0.3811	0.3827	0.3843	0.3859	0.3875	0.3891	0.3907	67
23	0.3907	0.3923	0.3939	0.3955	0.3971	0.3987	0.4003	0.4019	0.4035	0.4051	0.4067	66
24	0.4067	0.4083	0.4099	0.4115	0.4131	0.4147	0.4163	0.4179	0.4195	0.4210	0.4226	65
25	0.4226	0.4242	0.4258	0.4274	0.4289	0.4305	0.4321	0.4337	0.4352	0.4368	0.4384	64
26	0.4384	0.4399	0.4415	0.4431	0.4446	0.4462	0.4478	0.4493	0.4509	0.4524	0.4540	63
27	0.4540	0.4555	0.4571	0.4586	0.4602	0.4617	0.4633	0.4648	0.4664	0.4679	0.4695	62
28	0.4695	0.4710	0.4726	0.4741	0.4756	0.4772	0.4787	0.4802	0.4818	0.4833	0.4848	61
29	0.4848	0.4863	0.4879	0.4894	0.4909	0.4924	0.4939	0.4955	0.4970	0.4985	0.5000	60°
30°	0.5000	0.5015	0.5030	0.5045	0.5060	0.5075	0.5090	0.5105	0.5120	0.5135	0.5150	59
31	0.5150	0.5165	0.5180	0.5195	0.5210	0.5225	0.5240	0.5255	0.5270	0.5284	0.5299	58
32	0.5299	0.5314	0.5329	0.5344	0.5358	0.5373	0.5388	0.5402	0.5417	0.5432	0.5446	57
33	0.5446	0.5461	0.5476	0.5490	0.5505	0.5519	0.5534	0.5548	0.5563	0.5577	0.5592	56
34	0.5592	0.5606	0.5621	0.5635	0.5650	0.5664	0.5678	0.5693	0.5707	0.5721	0.5736	55
35	0.5736	0.5750	0.5764	0.5779	0.5793	0.5807	0.5821	0.5835	0.5850	0.5864	0.5878	54
36	0.5878	0.5892	0.5906	0.5920	0.5934	0.5948	0.5962	0.5976	0.5990	0.6004	0.6018	53
37	0.6018	0.6032	0.6046	0.6060	0.6074	0.6088	0.6101	0.6115	0.6129	0.6143	0.6157	52
38	0.6157	0.6170	0.6184	0.6198	0.6211	0.6225	0.6239	0.6252	0.6266	0.6280	0.6293	51
39	0.6293	0.6307	0.6320	0.6334	0.6347	0.6361	0.6374	0.6388	0.6401	0.6414	0.6428	50°
40°	0.6428	0.6441	0.6455	0.6468	0.6481	0.6494	0.6508	0.6521	0.6534	0.6547	0.6561	49
41	0.6561	0.6574	0.6587	0.6600	0.6613	0.6626	0.6639	0.6652	0.6665	0.6678	0.6691	48
42	0.6691	0.6704	0.6717	0.6730	0.6743	0.6756	0.6769	0.6782	0.6794	0.6807	0.6820	47
43	0.6820	0.6833	0.6845	0.6858	0.6871	0.6884	0.6896	0.6909	0.6921	0.6934	0.6947	46
44	0.6947	0.6959	0.6972	0.6984	0.6997	0.7009	0.7022	0.7034	0.7046	0.7059	0.7071	45
	°1.0	°0.9	°0.8	°0.7	°0.6	°0.5	°0.4	°0.3	°0.2	°0.1	°0.0	Deg.

NOTE: For cosines, use right-hand column of degrees and lower line of tenths.

Deg.	°0.0	°0.1	°0.2	°0.3	°0.4	°0.5	°0.6	°0.7	°0.8	°0.9	°1.0	
45	0.7071	0.7083	0.7096	0.7108	0.7120	0.7133	0.7145	0.7157	0.7169	0.7181	0.7193	44
46	0.7193	0.7206	0.7218	0.7230	0.7242	0.7254	0.7266	0.7278	0.7290	0.7302	0.7314	43
47	0.7314	0.7325	0.7337	0.7349	0.7361	0.7373	0.7385	0.7396	0.7408	0.7420	0.7431	42
48	0.7431	0.7443	0.7455	0.7466	0.7478	0.7490	0.7501	0.7513	0.7524	0.7536	0.7547	41
49	0.7547	0.7559	0.7570	0.7581	0.7593	0.7604	0.7615	0.7627	0.7638	0.7649	0.7660	40°
50°	0.7660	0.7672	0.7683	0.7694	0.7705	0.7716	9.7727	0.7738	0.7749	0.7760	0.7771	39
51	0.7771	0.7782	0.7793	0.7804	0.7815	0.7826	0.7837	0.7848	0.7859	0.7869	0.7880	38
52	0.7880	0.7891	0.7902	0.7912	0.7923	0.7934	0.7944	0.7955	0.7965	0.7976	0.7986	37
53	0.7986	0.7997	0.8007	0.8018	0.8028	0.8039	0.8049	0.8059	0.8070	0.8080	0.8090	36
54	0.8090	0.8100	0.8111	0.8121	0.8131	0.8141	0.8151	0.8161	0.8171	0.8181	0.8192	35
55	0.8192	0.8202	0.8211	0.8221	0.8231	0.8241	0.8251	0.8261	0.8271	0.8281	0.8290	34
56	0.8290	0.8300	0.8310	0.8320	0.8329	0.8339	0.8348	0.8358	0.8368	0.8377	0.8387	33
57	0.8387	0.8396	0.8406	0.8415	0.8425	0.8434	0.8443	0.8453	0.8462	0.8471	0.8480	32
58	0.8480	0.8490	0.8499	0.8508	0.8517	0.8526	0.8536	0.8545	0.8554	0.8563	0.8572	31
59	0.8572	0.8581	0.8590	0.8599	0.8607	0.8616	0.8625	0.8634	0.8643	0.8652	0.8660	30°
60°	0.8660	0.8669	0.8678	0.8686	0.8695	0.8704	0.8712	0.8721	0.8729	0.8738	0.8746	29
61	0.8746	0.8755	0.8763	0.8771	0.8780	0.8788	0.8796	0.8805	0.8813	0.8821	0.8829	28
62	0.8829	0.8838	0.8846	0.8854	0.8862	0.8870	0.8878	0.8886	0.8894	0.8902	0.8910	27
63	0.8910	0.8918	0.8926	0.8934	0.8942	0.8949	0.8957	0.8965	0.8973	0.8980	0.8988	26
64	0.8988	0.8996	0.9003	0.9011	0.9018	0.9026	0.9033	0.9041	0.9048	0.9056	0.9063	25
65	0.9063	0.9070	0.9078	0.9085	0.9092	0.9100	0.9107	0.9114	0.9121	0.9128	0.9135	24
66	0.9135	0.9143	0.9150	0.9157	0.9164	0.9171	0.9178	0.9184	0.9191	0.9198	0.9205	23
67	0.9205	0.9212	0.9219	0.9225	0.9232	0.9239	0.9245	0.9252	0.9259	0.9265	0.9272	22
68	0.9272	0.9278	0.9285	0.9291	0.9298	0.9304	0.9311	0.9317	0.9323	0.9330	0.9336	21
69	0.9336	0.9342	0.9348	0.9354	0.9361	0.9367	0.9373	0.9379	0.9385	0.9391	0.9397	20°
70°	0.9397	0.9403	0.9409	0.9415	0.9421	0.9426	0.9432	[illegible]	0.9444	0.9449	0.9455	19
71	0.9455	0.9461	0.9466	0.9472	0.9478	0.9483	0.9489	0.9494	0.9500	0.9505	0.9511	18
72	0.9511	0.9516	0.9521	0.9527	0.9532	0.9537	0.9542	0.9548	0.9553	0.9558	0.9563	17
73	0.9563	0.9568	0.9573	0.9578	0.9583	0.9588	0.9593	0.9598	0.9603	0.9608	0.9613	16
74	0.9613	0.9617	0.9622	0.9627	0.9632	0.9636	0.9641	0.9646	0.9650	0.9655	0.9659	15
75	0.9659	0.9664	0.9668	0.9673	0.9677	0.9681	0.9686	0.9690	0.9694	0.9699	0.9703	14
76	0.9703	0.9707	0.9711	0.9715	0.9720	0.9724	0.9728	0.9732	0.9736	0.9740	0.9744	13
77	0.9744	0.9748	0.9751	0.9755	0.9759	0.9763	0.9767	0.9770	0.9774	0.9778	0.9781	12
78	0.9781	0.9785	0.9789	0.9792	0.9796	0.9799	0.9803	0.9806	0.9810	0.9813	0.9816	11
79	0.9816	0.9820	0.9823	0.9826	0.9829	0.9833	0.9836	0.9839	0.9842	0.9845	0.9848	10°
80°	0.9848	0.9851	0.9854	0.9857	0.9860	0.9863	0.9866	0.9869	0.9871	0.9874	0.9877	9
81	0.9877	0.9880	0.9882	0.9885	0.9888	0.9890	0.9893	0.9895	0.9898	0.9900	0.9903	8
82	0.9903	0.9905	0.9907	0.9910	0.9912	0.9914	0.9917	0.9919	0.9921	0.9923	0.9925	7
83	0.9925	0.9928	0.9930	0.9932	0.9934	0.9936	0.9938	0.9940	0.9942	0.9943	0.9945	6
84	0.9945	0.9947	0.9949	0.9951	0.9952	0.9954	0.9956	0.9957	0.9959	0.9960	0.9962	5
85	0.9962	0.9963	0.9965	0.9966	0.9968	0.9969	0.9971	0.9972	0.9973	0.9974	0.9976	4
86	0.9976	0.9977	0.9978	0.9979	0.9980	0.9981	0.9982	0.9983	0.9984	0.9985	0.9986	3
87	0.9986	0.9987	0.9988	0.9989	0.9990	0.9990	0.9991	0.9992	0.9993	0.9993	0.9994	2
88	0.9994	0.9995	0.9995	0.9996	0.9996	0.9997	0.9997	0.9997	0.9998	0.9998	0.9998	1
89	0.9998	0.9999	0.9999	0.9999	0.9999	1.0000	1.0000	1.0000	1.0000	1.0000	1.0000	0°
	°1.0	°0.9	°0.8	°0.7	°0.6	°0.5	°0.4	°0.3	°0.2	°0.1	°0.0	Deg.

Appendix B Trigonometric Functions—Natural Tangents and Cotangents

NOTE: For cotangents, use right-hand column of degrees and lower line of tenths.

Deg.	°0.0	°0.1	°0.2	°0.3	°0.4	°0.5	°0.6	°0.7	°0.8	°0.9	°1.0	
0°	0.0000	0.0017	0.0035	0.0052	0.0070	0.0087	0.0105	0.0122	0.0140	0.0157	0.0175	89
1	0.0175	0.0192	0.0209	0.0227	0.0244	0.0262	0.0279	0.0297	0.0314	0.0332	0.0349	88
2	0.0349	0.0367	0.0384	0.0402	0.0419	0.0437	0.0454	0.0472	0.0489	0.0507	0.0524	87
3	0.0524	0.0542	0.0559	0.0577	0.0594	0.0612	0.0629	0.0647	0.0664	0.0682	0.0699	86
4	0.0699	0.0717	0.0734	0.0752	0.0769	0.0787	0.0805	0.0822	0.0840	0.0857	0.0875	85
5	0.0875	0.0892	0.0910	0.0928	0.0945	0.0963	0.0981	0.0998	0.1016	0.1033	0.1051	84
6	0.1051	0.1069	0.1086	0.1104	0.1122	0.1139	0.1157	0.1175	0.1192	0.1210	0.1228	83
7	0.1228	0.1246	0.1263	0.1281	0.1299	0.1317	0.1334	0.1352	0.1370	0.1388	0.1405	82
8	0.1405	0.1423	0.1441	0.1459	0.1477	0.1495	0.1512	0.1530	0.1548	0.1566	0.1584	81
9	0.1584	0.1602	0.1620	0.1638	0.1655	0.1673	0.1691	0.1709	0.1727	0.1745	0.1763	80°
10°	0.1763	0.1781	0.1799	0.1817	0.1835	0.1853	0.1871	0.1890	0.1908	0.1926	0.1944	79
11	0.1944	0.1962	0.1980	0.1998	0.2016	0.2035	0.2053	0.2071	0.2089	0.2107	0.2126	78
12	0.2126	0.2144	0.2162	0.2180	0.2199	0.2217	0.2235	0.2254	0.2272	0.2290	0.2309	77
13	0.2309	0.2327	0.2345	0.2364	0.2382	0.2401	0.2419	0.2438	0.2456	0.2475	0.2493	76
14	0.2493	0.2512	0.2530	0.2549	0.2568	0.2586	0.2605	0.2623	0.2642	0.2661	0.2679	75
15	0.2679	0.2698	0.2717	0.2736	0.2754	0.2773	0.2792	0.2811	0.2830	0.2849	0.2867	74
16	0.2867	0.2886	0.2905	0.2924	0.2943	0.2962	0.2981	0.3000	0.3019	0.3038	0.3057	73
17	0.3057	0.3076	0.3096	0.3115	0.3134	0.3153	0.3172	0.3191	0.3211	0.3230	0.3249	72
18	0.3249	0.3269	0.3288	0.3307	0.3327	0.3346	0.3365	0.3385	0.3404	0.3424	0.3443	71
19	0.3443	0.3463	0.3482	0.3502	0.3522	0.3541	0.3561	0.3581	0.3600	0.3620	0.3640	70°
20°	0.3640	0.3659	0.3679	0.3699	0.3719	0.3739	0.3759	0.3779	0.3799	0.3819	0.3839	69
21	0.3839	0.3859	0.3879	0.3899	0.3919	0.3939	0.3959	0.3979	0.4000	0.4020	0.4040	68
22	0.4040	0.4061	0.4081	0.4101	0.4122	0.4142	0.4163	0.4183	0.4204	0.4224	0.4245	67
23	0.4245	0.4265	0.4286	0.4307	0.4327	0.4348	0.4369	0.4390	0.4411	0.4431	0.4452	66
24	0.4452	0.4473	0.4494	0.4515	0.4536	0.4557	0.4578	0.4599	0.4621	0.4642	0.4663	65
25	0.4663	0.4684	0.4706	0.4727	0.4748	0.4770	0.4791	0.4813	0.4834	0.4856	0.4877	64
26	0.4877	0.4899	0.4921	0.4942	0.4964	0.4986	0.5008	0.5029	0.5051	0.5073	0.5095	63
27	0.5095	0.5117	0.5139	0.5161	0.5184	0.5206	0.5228	0.5250	0.5272	0.5295	0.5317	62
28	0.5317	0.5340	0.5362	0.5384	0.5407	0.5430	0.5452	0.5475	0.5498	0.5520	0.5543	61
29	0.5543	0.5566	0.5589	0.5612	0.5635	0.5658	0.5681	0.5704	0.5727	0.5750	0.5774	60°
30°	0.5774	0.5797	0.5820	0.5844	0.5867	0.5890	0.5914	0.5938	0.5961	0.5985	0.6009	59
31	0.6009	0.6032	0.6056	0.6080	0.6104	0.6128	0.6152	0.6176	0.6200	0.6224	0.6249	58
32	0.6249	0.6273	0.6297	0.6322	0.6346	0.6371	0.6395	0.6420	0.6445	0.6469	0.6494	57
33	0.6494	0.6519	0.6544	0.6569	0.6594	0.6619	0.6644	0.6669	0.6694	0.6720	0.6745	56
34	0.6745	0.6771	0.6796	0.6822	0.6847	0.6873	0.6899	0.6924	0.6950	0.6976	0.7002	55
35	0.7002	0.7028	0.7054	0.7080	0.7107	0.7133	0.7159	0.7186	0.7212	0.7239	0.7265	54
36	0.7265	0.7292	0.7319	0.7346	0.7373	0.7400	0.7427	0.7454	0.7481	0.7508	0.7536	53
37	0.7536	0.7563	0.7590	0.7618	0.7646	0.7673	0.7701	0.7729	0.7757	0.7785	0.7813	52
38	0.7813	0.7841	0.7869	0.7898	0.7926	0.7954	0.7983	0.8012	0.8040	0.8069	0.8098	51
39	0.8098	0.8127	0.8156	0.8185	0.8214	0.8243	0.8273	0.8302	0.8332	0.8361	0.8391	50°
40°	0.8391	0.8421	0.8451	0.8481	0.8511	0.8541	0.8571	0.8601	0.8632	0.8662	0.8693	49
41	0.8693	0.8724	0.8754	0.8785	0.8816	0.8847	0.8878	0.8910	0.8941	0.8972	0.9004	48
42	0.9004	0.9036	0.9067	0.9099	0.9131	0.9163	0.9195	0.9228	0.9260	0.9293	0.9325	47
43	0.9325	0.9358	0.9391	0.9424	0.9457	0.9490	0.9523	0.9556	0.9590	0.9623	0.9657	46
44	0.9657	0.9691	0.9725	0.9759	0.9793	0.9827	0.9861	0.9896	0.9930	0.9965	1.0000	45
	°1.0	°0.9	°0.8	°0.7	°0.6	°0.5	°0.4	°0.3	°0.2	°0.1	°0.0	Deg.

NOTE: For cotangents, use right-hand column of degrees and lower line of tenths.

Deg.	°0.0	°0.1	°0.2	°0.3	°0.4	°0.5	°0.6	°0.7	°0.8	°0.9	°1.0	
45	1.0000	1.0035	1.0070	1.0105	1.0141	1.0176	1.0212	1.0247	1.0283	1.0319	1.0355	44
46	1.0355	1.0392	1.0428	1.0464	1.0501	1.0538	1.0575	1.0612	1.0649	1.0686	1.0724	43
47	1.0724	1.0761	1.0799	1.0837	1.0875	1.0913	1.0951	1.0990	1.1028	1.1067	1.1106	42
48	1.1106	1.1145	1.1184	1.1224	1.1263	1.1303	1.1343	1.1383	1.1423	1.1463	1.1504	41
49	1.1504	1.1544	1.1585	1.1626	1.1667	1.1708	1.1750	1.1792	1.1833	1.1875	1.1918	40°
50°	1.1918	1.1960	1.2002	1.2045	1.2088	1.2131	1.2174	1.2218	1.2261	1.2305	1.2349	39
51	1.2349	1.2393	1.2437	1.2482	1.2527	1.2572	1.2617	1.2662	1.2708	1.2753	1.2799	38
52	1.2799	1.2846	1.2892	1.2938	1.2985	1.3032	1.3079	1.3127	1.3175	1.3222	1.3270	37
53	1.3270	1.3319	1.3367	1.3416	1.3465	1.3514	1.3564	1.3613	1.3663	1.3713	1.3764	36
54	1.3764	1.3814	1.3865	1.3916	1.3968	1.4019	1.4071	1.4124	1.4176	1.4229	1.4281	35
55	1.4281	1.4335	1.4388	1.4442	1.4496	1.4550	1.4605	1.4659	1.4715	1.4770	1.4826	34
56	1.4826	1.4882	1.4938	1.4994	1.5051	1.5108	1.5166	1.5224	1.5282	1.5340	1.5399	33
57	1.5399	1.5458	1.5517	1.5577	1.5637	1.5697	1.5757	1.5818	1.5880	1.5941	1.6003	32
58	1.6003	1.6066	1.6128	1.6191	1.6255	1.6319	1.6383	1.6447	1.6512	1.6577	1.6643	31
59	1.6643	1.6709	1.6775	1.6842	1.6909	1.6977	1.7045	1.7113	1.7182	1.7251	1.7321	30°
60°	1.7321	1.7391	1.7461	1.7532	1.7603	1.7675	1.7747	1.7820	1.7893	1.7966	1.8040	29
61	1.8040	1.8115	1.8190	1.8265	1.8341	1.8418	1.8495	1.8572	1.8650	1.8728	1.8807	28
62	1.8807	1.8887	1.8967	1.9047	1.9128	1.9210	1.9292	1.9375	1.9458	1.9542	1.9626	27
63	1.9626	1.9711	1.9797	1.9883	1.9970	2.0057	2.0145	2.0233	2.0323	2.0413	2.0503	26
64	2.0503	2.0594	2.0686	2.0778	2.0872	2.0965	2.1060	2.1155	2.1251	2.1348	2.1445	25
65	2.1445	2.1543	2.1642	2.1742	2.1842	2.1943	2.2045	2.2148	2.2251	2.2355	2.2460	24
66	2.2460	2.2566	2.2673	2.2781	2.2889	2.2998	2.3109	2.3220	2.3332	2.3445	2.3559	23
67	2.3559	2.3673	2.3789	2.3906	2.4023	2.4142	2.4262	2.4383	2.4504	2.4627	2.4751	22
68	2.4751	2.4876	2.5002	2.5129	2.5257	2.5386	2.5517	2.5649	2.5782	2.5916	2.6051	21
69	2.6051	2.6187	2.6325	2.6464	2.6605	2.6746	2.6889	2.7034	2.7179	2.7326	2.7475	20°
70°	2.7475	2.7625	2.7776	2.7929	2.8083	2.8239	2.8397	2.8556	2.8716	2.8878	2.9042	19
71	2.9042	2.9208	2.9375	2.9544	2.9714	2.9887	3.0061	3.0237	3.0415	3.0595	3.0777	18
72	3.0777	3.0961	3.1146	3.1334	3.1524	3.1716	3.1910	3.2106	3.2305	3.2506	3.2709	17
73	3.2709	3.2914	3.3122	3.3332	3.3544	3.3759	3.3977	3.4197	3.4420	3.4646	3.4874	16
74	3.4874	3.5105	3.5339	3.5576	3.5816	3.6059	3.6305	3.6554	3.6806	3.7062	3.7321	15
75	3.7321	3.7583	3.7848	3.8118	3.8391	3.8667	3.8947	3.9232	3.9520	3.9812	4.0108	14
76	4.0108	4.0408	4.0713	4.1022	4.1335	4.1653	4.1976	4.2303	4.2635	4.2972	4.3315	13
77	4.3315	4.3662	4.4015	4.4374	4.4737	4.5107	4.5483	4.5864	4.6252	4.6646	4.7046	12
78	4.7046	4.7453	4.7867	4.8288	4.8716	4.9152	4.9594	5.0045	5.0504	5.0970	5.1446	11
79	5.1446	5.1929	5.2422	5.2924	5.3435	5.3955	5.4486	5.5026	5.5578	5.6140	5.6713	10°
80°	5.6713	5.7297	5.7894	5.8502	5.9124	5.9758	6.0405	6.1066	6.1742	6.2432	6.3138	9
81	6.3138	6.3859	6.4596	6.5350	6.6122	6.6912	6.7720	6.8548	6.9395	7.0264	7.1154	8
82	7.1154	7.2066	7.3002	7.3962	7.4947	7.5958	7.6996	7.8062	7.9158	8.0285	8.1443	7
83	8.1443	8.2636	8.3863	8.5126	8.6427	8.7769	8.9152	9.0579	9.2052	9.3572	9.5144	6
84	9.5144	9.677	9.845	10.02	10.20	10.39	10.58	10.78	10.99	11.20	11.43	5
85	11.43	11.66	11.91	12.16	12.43	12.71	13.00	13.30	13.62	13.95	14.30	4
86	14.30	14.67	15.06	15.46	15.89	16.35	16.83	17.34	17.89	18.46	19.08	3
87	19.08	19.74	20.45	21.20	22.02	22.90	23.86	24.90	26.03	27.27	28.64	2
88	28.64	30.14	31.82	33.69	35.80	38.19	40.92	44.07	47.74	52.08	57.29	1
89	57.29	63.66	71.62	81.85	95.49	114.6	143.2	191.0	286.5	573.0		0°
	°1.0	°0.9	°0.8	°0.7	°0.6	°0.5	°0.4	°0.3	°0.2	°0.1	°0.0	Deg.

Appendix C Alternating Current Formulas

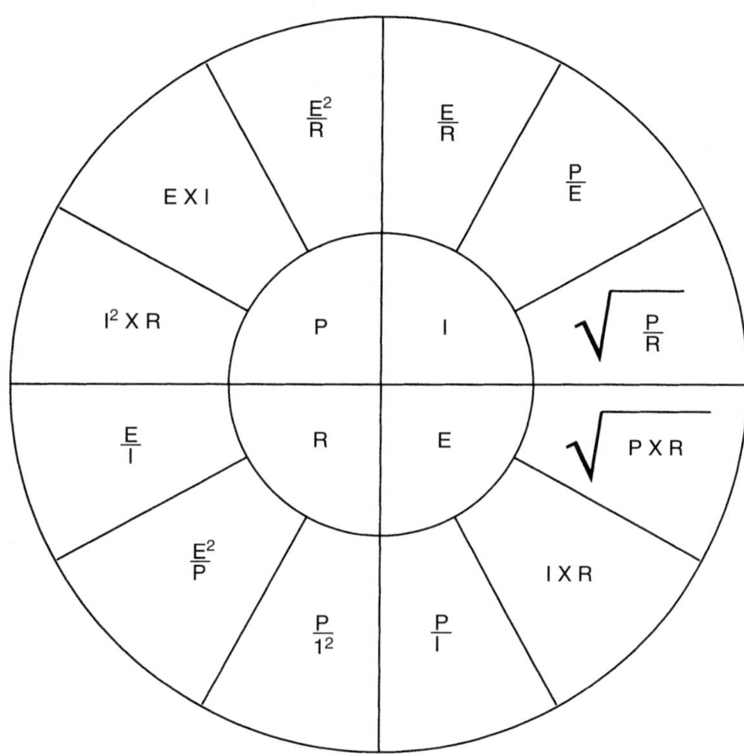

Instantaneous and Maximum Values

The instantaneous value of voltage and current for a sine wave is equal to the peak, or maximum, value of the wave form times the sine of the angle. For example, a wave form has a peak value of 300 V. What is the voltage at an angle of 22°?

$$\mathbf{E_{(INST)} = E_{(MAX)} \times SIN\angle}$$

$$\mathbf{E_{(INST)} = 300 \times 0.3746\ (SIN\ of\ 22°)}$$

$$\mathbf{E_{(INST)} = 112.328\ V}$$

$$\mathbf{E_{MAX} = \frac{E_{INST}}{SIN\angle}}$$

$$\mathbf{SIN\angle = \frac{E_{INST}}{E_{MAX}}}$$

Changing Peak, RMS, and Average Values

To change	To	Multiply by
Peak	RMS	0.707
Peak	Average	0.637
Peak	Peak-to-Peak	2
RMS	Peak	1.414
Average	Peak	1.567
RMS	Average	0.9
Average	RMS	1.111

Pure Resistive Circuit

$E = I \times R$ $\qquad I = \dfrac{E}{R}$ $\qquad R = \dfrac{E}{I}$ $\qquad P = E \times I$

$E = \dfrac{P}{I}$ $\qquad I = \dfrac{P}{R}$ $\qquad R = \dfrac{E^2}{P}$ $\qquad P = I^2 \times R$

$E = \sqrt{P \times R}$ $\qquad I = \sqrt{\dfrac{P}{R}}$ $\qquad R = \dfrac{P}{I^2}$ $\qquad P = \dfrac{E^2}{R}$

Series Resistive Circuits

$R_T = R_1 + R_2 + R_3$

$I_T = I_1 = I_2 = I_3$

$E_T = E_1 + E_2 + E_3$

$P_T = P_1 + P_2 + P_3$

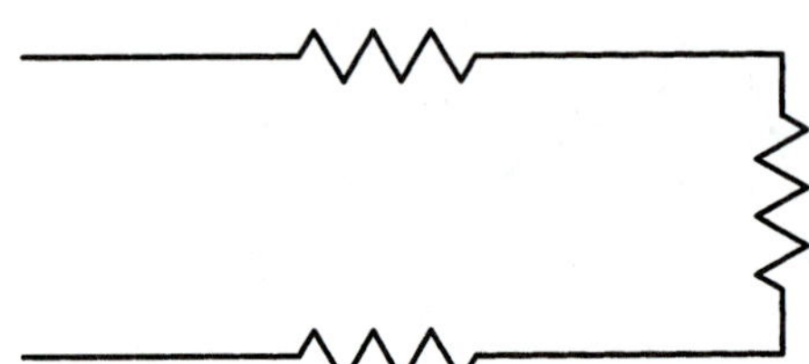

Parallel Resistive Circuits

$$R_T = \frac{1}{\dfrac{1}{R_1} + \dfrac{1}{R_2} + \dfrac{1}{R_3}}$$

$$R_T = \frac{R_1 \times R_2}{R_1 + R_2}$$

$I_T = I_1 + I_2 + I_3$

$E_T = E_1 = E_2 = E_3$

$P_T = P_1 + P_2 + P_3$

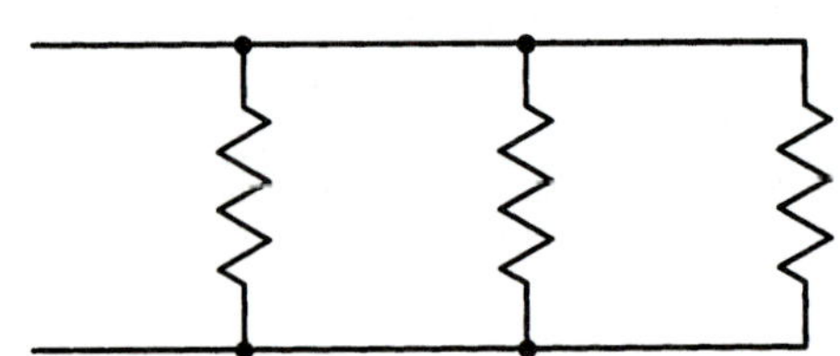

Pure Inductive Circuits

In a pure inductive circuit, the current lags the voltage by 90°. Therefore, there is no true power, or watts, and the power factor is 0. VARs is the inductive equivalent of watts.

$L = \dfrac{X_L}{2\pi F}$ $\qquad X_L = 2\pi FL$

$E_L = I_L \times X_L$ $\qquad I_L = \dfrac{E_L}{X_L}$ $\qquad X_L = \dfrac{E_L}{I_L}$ $\qquad VARs_L = E_L \times I_L$

$E_L = \sqrt{VARs_L \times X_L}$ $\qquad I_L = \dfrac{VARs_L}{E_L}$ $\qquad X_L = \dfrac{E_L^2}{VARs_L}$ $\qquad VARs_L = I_L^2 \times X_L$

$E_L = \dfrac{VARs_L}{I_L}$ $\qquad I_L = \sqrt{\dfrac{VARs_L}{X_L}}$ $\qquad X_L = \dfrac{VARs}{I_L^2}$ $\qquad VARs_L = \dfrac{E_L^2}{X_L}$

Series Inductive Circuits

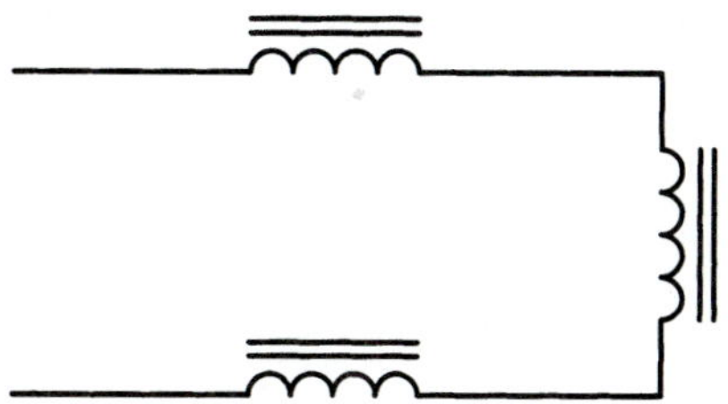

$E_{LT} = E_{L1} + E_{L2} + E_{L3}$

$X_{LT} = X_{L1} + X_{L2} + X_{L3}$

$I_{LT} = I_{L1} = I_{L2} = I_{L3}$

$VARs_{LT} = VARs_{L1} + VARs_{L2} + VARs_{L3}$

$L_T = L_1 + L_2 + L_3$

Parallel Inductive Circuits

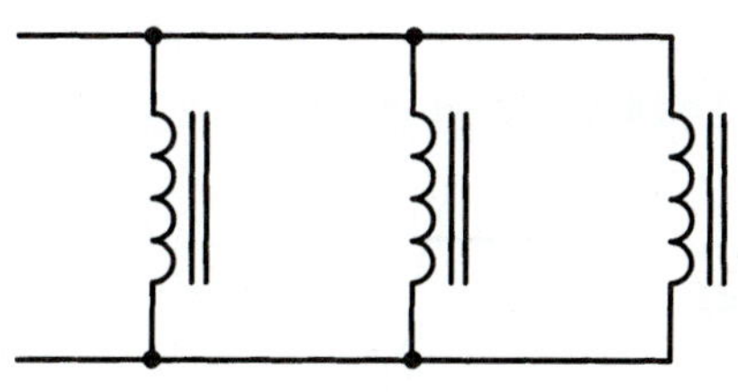

$E_{LT} = E_{L1} = E_{L2} = E_{L3}$

$I_{LT} = I_{L1} + I_{L2} + I_{L3}$

$$X_{LT} = \frac{X_{L1} \times X_{L2}}{X_{L1} + X_{L2}}$$

$$X_{LT} = \frac{1}{\frac{1}{X_{L1}} + \frac{1}{X_{L2}} + \frac{1}{X_{L3}}} \qquad L_T = \frac{1}{\frac{1}{L_1} + \frac{1}{L_2} + \frac{1}{L_3}} \qquad L_T = \frac{L_1 \times L_2}{L_1 + L_2}$$

$VARs_{LT} = VARs_{L1} + VARs_{L2} + VARs_{L3}$

Pure Capacitive Circuits

In a pure capacitive circuit, the current leads the voltage by 90°. For this reason there is no true power, or watts, and no power factor. VARs is the equivalent of watts in a pure capacitive circuit

The value of C in the formula for finding capacitance is in farads and must be changed into the capacitive units being used.

$$C = \frac{1}{2\pi F X_c} \qquad X_c = \frac{1}{2\pi FC}$$

$$E_c = I_c \times X_c \qquad I_c = \frac{E_c}{X_c} \qquad X_c = \frac{E_c}{I_c} \qquad VARs_c = E_c \times I_c$$

$$E_c = \sqrt{VARs_c \times X_c} \qquad I_c \frac{VARs_c}{E_c} \qquad X_c = \frac{E_c^2}{VARs_c} \qquad VARs_c = I_c^2 \times X_c$$

$$E_c = \frac{VARs_c}{I_c} \qquad I_c = \sqrt{\frac{VARs_c}{X_c}} \qquad X_c = \frac{VARs_c}{I_c^2} \qquad VARs_c = \frac{E_c^2}{X_c}$$

Series Capacitive Circuits

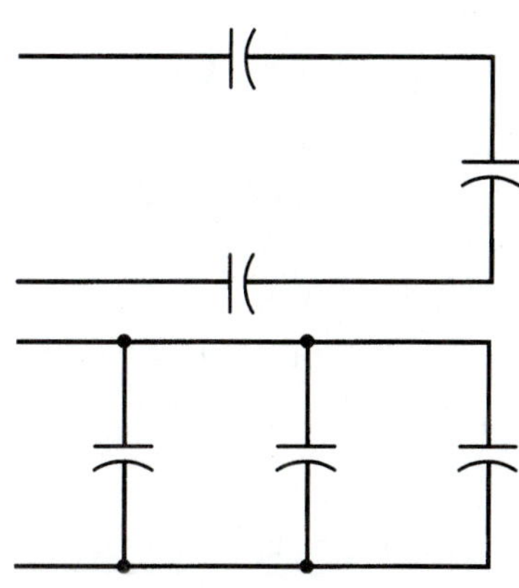

$E_{CT} = E_{C1} + E_{C2} + E_{C3}$

$I_{CT} = I_{C1} = I_{C2} = I_{C3}$

$X_{CT} = X_{C1} + X_{C2} + X_{C3}$

$VARs_{CT} = VARs_{C1} + VARs_{C2} + VARs_{C3}$

$$C_T = \frac{1}{\frac{1}{C_1} + \frac{1}{C_2} + \frac{1}{C_3}} \qquad C_T = \frac{C_1 \times C_2}{C_1 + C_2}$$

Parallel Capacitive Circuits

$E_{CT} = E_{C1} = E_{C2} = E_{C3}$

$$X_{CT} = \frac{1}{\frac{1}{X_{C1}} + \frac{1}{X_{C2}} + \frac{1}{X_{C3}}}$$

$$X_{CT} = \frac{X_{C1} \times X_{C2}}{X_{C1} + X_{C2}}$$

$I_{CT} = I_{C1} + I_{C2} + I_{C3}$

$C_T = C_1 + C_2 + C_3$

$VARs_{CT} = VARs_{C1} + VARs_{C2} + VARs_{C3}$

Resistive-Inductive Series Circuits

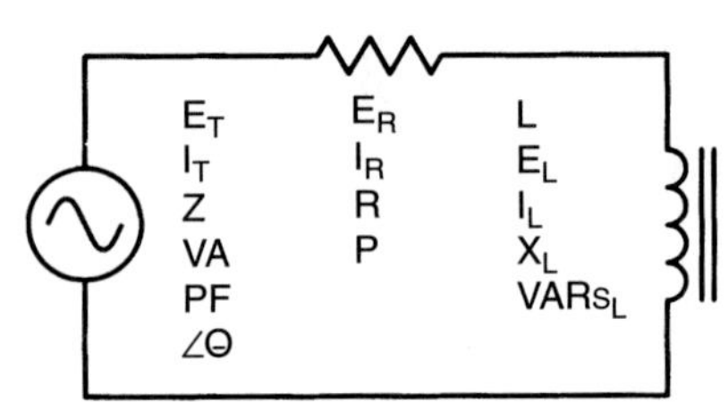

To find values for the resistor, use the formulas in the Pure Resistive section.
To find values for the inductor, use the formulas in the Pure Inductive section.

$E_T = \sqrt{E_R^2 + E_L^2}$

$E_T = I_T \times Z$

$$E_T = \frac{VA}{I_T}$$

$$E_T = \frac{E_R}{PF}$$

$$PF = \frac{R}{Z}$$

$$PF = \frac{P}{VA}$$

$$PF = \frac{E_R}{E_T}$$

$PF = COS\angle\theta$

$R = \sqrt{Z^2 - X_L^2}$

$$R = \frac{E_R}{I_R}$$

$$R = \frac{E_R^2}{P}$$

$$R = \frac{P}{I_R^2}$$

$R = Z \times PF$

$VARs_L = \sqrt{VA^2 - P^2}$

$Z = \sqrt{R^2 + X_L^2}$

$$Z = \frac{E_T}{I_T}$$

$$Z = \frac{VA}{I_T^2}$$

$$Z = \frac{R}{PF}$$

$P = E_R \times I_R$

$P = \sqrt{VA^2 - VARs_L^2}$

$$P = \frac{E_R^2}{R}$$

$P = I_R^2 \times R$

$P = VA \times PF$

$E_L = I_L \times X_L$

$E_L = \sqrt{E_T^2 - E_L^2}$

$E_L = \sqrt{VARs_L \times X_L}$

$$E_L = \frac{VARs_L}{I_L}$$

$VARs_L = E_L \times I_L$

$VA = E_T \times I_T$

$VA = I_T^2 \times Z$

$$VA = \frac{E_T^2}{Z}$$

$VA = \sqrt{P^2 + VARs_L^2}$

$$Z = \frac{E_T^2}{VA}$$

$E_R = I_R \times R$

$E_R = \sqrt{P \times R}$

$$E_R = \frac{P}{I_R}$$

$E_R = \sqrt{E_T^2 - E_L^2}$

$E_R = E_T \times PF$

$I_L = I_R = I_T$

$$I_L = \frac{E_L}{X_L}$$

$$I_L = \frac{VARs_L}{E_L}$$

$$I_L = \sqrt{\frac{VARs_L}{X_L}}$$

$$VARs_L = \frac{E_L^2}{X_L}$$

$$L = \frac{X_L}{2\pi F}$$

$I_T = I_R = I_L$

$$I_T = \frac{E_T}{Z}$$

$$I_T = \frac{VA}{E_T}$$

$$VA = \frac{P}{PF}$$

$I_R = I_T = I_L$

$$I_R = \frac{E_R}{R}$$

$$I_R = \frac{P}{E_R}$$

$$I_R = \sqrt{\frac{P}{R}}$$

$X_L = \sqrt{Z^2 - R^2}$

$$X_L = \frac{E_L}{I_L}$$

$$X_L = \frac{E_L^2}{VARs_L}$$

$$X_L = \frac{VARs_L}{I_L^2}$$

$X_L = 2\pi FL$

$VARs_L = I_L^2 \times X_L$

J

L

M

N

O

P

T

U

V

W